高 等 学 校 教 材

微积分（英文版）
（第2卷）

CALCULUS （Ⅱ）

张宇　黄艳　主编

哈尔滨工业大学出版社
HARBIN INSTITUTE OF TECHNOLOGY PRESS

内容简介

本书为《微积分》一书的第二卷,适用于工科院校非数学专业本科新生,亦可作为工程技术人员的参考书籍.本卷包含四个章节,内容涵盖多元函数微分学,多元函数积分学,第二型曲线积分、第二型曲面积分及无穷级数.本书包含大量例题及习题.

图书在版编目(CIP)数据

微积分.2＝Calculus.2:外文、英文/张宇,黄艳主编.—哈尔滨:哈尔滨工业大学出版社,2016.3
ISBN 978-7-5603-5896-3

Ⅰ.①微…　Ⅱ.①张…②黄…　Ⅲ.①微积分-高等学校-教材-英文　Ⅳ.①O172

中国版本图书馆 CIP 数据核字(2016)第 052472 号

策划编辑	刘培杰　张永芹	
责任编辑	张永芹　王勇钢	
封面设计	孙茵艾	
出版发行	哈尔滨工业大学出版社	
社　　址	哈尔滨市南岗区复华四道街 10 号　邮编 150006	
传　　真	0451-86414749	
网　　址	http://hitpress.hit.edu.cn	
印　　刷	哈尔滨市工大节能印刷厂	
开　　本	787mm×960mm　1/16　印张 19.25　字数 370 千字	
版　　次	2016 年 3 月第 1 版　2016 年 3 月第 1 次印刷	
书　　号	ISBN 978-7-5603-5896-3	
定　　价	48.00 元	

(如因印装质量问题影响阅读,我社负责调换)

Preface

Mathematics is one of the most important and wide applied science. Calculus is the branch of mathematics which studies the change of quantities. Modern calculus was developed in 17th century Europe by Newton and Leibniz. It is one of the most fundamental and well studied branch of mathematics. Calculus has comprehensive application in Engineering, economics, business and social and life sciences. Calculus is a part of the modern mathematics education and is important to the students for their future career pursuing.

There are a lot of excellent calculus text books in Chinese which are suitable for the newly college students major in engineering, economics and so on. The authors of this book are very experienced in teaching college calculus, especially in teaching in English for classy students and oversea students. The authors had some trouble in finding a suitable text book in English. Although there are some famous and popular calculus textbooks in English, but because of the uniqueness of our education system, none of them fits our requirement. The publishment of this book will fill in this blank.

This book is dedicate to college freshmen major in engineering, business, economic and so on. It can also be used as a reference for technicians. Some of the key features of this book are:

1. Abundant theories. This book not only contains necessary theories for students of technology, but also contains some classic theories of mathematical analysis for science students. It gives the students more solid foundation of mathematics.

2. Comprehensive examples and exercises. This book contains plentiful of examples and exercises. The examples and exercises are carefully graded, progressing from basic conceptual exercises and skill-development problems to more challenging problems involving applications and proofs. Many of the exercises and examples are related to real-world phenomena.

3. Easy to look up. This book is well organised and easy to look up for theories, it can be used as a reference.

During the preparation of this book, a lot of people gave us a great deal of help. Our colleagues who are all very experienced gave us a lot of helpful suggestions, the editors offered us a lot of assistance. Great thanks to all of these people who helped us to make this book possible.

Authors
January 15th, 2016

Contents

Chapter 8 Differential Calculus of Multivariable Functions ········· 1

    8.1 Limits and Continuity of Multivariable Functions ············· 1

    8.2 Partial Derivatives and Higher-Order Partial Derivatives ········· 8

    8.3 Linear Approximations and Total Differentials ··············· 15

    8.4 The Chain Rule ·· 21

    8.5 Implicit Differentiation ··· 26

    8.6 Applications of Partial Derivatives to Analytic Geometry ········ 35

    8.7 Extreme Values of Functions of Several Variables ············· 41

    8.8 Directional Derivatives and The Gradient Vector ·············· 53

    8.9 Examples ·· 57

    Exercises 8 ··· 61

Chapter 9 Multiple Integrals ·· 74

    9.1 Double Integrals ·· 74

    9.2 Calculating Double Integrals ··································· 78

    9.3 Calculating Triple Integrals ····································· 89

    9.4 Concepts and Calculations of The First Type Curve Integral ······ 101

    9.5 The First Type Surface Integral ································ 106

    9.6 Application of Integrals ·· 111

    9.7 Examples ·· 114

    Exercises 9 ··· 119

Chapter 10 The Second Type Curve Integral, Surface Integral,

          and Vector Field ·· 131

    10.1 The Second Type Curve Integral ······························ 131

10.2	The Green's Theorem	140
10.3	Conditions for Plane Curve Integrals Being Independent of Path, Conservative Fields	146
10.4	The Second Type Surface Integral	154
10.5	The Gauss Formula, The Flux and Divergence	162
10.6	The Stokes' Theorem, Circulation and Curl	170
10.7	Examples	177
	Exercises 10	183

Chapter 11 Infinite Series 197

11.1	Convergence and Divergence of Infinite Series	198
11.2	The Discriminances for Convergence and Divergence of Infinite Series with Positive Terms	205
11.3	Series With Arbitrary Terms, Absolute Convergence	213
11.4*	The Discriminances for Convergence of Improper Integral, Γ Function	218
11.5	Series with Function Terms, Uniform Convergence	223
11.6	Power Series	231
11.7	Expanding Functions as Power Series	240
11.8	Some Applications of The Power Series	253
11.9	Fourier Series	257
11.10	Examples	273
	Exercises 11	277

Appendix IV Change of Variables in Multiple Integrals 293

Appendix V Radius of Convergence of Power Series 300

Chapter 8 Differential Calculus of Multivariable Functions

Functions with two or more independent variables appear more often in science than functions of a single variable. In this chapter we extend the basic ideas of one variable differential calculus to such functions. With functions of several independent variables, we work with partial derivatives, which, in turn, give rise to directional derivatives and the gradient, some fundamental concepts in calculus. Partial derivatives allow us to find maximum and minimum values of multivariable functions. We define tangent planes, rather than tangent lines, that allow us to make linear approximations.

8.1 Limits and Continuity of Multivariable Functions

8.1.1 The n-Dimensional Space

If we introduce the rectangular coordinate system in space, then we have a one-to-one correspondence between points P in space and ordered triples (x,y,z). The set of all points described by the triples (x,y,z) is called the three-dimensional space. We denote this space by $\mathbf{R}^3$. Generally, we can consider an ordered n-tuple of real numbers $(x_1, x_2, \cdots, x_n)$ for any integer $n \geqslant 1$. Such an n-tuple is called an n-dimensional point, the individual numbers $x_1, x_2, \cdots, x_n$ being referred to as coordinates or components of the point. The set of all n-dimensional points is called the n-dimensional space. We denote this space by $\mathbf{R}^n$. We shall usually denote points by capital letters A, $B, C, \cdots$, and components by the corresponding small letters $a, b, c, \cdots$

The familiar formula for the distance between two points in $\mathbf{R}^3$ is easily extended to the n-dimentional formula. The distance $\rho(A,B)$ between the points $A(a_1, a_2, \cdots, a_n)$ and $B(b_1, b_2, \cdots, b_n)$ in $\mathbf{R}^n$ is

$$\rho(A,B) = \sqrt{(a_1 - b_1)^2 + (a_2 - b_2)^2 + \cdots + (a_n - b_n)^2}$$

Let $P_0 \in \mathbf{R}^n$ and $\delta > 0$, a set $\{P \mid \rho(P,P_0) < \delta, P \in \mathbf{R}^n\}$ is called a δ-neighborhood of the point P_0, denoted by $U_\delta(P_0)$ or $U(P_0, \delta)$. If we don't care about the size of δ, we often use "neighborhood" instead of "δ-neighborhood" of P_0, denoted by $U(P_0)$.

A point P_0 in a set E in $\mathbf{R}^n$ is an interior point of E if there exists a number $\delta > 0$ such that the δ-neighborhood $U_\delta(P_0)$ lies entirely in E. A point P_0 is a boundary point of E if every δ-neighborhood $U_\delta(P_0)$ contains points that lie outside of E as well as points that lie in E (Fig. 8.1). An interior point is necessarily a point of E. A boundary point of E needs not belong to E.

The interior points of a set make up the interior of the set. The set's boundary points make up its boundary. A set is open if it consists entirely of interior points. A set is closed if it contains all of its boundary points. A set is connected if every point can be connected to every other point by a smooth curve that lies entirely in the set. A connected open set is called an open region. The union of an open region and its boundary is called a closed region.

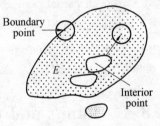

Fig. 8.1

A region in $\mathbf{R}^n$ is bounded if it lies inside a neighborhood of fixed radius. A region is unbounded if it is not bounded. For example, the set $\{(x,y,z) \mid x^2 + y^2 + z^2 \leqslant 1\}$ is a bounded region in $\mathbf{R}^3$, but the set $\{(x,y) \mid x + y > 0\}$ is an unbounded region in $\mathbf{R}^2$.

8.1.2 Functions of Several Variables

Many functions depend on more than one independent variable. The volume V of a right circular cylinder depends on its radius r and its height h. In fact, we know that $V = \pi r^2 h$. The temperature T at a point on the surface of the earth depends on the longitude x and latitude y of the point and on the time t, so we could write $T = f(x,y,t)$.

Definition 8.1 (Functions of Two Variables) A function f of two variables is a rule that assigns to each ordered pair of real numbers (x,y) in a set D a unique real number z, denoted by

Chapter 8 Differential Calculus of Multivariable Functions

$$z = f(x,y), (x,y) \in D$$

where x and y are called the independent variables and z is called the dependent variable. The set D is the domain of f and its range is the set of the values that f takes on, that is, $\{f(x,y) \mid (x,y) \in D\}$.

In general, a function f of n variables is a rule that assigns to each n-tuple $(x_1, x_2, \cdots, x_n)$ of real numbers a unique real number $u = f(x_1, x_2, \cdots, x_n)$. The variables x_1 to x_n are the independent variables and u is the dependent variable.

As with functions of one variable, a function of several variables may have a domain that is restricted by the context of the problem. For example, if the independent variables correspond to price or length or population, they take only nonnegative values, even though the associated function may be defined for negative values of the variables. If not stated otherwise, the domain D is the set of points for which the function is defined.

Example 1 The domain of the function $z = \ln(x + y)$ is $\{(x,y) \mid x + y > 0\}$, which is the set of the points that lie above the line $y = -x$. The domain is an unbounded open region in $\mathbf{R}^2$ (Fig. 8.2).

Example 2 The domain of the function $z = \dfrac{\sqrt{2x - x^2 - y^2}}{\sqrt{x^2 + y^2 - 1}}$ is

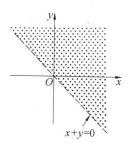

Fig. 8.2

$$\{(x,y) \mid (x-1)^2 + y^2 \leqslant 1,\ x^2 + y^2 > 1\}$$

which is the set of the points on or within the circle $(x-1)^2 + y^2 = 1$ and outside the circle $x^2 + y^2 = 1$. The domain is a bounded set in $\mathbf{R}^2$ (Fig. 8.3).

Example 3 The domain of the function $u = \sqrt{z - x^2 - y^2} + \arcsin(x^2 + y^2 + z^2)$ is

$$\{(x,y,z) \mid x^2 + y^2 \leqslant z,\ x^2 + y^2 + z^2 \leqslant 1\}$$

which is the set of the points that lie on or above the paraboloid $z = x^2 + y^2$ and on or within the sphere $x^2 + y^2 + z^2 = 1$. The domain is a bounded closed region in $\mathbf{R}^3$ (Fig. 8.4).

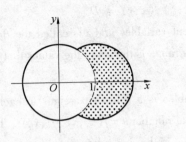 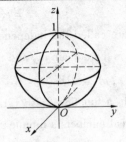

Fig. 8.3　　　　　　　Fig. 8.4

The graph of a function f of two variables is the set of points (x,y,z) that satisfy the equation $z = f(x,y)$ i. e. the surface $z = f(x,y)$. But for functions of three or more independent variables, no geometric picture is available.

8.1.3　Limits and Continuity

This subsection deals with limits and continuity for multivariable functions. There are a number of differences between the calculus of one and two variables. However, the calculus of functions of three or more variables differs only slightly from that of functions of two variables. The study here will be limited largely to functions of two variables.

The following definition of limit for functions of two variables is analogous to the limit definition for functions of one variable.

Definition 8.2　Let f be a function of two variables whose domain D includes points arbitrarily close to $P_0(x_0, y_0)$. Then we say that the limit of $f(x,y)$ as $P(x,y)$ approaches $P_0(x_0, y_0)$ is A and we write

$$\lim_{(x,y)\to(x_0,y_0)} f(x,y) = \lim_{P\to P_0} f(P) = A$$

if for every number $\varepsilon > 0$, there is a corresponding number $\delta > 0$ such that

$$|f(x,y) - A| < \varepsilon$$

whenever $(x,y) \in D$ and $0 < \rho(P, P_0) = \sqrt{(x - x_0)^2 + (y - y_0)^2} < \delta$.

Other notations for the limit in Definition 8.2 are

$$\lim_{\substack{x\to x_0 \\ y\to y_0}} f(x,y) = A \quad \text{and} \quad f(x,y) \to A \text{ as } (x,y) \to (x_0, y_0)$$

Example 4　Prove that $\lim\limits_{(x,y)\to(0,0)} (x^2 + y^2) \sin \dfrac{1}{xy} = 0$.

Solution　Let $\varepsilon > 0$ be given. We want to find $\delta > 0$ such that

Chapter 8 Differential Calculus of Multivariable Functions

$$\left|(x^2+y^2)\sin\frac{1}{xy}-0\right|<\varepsilon \quad \text{whenever} \quad 0<\sqrt{x^2+y^2}<\delta\,(\,xy\neq 0\,)$$

But

$$\left|(x^2+y^2)\sin\frac{1}{xy}-0\right|=(x^2+y^2)\left|\sin\frac{1}{xy}\right|\leqslant x^2+y^2$$

Thus, if we choose $\delta=\sqrt{\varepsilon}$ and let $0<\sqrt{x^2+y^2}<\delta\,(\,xy\neq 0\,)$, then

$$\left|(x^2+y^2)\sin\frac{1}{xy}-0\right|\leqslant x^2+y^2<\delta^2=\varepsilon$$

Thus, by Definition 8.2

$$\lim_{(x,y)\to(0,0)}(x^2+y^2)\sin\frac{1}{xy}=0$$

The condition $\rho(P,P_0)<\delta$ in Definition 8.2 means that the distance between $P(x,y)$ and $P_0(x_0,y_0)$ is less than δ as P approaches P_0 from all possible directions. Therefore, the limit exists only if $f(x,y)$ approaches A as P approaches P_0 along all possible paths in the domain of f. Thus, if we can find two different paths of approach along which the function $f(x,y)$ has different limits, then it follows that $\lim\limits_{(x,y)\to(x_0,y_0)} f(x,y)$ does not exist.

Example 5 Show that $\lim\limits_{(x,y)\to(0,0)}\dfrac{xy^2}{x^2+y^4}$ does not exist.

Solution Let $f(x,y)=\dfrac{xy^2}{x^2+y^4}$. First let $(x,y)\to(0,0)$ along any nonvertical line through the origin. Then $y=kx$, where k is the slope, and

$$f(x,y)=f(x,kx)=\frac{x\,(kx)^2}{x^2+(kx)^4}=\frac{k^2x}{1+k^4x^2}$$

So

$$\lim_{\substack{(x,y)\to(0,0)\\ y=kx}} f(x,y)=\lim_{x\to 0} f(x,kx)=\lim_{x\to 0}\frac{k^2x}{1+k^4x^2}=0$$

We now let $(x,y)\to(0,0)$ along the y-axis. Then $x=0$ and $f(0,y)=0$. So

$$\lim_{\substack{(x,y)\to(0,0)\\ x=0}} f(x,y)=\lim_{y\to 0} f(0,y)=\lim_{y\to 0} 0=0$$

Thus, we have obtained identical limits along every lines through the origin. But that does not show that the given limit is 0. If we now let $(x,y)\to(0,0)$ along the parabola $x=y^2$, then we have

$$f(x,y)=f(y^2,y)=\frac{y^2\cdot y^2}{(y^2)^2+y^4}=\frac{1}{2}$$

So
$$\lim_{\substack{(x,y)\to(0,0)\\ x=y^2}} f(x,y) = \lim_{y\to 0} f(y^2,y) = \lim_{y\to 0} \frac{1}{2} = \frac{1}{2}$$

Since different paths lead to different limiting values, the given limit does not exist.

Just as for functions of one variable, the calculation of limits for functions of two (or more) variables can be greatly simplified by the use of properties of limits. The limit laws for functions of one variable can be extended to functions of two (or more) variables. The limit of a sum is the sum of the limits, the limit of a product is the product of the limits, and so on. The Squeeze Theorem also holds.

Example 6 $\lim\limits_{\substack{x\to 0\\ y\to a}} \dfrac{\sin(xy)}{x} = \lim\limits_{\substack{x\to 0\\ y\to a}} y \cdot \dfrac{\sin(xy)}{xy} = a \cdot 1 = a.$

Example 7 Evaluate $\lim\limits_{\substack{x\to 0\\ y\to 0}} \dfrac{2xy^2}{x^2 + y^2 - y^4}$.

Solution We use polar coordinates to find this limit. Let $x = r\cos\theta, y = r\sin\theta$. Then $(x,y)\to(0,0)$ is equivalent to $r\to 0$ and
$$\frac{2xy^2}{x^2 + y^2 - y^4} = \frac{2r\cos\theta \sin^2\theta}{1 - r^2 \sin^4\theta}$$

Since $\left|\dfrac{2r\cos\theta \sin^2\theta}{1 - r^2 \sin^4\theta}\right| < \dfrac{2r}{1 - r^2}$ ($0 < r < 1$) and $\lim\limits_{r\to 0}\dfrac{2r}{1 - r^2} = 0$, by the Squeeze Theorem we have
$$\lim_{\substack{x\to 0\\ y\to 0}} \frac{2xy^2}{x^2 + y^2 - y^4} = \lim_{r\to 0} \frac{2r\cos\theta \sin^2\theta}{1 - r^2 \sin^4\theta} = 0$$

The definition of continuity for functions of two variables is essentially the same as for functions of one variable.

Definition 8.3 Let f be a function of two variables and assume that f is defined at the point $P_0(x_0,y_0)$ and there are points $P(x,y)$ in the domain of f arbitrarily close to (x_0,y_0). Then f is continuous at (x_0,y_0) if
$$\lim_{(x,y)\to(x_0,y_0)} f(x,y) = f(x_0,y_0)$$

If we write $\Delta z = f(P) - f(P_0)$ and $\rho = \rho(P,P_0)$, then the function $z = f(P)$ being continuous at $P_0(x_0,y_0)$ is equivalent to
$$\lim_{\rho\to 0} \Delta z = 0$$

We say that f is continuous on a set E if f is continuous at each point of E. A func-

Chapter 8 Differential Calculus of Multivariable Functions

tion f is continuous if it is continuous at every point of its domain.

The intuitive meaning of continuity is that if the point (x,y) changes by a small amount, then the value of $f(x,y)$ changes by a small amount. This means that a surface that is the graph of a continuous function has no hole or break.

Just as for functions of one variable, the sums, differences, products, quotients, and compositions of continuous functions of two (or more) variables are continuous on their domains.

A function of several variables built from a finite number of basic elementary functions of each independent variable through combinations and compositions, which may be represented by a single formula, is called an elementary function of several variables. Elementary functions of several variables are continuous on the interior of its domain.

For example, the function $f(x,y) = \dfrac{xy}{1 + x^2 + y^2}$ is continuous on $\mathbf{R}^2$; the function $f(x,y) = \dfrac{xy^2}{x^2 + y^4}$ is continuous except at $(0,0)$; the function $f(x,y) = \sin\dfrac{1}{1 - x^2 - y^2}$ is continuous except on the circle $x^2 + y^2 = 1$.

Example 8 Determine the points at which the following function

$$f(x,y) = \begin{cases} (1 + x^2 + y^2)^{\frac{1}{x^2+y^2}}, & (x,y) \neq (0,0) \\ e, & (x,y) = (0,0) \end{cases}$$

is continuous.

Solution The function f is continuous at any point $(x,y) \neq (0,0)$ since it is equal to an elementary function there. Also, we have

$$\lim_{(x,y)\to(0,0)} f(x,y) = \lim_{(x,y)\to(0,0)} (1 + x^2 + y^2)^{\frac{1}{x^2+y^2}} = e = f(0,0)$$

Therefore, f is continuous at $(0,0)$, and so it is continuous on $\mathbf{R}^2$.

Example 9 Where is the following function

$$f(x,y) = \begin{cases} \dfrac{xy}{x^2 + y^2}, & (x,y) \neq (0,0) \\ 0, & (x,y) = (0,0) \end{cases}$$

continuous?

Solution The function f is continuous for $(x,y) \neq (0,0)$ because it is equal to an elementary function there.

At $(0,0)$, the value of f is defined, but f has no limit as $(x,y) \to (0,0)$. In fact, for every value of k, if we let $(x,y) \to (0,0)$ along the line $y = kx$, then we have

$$\lim_{\substack{(x,y)\to(0,0)\\y=kx}} f(x,y) = \lim_{x\to 0} f(x,kx) = \lim_{x\to 0} \frac{x(kx)}{x^2 + (kx)^2} = \frac{k}{1+k^2}$$

This limit changes with k, so the limit $\lim_{(x,y)\to(0,0)} f(x,y)$ fails to exist, and the function f is not continuous at $(0,0)$.

Therefore, f is continuous on $\mathbf{R}^2$ except $(0,0)$.

Continuous functions of several variables on closed bounded sets have the following important properties:

1. If a function is continuous on a closed bounded set, then it attains an absolute maximum value and an absolute minimum value at some points in that set.

2. If a function is continuous on a closed bounded set, then it must take all values between its absolute minimum and absolute maximum values on that set.

8.2 Partial Derivatives and Higher-Order Partial Derivatives

8.2.1 Partial Derivatives

Derivatives may be defined for functions of several variables with respect to any of the independent variables. The resulting derivatives are called partial derivatives.

If f is a function of two variables x and y, suppose that we let only x vary while keeping y fixed, say $y = y_0$, where y_0 is a constant. If the function of one variable x, namely, $g(x) = f(x,y_0)$, has a derivative at x_0, then by the definition of a derivative, we have

$$g'(x_0) = \lim_{\Delta x \to 0} \frac{g(x_0 + \Delta x) - g(x_0)}{\Delta x} = \lim_{\Delta x \to 0} \frac{f(x_0 + \Delta x, y_0) - f(x_0, y_0)}{\Delta x}$$

We define the partial derivative of f with respect to x at the point (x_0, y_0) as the ordinary derivative of g at x_0.

Definition 8.4 The partial derivative of f with respect to x at the point (x_0, y_0), denoted by $f'_x(x_0, y_0)$, is

Chapter 8 Differential Calculus of Multivariable Functions

$$f'_x(x_0,y_0) = \lim_{\Delta x \to 0} \frac{f(x_0 + \Delta x, y_0) - f(x_0,y_0)}{\Delta x}$$

provided this limit exists.

Similarly, the partial derivative of f with respect to y at the point (x_0, y_0), denoted by $f'_y(x_0, y_0)$, is defined by

$$f'_y(x_0,y_0) = \lim_{\Delta y \to 0} \frac{f(x_0, y_0 + \Delta y) - f(x_0,y_0)}{\Delta y}$$

provided this limit exists.

If we now let the point (x_0, y_0) vary, then f'_x and f'_y become functions of two variables.

Definition 8.5 (Partial Derivative Functions) If f is a function of two variables, its partial derivatives are the functions f'_x and f'_y defined by

$$f'_x(x,y) = \lim_{\Delta x \to 0} \frac{f(x + \Delta x, y) - f(x,y)}{\Delta x}$$

$$f'_y(x,y) = \lim_{\Delta y \to 0} \frac{f(x, y + \Delta y) - f(x,y)}{\Delta y}$$

provided these limits exist.

The partial derivative of f with respect to x at the point (x_0, y_0) is the value of the function $f'_x(x,y)$ at (x_0, y_0), that is

$$f'_x(x_0,y_0) = f'_x(x,y)\big|_{(x_0,y_0)}$$

There are many alternative notations for partial derivatives. If $z = f(x,y)$, we write

$$f'_x(x,y) = z'_x = f'_1(x,y) = \frac{\partial f}{\partial x} = \frac{\partial z}{\partial x} = \frac{\partial}{\partial x}f(x,y)$$

$$f'_y(x,y) = z'_y = f'_2(x,y) = \frac{\partial f}{\partial y} = \frac{\partial z}{\partial y} = \frac{\partial}{\partial y}f(x,y)$$

In general, if u is a function of n variables, $u = f(x_1, x_2, \cdots, x_n)$, its partial derivatives with respect to the ith variable x_i is

$$\frac{\partial u}{\partial x_i} = \lim_{\Delta x_i \to 0} \frac{f(x_1, \cdots, x_{i-1}, x_i + \Delta x_i, x_{i+1}, \cdots, x_n) - f(x_1, \cdots, x_i, \cdots, x_n)}{\Delta x_i}$$

All the rules and results for ordinary derivatives can be used to compute partial derivatives. Specifically, to compute $f'_x(x,y)$, we treat y as a constant and take an ordinary derivative with respect to x. Similarly, to compute $f'_y(x,y)$, we treat x as a constant and take an ordinary derivative with respect to y.

Calculus(II)

Example 1 Find the two partial derivatives of the function $z = x^2y + \sin y$ at the point $(1,0)$.

Solution Holding y constant and differentiating with respect to x, we get

$$\frac{\partial z}{\partial x} = \frac{\partial}{\partial x}(x^2y + \sin y) = 2xy$$

and so

$$\left.\frac{\partial z}{\partial x}\right|_{(1,0)} = 2xy\bigg|_{\substack{x=1\\y=0}} = 0$$

Holding x constant and differentiating with respect to y, we get

$$\frac{\partial z}{\partial y} = \frac{\partial}{\partial y}(x^2y + \sin y) = x^2 + \cos y$$

and so

$$\left.\frac{\partial z}{\partial y}\right|_{(1,0)} = (x^2 + \cos y)\bigg|_{\substack{x=1\\y=0}} = 2$$

Example 2 Find the three partial derivatives of the function $f(x,y,z) = (z - e^{xy})\sin \ln x^2$ at the point $(1,0,2)$.

Solution It is simpler to calculate the partial derivatives as follows

$$f'_x(1,0,2) = \frac{d}{dx}f(x,0,2)\bigg|_{x=1} = \frac{d}{dx}(\sin \ln x^2)\bigg|_{x=1} = \frac{2}{x}\cos \ln x^2\bigg|_{x=1} = 2$$

$$f'_y(1,0,2) = \frac{d}{dy}f(1,y,2)\bigg|_{y=0} = \frac{d}{dy}(0)\bigg|_{y=0} = 0$$

$$f'_z(1,0,2) = \frac{d}{dz}f(1,0,z)\bigg|_{z=2} = \frac{d}{dz}(0)\bigg|_{z=2} = 0$$

Example 3 Find $\dfrac{\partial z}{\partial x}$ and $\dfrac{\partial z}{\partial y}$ if $z = x^y$ ($x > 0$).

Solution The two partial derivatives are

$$\frac{\partial z}{\partial x} = yx^{y-1}, \quad \frac{\partial z}{\partial y} = x^y \ln x$$

Example 4 If resistors of R_1, R_2 and R_3 ohms are connected in parallel to make an R-ohm resistor, the value of R can be found from the equation

$$\frac{1}{R} = \frac{1}{R_1} + \frac{1}{R_2} + \frac{1}{R_3}$$

Find the value of $\dfrac{\partial R}{\partial R_2}$ when $R_1 = 30$, $R_2 = 45$ and $R_3 = 90$ ohms.

Chapter 8 Differential Calculus of Multivariable Functions

Solution We have
$$R = \left(\frac{1}{R_1} + \frac{1}{R_2} + \frac{1}{R_3}\right)^{-1}$$

To find $\dfrac{\partial R}{\partial R_2}$, we treat R_1 and R_3 as constants and differentiate with respect to R_2

$$\frac{\partial R}{\partial R_2} = -\left(\frac{1}{R_1} + \frac{1}{R_2} + \frac{1}{R_3}\right)^{-2}\left(-\frac{1}{R_2^2}\right) = \frac{R^2}{R_2^2}$$

When $R_1 = 30$, $R_2 = 45$ and $R_3 = 90$, $R = 15$ and so
$$\frac{\partial R}{\partial R_2} = \frac{15^2}{45^2} = \frac{1}{9}$$

Example 5 The gas law for a fixed mass m of an ideal gas at absolute temperature T, pressure P, and volume V is $PV = mRT$, where R is the gas constant. Show that
$$\frac{\partial P}{\partial V}\frac{\partial V}{\partial T}\frac{\partial T}{\partial P} = -1$$

Solution Since
$$P = \frac{mRT}{V},\; V = \frac{mRT}{P},\; T = \frac{PV}{mR}$$

we have
$$\frac{\partial P}{\partial V} = -\frac{mRT}{V^2},\; \frac{\partial V}{\partial T} = \frac{mR}{P},\; \frac{\partial T}{\partial P} = \frac{V}{mR}$$

Therefore
$$\frac{\partial P}{\partial V}\frac{\partial V}{\partial T}\frac{\partial T}{\partial P} = \left(-\frac{mRT}{V^2}\right)\left(\frac{mR}{P}\right)\left(\frac{V}{mR}\right) = -\frac{mRT}{PV} = -1$$

Example 5 tells us that, unlike the derivative $\dfrac{dy}{dx}$ of the function $y = f(x)$, the partial derivatives $\dfrac{\partial z}{\partial x}$ and $\dfrac{\partial z}{\partial y}$ of the function $z = f(x,y)$ can not be interpreted as ratios of differentials.

A function $f(x,y)$ can have partial derivatives with respect to both x and y at a point without being continuous there. This is different from functions of a single variable, where the existence of a derivative implies continuity.

Example 6 Let $f(x,y) = \begin{cases} 0, xy \neq 0 \\ 1, xy = 0 \end{cases}$. (a) Prove that f is not continuous at the origin. (b) Show that both partial derivatives of f exist at the origin.

Solution   (a) Since $f(0,0) = 1$, but
$$\lim_{\substack{(x,y) \to (0,0) \\ y=x}} f(x,y) = \lim_{x \to 0} f(x,x) = \lim_{x \to 0} 0 = 0$$
then f is not continuous at $(0,0)$.

(b) By the definition of partial derivatives, we obtain
$$f'_x(0,0) = \lim_{\Delta x \to 0} \frac{f(0 + \Delta x, 0) - f(0,0)}{\Delta x} = \lim_{\Delta x \to 0} \frac{1-1}{\Delta x} = 0$$
$$f'_y(0,0) = \lim_{\Delta y \to 0} \frac{f(0, 0 + \Delta y) - f(0,0)}{\Delta y} = \lim_{\Delta y \to 0} \frac{1-1}{\Delta y} = 0$$
Then, the two partial derivatives of f exist at the origin.

Interpretations of Partial Derivatives

If (x_0, y_0) is a point in the domain of a function f, then the vertical plane $y = y_0$ intersects the surface $z = f(x,y)$ in a curve C_x. This curve is the graph of the function $z = f(x, y_0)$ in the plane $y = y_0$, so the slope of its tangent T_x at $P_0(x_0, y_0, f(x_0, y_0))$ is $f'_x(x_0, y_0)$. Likewise, the vertical plane $x = x_0$ intersects the surface $z = f(x,y)$ in a curve C_y. This curve is the graph of the function $z = f(x_0, y)$ in the plane $x = x_0$, so the slope of its tangent T_y at $P_0(x_0, y_0, f(x_0, y_0))$ is $f'_y(x_0, y_0)$ (Fig. 8.5).

Partial derivatives can also be interpreted as rates of change. If $z = f(x,y)$, then $\dfrac{\partial z}{\partial x}$ represents the rate of change of z with respect to x when y is fixed. Similarly, $\dfrac{\partial z}{\partial y}$ represents the rate of change of z with respect to y when x is fixed.

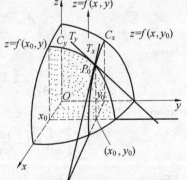

Fig. 8.5

8.2.2 Higher-Order Partial Derivatives

If f is a function of two variables, then its partial derivatives $\dfrac{\partial f}{\partial x}$ and $\dfrac{\partial f}{\partial y}$ are also functions of two variables, so we can consider their partial derivatives $\dfrac{\partial}{\partial x}\left(\dfrac{\partial f}{\partial x}\right)$, $\dfrac{\partial}{\partial y}\left(\dfrac{\partial f}{\partial x}\right)$,

Chapter 8 Differential Calculus of Multivariable Functions

$\dfrac{\partial}{\partial x}\left(\dfrac{\partial f}{\partial y}\right)$, and $\dfrac{\partial}{\partial y}\left(\dfrac{\partial f}{\partial y}\right)$, which are called the second (second-order) partial derivatives of f. If $z = f(x,y)$, then these derivatives are usually denoted by

$$\dfrac{\partial}{\partial x}\left(\dfrac{\partial f}{\partial x}\right) = \dfrac{\partial^2 f}{\partial x^2} = f''_{xx}(x,y) = f''_{11}(x,y) = \dfrac{\partial^2 z}{\partial x^2} = z''_{xx}$$

$$\dfrac{\partial}{\partial y}\left(\dfrac{\partial f}{\partial x}\right) = \dfrac{\partial^2 f}{\partial x \partial y} = f''_{xy}(x,y) = f''_{12}(x,y) = \dfrac{\partial^2 z}{\partial x \partial y} = z''_{xy}$$

$$\dfrac{\partial}{\partial x}\left(\dfrac{\partial f}{\partial y}\right) = \dfrac{\partial^2 f}{\partial y \partial x} = f''_{yx}(x,y) = f''_{21}(x,y) = \dfrac{\partial^2 z}{\partial y \partial x} = z''_{yx}$$

$$\dfrac{\partial}{\partial y}\left(\dfrac{\partial f}{\partial y}\right) = \dfrac{\partial^2 f}{\partial y^2} = f''_{yy}(x,y) = f''_{22}(x,y) = \dfrac{\partial^2 z}{\partial y^2} = z''_{yy}$$

Notice the order in which the derivatives are taken: the notation $\dfrac{\partial^2 f}{\partial x \partial y}$ means that we first differentiate with respect to x and then with respect to y, whereas in computing $\dfrac{\partial^2 f}{\partial y \partial x}$ the order is reversed.

Example 7 Find the second partial derivatives of the function $z = \ln(x^2 + y)$.

Solution First, we compute

$$\dfrac{\partial z}{\partial x} = \dfrac{\partial}{\partial x}(\ln(x^2 + y)) = \dfrac{2x}{x^2 + y}$$

$$\dfrac{\partial z}{\partial y} = \dfrac{\partial}{\partial y}(\ln(x^2 + y)) = \dfrac{1}{x^2 + y}$$

For the second partial derivatives, we have

$$\dfrac{\partial^2 z}{\partial x^2} = \dfrac{\partial}{\partial x}\left(\dfrac{2x}{x^2 + y}\right) = \dfrac{2(x^2 + y) - (2x)(2x)}{(x^2 + y)^2} = \dfrac{2(y - x^2)}{(x^2 + y)^2}$$

$$\dfrac{\partial^2 z}{\partial x \partial y} = \dfrac{\partial}{\partial y}\left(\dfrac{2x}{x^2 + y}\right) = (2x)\dfrac{-1}{(x^2 + y)^2} = \dfrac{-2x}{(x^2 + y)^2}$$

$$\dfrac{\partial^2 z}{\partial y \partial x} = \dfrac{\partial}{\partial x}\left(\dfrac{1}{x^2 + y}\right) = \dfrac{-1}{(x^2 + y)^2}(2x) = \dfrac{-2x}{(x^2 + y)^2}$$

$$\dfrac{\partial^2 z}{\partial y^2} = \dfrac{\partial}{\partial y}\left(\dfrac{1}{x^2 + y}\right) = \dfrac{-1}{(x^2 + y)^2}$$

Notice that the two mixed partial derivatives in Example 7 are equal, that is, $\dfrac{\partial^2 z}{\partial x \partial y} = \dfrac{\partial^2 z}{\partial y \partial x}$. This is not just a coincidence. They must be equal whenever $\dfrac{\partial^2 z}{\partial x \partial y}$ and $\dfrac{\partial^2 z}{\partial x \partial y}$

$\dfrac{\partial^2 z}{\partial y \partial x}$ are continuous, as stated in the following theorem.

Theorem 8.1 (The Mixed Derivative Theorem) Assume that f is defined on a open region D of $\mathbf{R}^2$, and $f''_{xy}(x,y)$ and $f''_{yx}(x,y)$ are continuous throughout D. Then $f''_{xy}(x,y) = f''_{yx}(x,y)$ at all points of D.

You can find a proof of Theorem 8.1 in advanced textbooks.

Partial derivatives of order three or higher can also be defined. As with second derivatives, the order of differentiation is immaterial as long as the derivatives through the order in question are continuous. For example

$$\frac{\partial}{\partial y}\left(\frac{\partial^2 f}{\partial x \partial y}\right) = \frac{\partial^3 f}{\partial x \partial y^2} = f'''_{xyy}(x,y) = f'''_{122}(x,y) = \frac{\partial^3 z}{\partial x \partial y^2} = z'''_{xyy}$$

and $f'''_{xyy}(x,y) = f'''_{yxy}(x,y) = f'''_{yyx}(x,y)$ if these functions are continuous.

Indeed there are functions whose mixed partial derivatives are not equal.

Example 8 Let

$$f(x,y) = \begin{cases} xy\dfrac{x^2 - y^2}{x^2 + y^2}, & (x,y) \neq (0,0) \\ 0, & (x,y) = (0,0) \end{cases}$$

Show that both $f''_{xy}(0,0)$ and $f''_{yx}(0,0)$ exist and $f''_{xy}(0,0) \neq f''_{yx}(0,0)$.

Solution When $(x,y) \neq (0,0)$, we have

$$f'_x(x,y) = \frac{\partial}{\partial x}\left(xy\frac{x^2 - y^2}{x^2 + y^2}\right) = \frac{y(x^4 + 4x^2 y^2 - y^4)}{(x^2 + y^2)^2}$$

$$f'_y(x,y) = \frac{\partial}{\partial y}\left(xy\frac{x^2 - y^2}{x^2 + y^2}\right) = \frac{x(x^4 - 4x^2 y^2 - y^4)}{(x^2 + y^2)^2}$$

and when $(x,y) = (0,0)$, we have

$$f'_x(0,0) = \lim_{\Delta x \to 0} \frac{f(\Delta x, 0) - f(0,0)}{\Delta x} = \lim_{\Delta x \to 0} \frac{0 - 0}{\Delta x} = 0$$

$$f'_y(0,0) = \lim_{\Delta y \to 0} \frac{f(0, \Delta y) - f(0,0)}{\Delta y} = \lim_{\Delta y \to 0} \frac{0 - 0}{\Delta y} = 0$$

Thus, we obtain

$$f'_x(x,y) = \begin{cases} \dfrac{y(x^4 + 4x^2 y^2 - y^4)}{(x^2 + y^2)^2}, & (x,y) \neq (0,0) \\ 0, & (x,y) = (0,0) \end{cases}$$

Chapter 8 Differential Calculus of Multivariable Functions

$$f'_y(x,y) = \begin{cases} \dfrac{x(x^4 - 4x^2y^2 - y^4)}{(x^2+y^2)^2}, & (x,y) \neq (0,0) \\ 0, & (x,y) = (0,0) \end{cases}$$

By the definition of partial derivatives, we get

$$f''_{xy}(0,0) = \lim_{\Delta y \to 0} \frac{f'_x(0,\Delta y) - f'_x(0,0)}{\Delta y} = \lim_{\Delta y \to 0} \frac{-\Delta y}{\Delta y} = -1$$

$$f''_{yx}(0,0) = \lim_{\Delta x \to 0} \frac{f'_y(\Delta x,0) - f'_y(0,0)}{\Delta x} = \lim_{\Delta x \to 0} \frac{\Delta x}{\Delta x} = 1$$

This shows that $f''_{xy}(0,0)$ and $f''_{yx}(0,0)$ exist and $f''_{xy}(0,0) \neq f''_{yx}(0,0)$.

8.3 Linear Approximations and Total Differentials

8.3.1 Linear Approximations

Recall that for a function of one variable, $y = f(x)$, if x changes from x_0 to $x_0 + \Delta x$, we defined the increment of y as

$$\Delta y = f(x_0 + \Delta x) - f(x_0)$$

In Chapter 3 we showed that if f is differentiable at x_0, then

$$\Delta y = f'(x_0)\Delta x + o(\Delta x)$$

Now consider a function of two variables, $z = f(x,y)$, and suppose that x changes from x_0 to $x_0 + \Delta x$ and y changes from y_0 to $y_0 + \Delta y$. Then the corresponding increment of z is

$$\Delta z = f(x_0 + \Delta x, y_0 + \Delta y) - f(x_0, y_0)$$

Thus, the increment Δz represents the change in the value of f when (x,y) changes from (x_0,y_0) to $(x_0+\Delta x, y_0+\Delta y)$. By analogy with the one variable case, we define the differentiability of a function of two variables as follows.

Definition 8.6 (Differentiability) A function $z = f(x,y)$ is differentiable at (x_0, y_0) if both $f'_x(x_0,y_0)$ and $f'_y(x_0,y_0)$ exist and the increment Δz can be expressed in the form

$$\Delta z = f'_x(x_0,y_0)\Delta x + f'_y(x_0,y_0)\Delta y + o(\rho)$$

where $\rho = \sqrt{(\Delta x)^2 + (\Delta y)^2}$.

A function is differentiable on an open set E if it is differentiable at every point of

E.

It's sometimes difficult to use Definition 8.6 directly to verify the differentiability of a function, but the following theorem provides a convenient sufficient condition for differentiability.

Theorem 8.2 (Continuity of Partial Derivatives Implies Differentiablity) If the partial derivatives f'_x and f'_y exist near (x_0, y_0) and are continuous at (x_0, y_0), then f is differentiable at (x_0, y_0).

Proof Suppose that $f'_x(x,y)$ and $f'_y(x,y)$ exist on a neighborhood $U(P_0)$ of the point $P_0(x_0, y_0)$ and let $(x_0 + \Delta x, y_0 + \Delta y)$ be any point in $U(P_0)$. We assume that Δx and Δy are small enough so that the line segment joining P_0 to the point $(x_0 + \Delta x, y_0)$ and the line segment joining the point $(x_0 + \Delta x, y_0)$ to $(x_0 + \Delta x, y_0 + \Delta y)$ lie in $U(P_0)$.

If $z = f(x,y)$, we may think of Δz as the sum of two increments
$$\Delta z = f(x_0 + \Delta x, y_0 + \Delta y) - f(x_0, y_0)$$
$$= [f(x_0 + \Delta x, y_0) - f(x_0, y_0)] + [f(x_0 + \Delta x, y_0 + \Delta y) - f(x_0 + \Delta x, y_0)]$$
where the change $f(x_0 + \Delta x, y_0) - f(x_0, y_0)$ is caused by changing x from x_0 to $x_0 + \Delta x$ while holding y equal to y_0 and the change $f(x_0 + \Delta x, y_0 + \Delta y) - f(x_0 + \Delta x, y_0)$ is caused by changing y from y_0 to $y_0 + \Delta y$ while holding x equal to $x_0 + \Delta x$.

On the closed interval of x-values joining x_0 to $x_0 + \Delta x$, the function $g(x) = f(x, y_0)$ is a differentiable (and hence continuous) function of x, with derivative $g'(x) = f'_x(x, y_0)$. By the Mean Value Theorem, there is an x-value ξ between x_0 and $x_0 + \Delta x$ at which
$$g(x_0 + \Delta x) - g(x_0) = g'(\xi) \Delta x$$
or
$$f(x_0 + \Delta x, y_0) - f(x_0, y_0) = f'_x(\xi, y_0) \Delta x$$

Similarly, $h(y) = f(x_0 + \Delta x, y)$ is a differentiable (and hence continuous) function of y on the closed y-interval joining y_0 and $y_0 + \Delta y$, with derivative $h'(y) = f'_y(x_0 + \Delta x, y)$. Therefore, there is a y-value η between y_0 and $y_0 + \Delta y$ at which
$$h(y_0 + \Delta y) - h(y_0) = h'(\eta) \Delta y$$
or
$$f(x_0 + \Delta x, y_0 + \Delta y) - f(x_0 + \Delta x, y_0) = f'_y(x_0 + \Delta x, \eta) \Delta y$$

Now, as Δx and $\Delta y \to 0$, we know that $\xi \to x_0$ and $\eta \to y_0$. Therefore, since f'_x and

Chapter 8 Differential Calculus of Multivariable Functions

f'_y are continuous at (x_0, y_0), we have
$$f'_x(\xi, y_0) = f'_x(x_0, y_0) + \alpha, \quad f'_y(x_0 + \Delta x, \eta) = f'_y(x_0, y_0) + \beta$$
where both α and β approach zero as Δx and $\Delta y \to 0$. Thus
$$\Delta z = f'_x(\xi, y_0)\Delta x + f'_y(x_0 + \Delta x, \eta)\Delta y$$
$$= [f'_x(x_0, y_0) + \alpha]\Delta x + [f'_y(x_0, y_0) + \beta]\Delta y$$
$$= f'_x(x_0, y_0)\Delta x + f'_y(x_0, y_0)\Delta y + \alpha\Delta x + \beta\Delta y$$
Since
$$\frac{|\alpha\Delta x + \beta\Delta y|}{\rho} \leq |\alpha|\frac{|\Delta x|}{\rho} + |\beta|\frac{|\Delta y|}{\rho} \leq |\alpha| + |\beta| \to 0$$
as Δx and $\Delta y \to 0$, where $\rho = \sqrt{(\Delta x)^2 + (\Delta y)^2}$, then $\alpha\Delta x + \beta\Delta y = o(\rho)$. Thus
$$\Delta z = f'_x(x_0, y_0)\Delta x + f'_y(x_0, y_0)\Delta y + o(\rho)$$
This shows that f is differentiable at (x_0, y_0).

But, when f is differentiable at the point (x_0, y_0), the partial derivatives f'_x and f'_y are not necessarily continuous at (x_0, y_0).

Example 1 Let
$$f(x,y) = \begin{cases} (x^2 + y^2)\sin\dfrac{1}{\sqrt{x^2 + y^2}}, & (x,y) \neq (0,0) \\ 0, & (x,y) = (0,0) \end{cases}$$
Show that f is differentiable at the origin, but its partial derivatives f'_x and f'_y are not continuous at the origin.

Solution By the definition of partial derivatives, we get
$$f'_x(0,0) = \lim_{\Delta x \to 0}\frac{f(\Delta x, 0) - f(0,0)}{\Delta x} = \lim_{\Delta x \to 0}\frac{(\Delta x)^2 \sin\frac{1}{|\Delta x|} - 0}{\Delta x} = \lim_{\Delta x \to 0}\Delta x \sin\frac{1}{|\Delta x|} = 0$$

$$f'_y(0,0) = \lim_{\Delta y \to 0}\frac{f(0,\Delta y) - f(0,0)}{\Delta y} = \lim_{\Delta y \to 0}\frac{(\Delta y)^2 \sin\frac{1}{|\Delta y|} - 0}{\Delta y} = \lim_{\Delta y \to 0}\Delta y \sin\frac{1}{|\Delta y|} = 0$$

Let $z = f(x,y)$. Then
$$\Delta z = f(\Delta x, \Delta y) - f(0,0) = [(\Delta x)^2 + (\Delta y)^2]\sin\frac{1}{\sqrt{(\Delta x)^2 + (\Delta y)^2}} = o(\rho)$$
where $\rho = \sqrt{(\Delta x)^2 + (\Delta y)^2}$. Hence
$$\Delta z = 0 \cdot \Delta x + 0 \cdot \Delta y + o(\rho) = f'_x(0,0)\Delta x + f'_y(0,0)\Delta y + o(\rho)$$

This shows that f is differentiable at $(0,0)$.

When $(x,y) \neq (0,0)$, we have

$$f'_x(x,y) = \frac{\partial}{\partial x}\left[(x^2+y^2)\sin\frac{1}{\sqrt{x^2+y^2}}\right] = 2x\sin\frac{1}{\sqrt{x^2+y^2}} - \frac{x}{\sqrt{x^2+y^2}}\sin\frac{1}{\sqrt{x^2+y^2}}$$

Since

$$\lim_{\substack{(x,y)\to(0,0)\\y=0}} f'_x(x,y) = \lim_{x\to 0} f'_x(x,0) = \lim_{x\to 0}\left(2x\sin\frac{1}{|x|} - \frac{x}{|x|}\cos\frac{1}{|x|}\right)$$

does not exist, $\lim_{(x,y)\to(0,0)} f'_x(x,y)$ does not exist, and so f'_x is not continuous at $(0,0)$.

Similarly, we can show that f'_y is not continuous at $(0,0)$.

The following theorem tells us that a function of two variables is continuous at every point where it is differentiable.

Theorem 8.3 (**Differentiablity Implies Continuity**) If a function f is differentiable at (x_0, y_0), then it is continuous at (x_0, y_0).

Proof If $z = f(x,y)$, then by the definition of differentiability, we have

$$\Delta z = f'_x(x_0,y_0)\Delta x + f'_y(x_0,y_0)\Delta y + o(\rho)$$

As $\rho \to 0$ (and hence $(\Delta x, \Delta y) \to (0,0)$), we have

$$\lim_{\rho \to 0} \Delta z = \lim_{(\Delta x, \Delta y)\to(0,0)} [f'_x(x_0,y_0)\Delta x + f'_y(x_0,y_0)\Delta y + o(\rho)] = 0$$

which implies continuity of f at (x_0, y_0).

Suppose that a surface S has equation $z = f(x,y)$, where f is differentiable. Then the tangent plane to the surface S at the point $P_0(x_0, y_0, f(x_0,y_0))$ is defined to be the plane that most closely approximates the surface S near the point P_0. By Definition 8.6, we have

$$f(x,y) = f(x_0,y_0) + f'_x(x_0,y_0)(x-x_0) + f'_y(x_0,y_0)(y-y_0) + $$
$$o(\sqrt{(x-x_0)^2 + (y-y_0)^2})$$

Then an equation of the tangent plane to the surface S at P_0 is

$$z = f(x_0,y_0) + f'_x(x_0,y_0)(x-x_0) + f'_y(x_0,y_0)(y-y_0)$$

The linear function whose graph is this tangent plane, that is

$$L(x,y) = f(x_0,y_0) + f'_x(x_0,y_0)(x-x_0) + f'_y(x_0,y_0)(y-y_0)$$

is called the linearization of f at the point (x_0, y_0) and the approximation

$$f(x,y) \approx f(x_0,y_0) + f'_x(x_0,y_0)(x-x_0) + f'_y(x_0,y_0)(y-y_0)$$

is called the linear approximation of f at (x_0, y_0).

Chapter 8 Differential Calculus of Multivariable Functions

Example 2 Show that the function $f(x,y) = x^y$ is differentiable at the point $(1,2)$ and find its linearization there. Then use it to approximate $(1.01)^{1.98}$.

Solution The partial derivatives are

$$f'_x(x,y) = yx^{y-1}, \quad f'_y(x,y) = x^y \ln x$$
$$f'_x(1,2) = 2, \quad f'_y(1,2) = 0$$

Both f'_x and f'_y are continuous near $(1,2)$, so f is differentiable at $(1,2)$ by Theorem 8.2. The linearization is

$$L(x,y) = f(1,2) + f'_x(1,2)(x-1) + f'_y(1,2)(y-2)$$
$$= 1 + 2(x-1) + 0 \cdot (y-2) = 2x - 1$$

The corresponding linear approximation is

$$x^y \approx 2x - 1$$

So

$$(1.01)^{1.98} \approx 2 \times 1.01 - 1 = 1.02$$

Differentiability and linear approximations can be defined in a similar manner for functions of more than two variables. For example, if $u = f(x,y,z)$, then f is differentiable at the point (x_0, y_0, z_0) if Δu can be expressed in the form

$$\Delta u = f'_x(x_0,y_0,z_0)\Delta x + f'_y(x_0,y_0,z_0)\Delta y + f'_z(x_0,y_0,z_0)\Delta z + o(\rho)$$

where $\rho = \sqrt{(\Delta x)^2 + (\Delta y)^2 + (\Delta z)^2}$, and the linearization of f at (x_0, y_0, z_0) is

$$L(x,y,z) = f'_x(x_0,y_0,z_0)(x-x_0) + f'_y(x_0,y_0,z_0)(y-y_0) + f'_z(x_0,y_0,z_0)(z-z_0)$$

8.3.2 Total Differentials

For a function of one variable, $y = f(x)$, the differential dx is defined to be an independent variable, and the differential of y is then defined as

$$dy = f'(x) dx$$

For a differentiable function of two variables, $z = f(x,y)$, we define the differentials dx and dy to be independent variables. By analogy with the one variable case we can define the differential of z.

Definition 8.7 Let $z = f(x,y)$ be a differentiable function. The differentials dx and dy are independent variables. The differential dz, also called the total differential, is defined in terms of dx and dy by the equation

$$dz = f'_x(x,y) dx + f'_y(x,y) dy$$

Sometimes the notation df is used in place of dz.

If we take $dx = \Delta x = x - x_0$ and $dy = \Delta y = y - y_0$, then the differential of z is
$$dz = f'_x(x_0, y_0)(x - x_0) + f'_y(x_0, y_0)(y - y_0)$$
Therefore, in the notation of differentials, the linear approximation of f at (x_0, y_0) can be written as
$$f(x,y) \approx f(x_0, y_0) + dz$$

Fig. 8.6 shows the geometric interpretation of the differential dz and the increment Δz: dz represents the change in height of the tangent plane, whereas Δz represents the change in height of the surface $z = f(x,y)$ when (x,y) changes from (x_0, y_0) to $(x_0 + \Delta x, y_0 + \Delta y)$.

Differentials can be defined in a similar manner for functions of more than two variables. For example, if $u = f(x,y,z)$ is a differentiable function, then the differential of u is defined by
$$du = f'_x(x,y,z)dx + f'_y(x,y,z)dy + f'_z(x,y,z)dz$$

Example 3 If $z = x^4 y^3 + 2x$, find the differential dz at the point $(1,2)$.

Solution Since

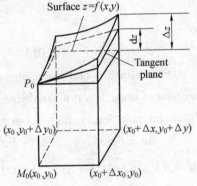

Fig. 8.6

$$\frac{\partial z}{\partial x} = 4x^3 y^3 + 2, \frac{\partial z}{\partial y} = 3x^4 y^2$$

are continuous on $\mathbf{R}^2$, the differential dz exists on $\mathbf{R}^2$ and

$$dz = \frac{\partial z}{\partial x}dx + \frac{\partial z}{\partial y}dy = (4x^3 y^3 + 2)dx + 3x^4 y^2 dy$$

Then the differential dz at $(1,2)$ is
$$dz\big|_{(1,2)} = 34dx + 12dy$$

Example 4 Find the differential du of the function $u = x + \sin\dfrac{y}{2} + e^{yz}$.

Solution Since

$$\frac{\partial u}{\partial x} = 1, \frac{\partial u}{\partial y} = \frac{1}{2}\cos\frac{y}{2} + ze^{yz}, \frac{\partial u}{\partial z} = ye^{yz}$$

are continuous on $\mathbf{R}^3$, the differential du exists on $\mathbf{R}^3$ and

Chapter 8 Differential Calculus of Multivariable Functions

$$du = \frac{\partial u}{\partial x}dx + \frac{\partial u}{\partial y}dy + \frac{\partial u}{\partial z}dz = dx + \left(\frac{1}{2}\cos\frac{y}{2} + ze^{yz}\right)dy + ye^{yz}dz$$

Example 5 The base radius and height of a right circular cone are measured as 10 cm and 25 cm, respectively, with a possible error in measurement of as much as 0.1 cm in each. Use differentials to estimate the maximum error in the calculated volume of the cone.

Solution If the base radius and height of the cone is r and h, then its volume is $V = \frac{\pi r^2 h}{3}$ and so

$$dV = \frac{\partial V}{\partial r}dr + \frac{\partial V}{\partial h}dh = \frac{2\pi rh}{3}dr + \frac{\pi r^2}{3}dh$$

Since each error is at most 0.1 cm, we have $|\Delta r| \leqslant 0.1$ and $|\Delta h| \leqslant 0.1$. To find the largest error in the volume, we therefore take $dr = 0.1$ and $dh = 0.1$ together with $r = 10$ and $h = 25$. This gives

$$\Delta V \approx dV = \frac{500\pi}{3}(0.1) + \frac{100\pi}{3}(0.1) = 20\pi$$

Thus, the maximum error in the calculated volume is about 20π cm^3 ≈ 63 cm^3. This may seem like a large error, but it is only about 2.4% of the volume of the cone.

Algebra Rules for Total Differentials

Algebra rules for differentials for functions of a single variable can be extended to functions of several variables. If u and v are differentiable functions of several variables, then

$$d(u + v) = du + dv$$
$$d(u - v) = du - dv$$
$$d(uv) = vdu + udv$$
$$d\left(\frac{u}{v}\right) = \frac{vdu - udv}{v^2} \quad (v \neq 0)$$

8.4 The Chain Rule

In this section, we combine ideas based on the basic Chain Rule with what we know about partial derivatives. Recall the basic Chain Rule: If y is a function of u and u

is a function of x, then $\dfrac{dy}{dx} = \dfrac{dy}{du}\dfrac{du}{dx}$. We first extend the Chain Rule to composite functions of the form $z = f(u,v)$, where u and v are functions of x and y.

Theorem 8.4 (The Chain Rule for Two Independent Variables and Two Intermediate Variables) Suppose that $z = f(u,v)$ is a differentiable function of u and v, where $u = u(x,y)$ and $v = v(x,y)$ both have partial derivatives with respect to x and y. Then z has partial derivatives with respect to x and y, given by the formulas

$$\frac{\partial z}{\partial x} = \frac{\partial z}{\partial u}\frac{\partial u}{\partial x} + \frac{\partial z}{\partial v}\frac{\partial v}{\partial x}$$

$$\frac{\partial z}{\partial y} = \frac{\partial z}{\partial u}\frac{\partial u}{\partial y} + \frac{\partial z}{\partial v}\frac{\partial v}{\partial y}$$

Proof Regard y as a constant and let x change from x to $x + \Delta x$. Then the change of Δx in x produces changes of Δu in u and Δv in v. These, in turn, produce a change of Δz in z. Since f is differentiable, we have

$$\Delta z = \frac{\partial z}{\partial u}\Delta u + \frac{\partial z}{\partial v}\Delta v + o(\rho)$$

where $\rho = \sqrt{(\Delta u)^2 + (\Delta v)^2}$. Dividing both sides of this equation by Δx, we get

$$\frac{\Delta z}{\Delta x} = \frac{\partial z}{\partial u}\frac{\Delta u}{\Delta x} + \frac{\partial z}{\partial v}\frac{\Delta v}{\Delta x} + \frac{o(\rho)}{\Delta x}$$

If we now let $\Delta x \to 0$, then $\Delta u \to 0$ and $\Delta v \to 0$ because both u and v have partial derivatives with respect to x and therefore $\rho \to 0$, so

$$\frac{\partial z}{\partial x} = \lim_{\Delta x \to 0}\frac{\Delta z}{\Delta x} = \frac{\partial z}{\partial u}\lim_{\Delta x \to 0}\frac{\Delta u}{\Delta x} + \frac{\partial z}{\partial v}\lim_{\Delta x \to 0}\frac{\Delta v}{\Delta x} + \lim_{\Delta x \to 0}\frac{o(\rho)}{\Delta x} = \frac{\partial z}{\partial u}\frac{\partial u}{\partial x} + \frac{\partial z}{\partial v}\frac{\partial v}{\partial x}$$

noticing that

$$\lim_{\Delta x \to 0}\frac{o(\rho)}{\Delta x} = \pm \lim_{\Delta x \to 0}\frac{o(\rho)}{\rho}\sqrt{\left(\frac{\Delta u}{\Delta x}\right)^2 + \left(\frac{\Delta v}{\Delta x}\right)^2} = 0 \cdot \sqrt{\left(\frac{\partial u}{\partial x}\right)^2 + \left(\frac{\partial v}{\partial x}\right)^2} = 0$$

Thus, the formula for $\dfrac{\partial z}{\partial x}$ is proved.

Similarly, we can prove the formula for $\dfrac{\partial z}{\partial y}$.

Now we consider the general situation in which a dependent variable is a function of n intermediate variables, each of which is, in turn, a function of m independent variables. Suppose that z is a differentiable function of n variables $u_1, u_2, \cdots, u_n$, and each u_j

Chapter 8 Differential Calculus of Multivariable Functions

is a function of m variables $x_1, x_2, \cdots, x_m$ which has partial derivative with respect to each x_i. Then z is a function of $x_1, x_2, \cdots, x_m$ and

$$\frac{\partial z}{\partial x_i} = \frac{\partial z}{\partial u_1}\frac{\partial u_1}{\partial x_i} + \frac{\partial z}{\partial u_2}\frac{\partial u_2}{\partial x_i} + \cdots + \frac{\partial z}{\partial u_n}\frac{\partial u_n}{\partial x_i}$$

for each $i = 1, 2, \cdots, m$. The proof is similar to that of Theorem 8.4.

When $m = 1$ in the general situation, each u_j is a function of one variable x_1 and therefore z is a function of the single variable x_1 and

$$\frac{dz}{dx_1} = \frac{\partial z}{\partial u_1}\frac{du_1}{dx_1} + \frac{\partial z}{\partial u_2}\frac{du_2}{dx_1} + \cdots + \frac{\partial z}{\partial u_n}\frac{du_n}{dx_1}$$

Example 1 If $z = e^u \sin v$, where $u = xy$ and $v = x + y$, find $\dfrac{\partial z}{\partial x}$ and $\dfrac{\partial z}{\partial y}$.

Solution Applying the Chain Rule, we get

$$\frac{\partial z}{\partial x} = \frac{\partial z}{\partial u}\frac{\partial u}{\partial x} + \frac{\partial z}{\partial v}\frac{\partial v}{\partial x} = (e^u \sin v)(y) + (e^u \cos v)(1) = e^{xy}[y\sin(x+y) + \cos(x+y)]$$

$$\frac{\partial z}{\partial y} = \frac{\partial z}{\partial u}\frac{\partial u}{\partial y} + \frac{\partial z}{\partial v}\frac{\partial v}{\partial y} = (e^u \sin v)(x) + (e^u \cos v)(1) = e^{xy}[x\sin(x+y) + \cos(x+y)]$$

Example 2 If $y = (\cos x)^{\sin x}$ ($\cos x > 0$), find $\dfrac{dy}{dx}$.

Solution The derivative of the function y may be found by Logarithmic Differentiation, but it is simpler to find the derivative by the Chain Rule. Let

$$u = \cos x, v = \sin x$$

Then

$$y = u^v$$

By the Chain Rule, we obtain

$$\frac{dy}{dx} = \frac{\partial y}{\partial u}\frac{du}{dx} + \frac{\partial y}{\partial v}\frac{dv}{dx} = (vu^{v-1})(-\sin x) + (u^v \ln u)(\cos x)$$

$$= (\cos x)^{1+\sin x}(\ln \cos x - \tan^2 x)$$

Example 3 If $z = \dfrac{y}{f(x^2 - y^2)}$ and f is differentiable, show that z satisfies the equation

$$\frac{1}{x}\frac{\partial z}{\partial x} + \frac{1}{y}\frac{\partial z}{\partial y} = \frac{z}{y^2}$$

Solution Let $t = x^2 - y^2$, then $z = g(y, t) = \dfrac{y}{f(t)}$. By the Chain Rule, we get

Calculus(Ⅱ)

$$\frac{\partial z}{\partial x} = \frac{\partial g}{\partial t}\frac{\partial t}{\partial x} = \left(y\frac{-1}{f^2(t)}f'(t)\right)(2x) = -\frac{2xyf'(t)}{f^2(t)}$$

$$\frac{\partial z}{\partial y} = \frac{\partial g}{\partial y}\frac{\partial y}{\partial y} + \frac{\partial g}{\partial t}\frac{\partial t}{\partial y} = \left(\frac{1}{f(t)}\right)(1) + \left(y\frac{-1}{f^2(t)}f'(t)\right)(-2y) = \frac{1}{f(t)} + \frac{2y^2 f'(t)}{f^2(t)}$$

Therefore

$$\frac{1}{x}\frac{\partial z}{\partial x} + \frac{1}{y}\frac{\partial z}{\partial y} = \frac{1}{x}\left(-\frac{2xyf'(t)}{f^2(t)}\right) + \frac{1}{y}\left(\frac{1}{f(t)} + \frac{2y^2 f'(t)}{f^2(t)}\right)$$

$$= -\frac{2yf'(t)}{f^2(t)} + \frac{1}{yf(t)} + \frac{2yf'(t)}{f^2(t)} = \frac{1}{yf(t)} = \frac{z}{y^2}$$

Example 4 If $u = f(x, xy, xyz)$, where f is differentiable, find $\dfrac{\partial u}{\partial x}$ and $\dfrac{\partial u}{\partial z}$.

Solution Let $s = xy$ and $t = xyz$, then $u = f(x, s, t)$. Using the Chain Rule, we get

$$\frac{\partial u}{\partial x} = \frac{\partial f}{\partial x}\frac{\partial x}{\partial x} + \frac{\partial f}{\partial s}\frac{\partial s}{\partial x} + \frac{\partial f}{\partial t}\frac{\partial t}{\partial x} = \frac{\partial f}{\partial x} + \frac{\partial f}{\partial s}\cdot y + \frac{\partial f}{\partial t}\cdot yz$$

$$\frac{\partial u}{\partial z} = \frac{\partial f}{\partial t}\frac{\partial t}{\partial z} = \frac{\partial f}{\partial t}\cdot xy$$

Notice that we can't write $\dfrac{\partial u}{\partial x}$ in place of $\dfrac{\partial f}{\partial x}$ since this will lead to confusion.

Example 5 If $z = F(x, y)$ has continuous second partial derivatives and $y = \varphi(x)$ has second derivative, find $\dfrac{d^2 z}{dx^2}$.

Solution By the Chain Rule, we have

$$\frac{dz}{dx} = \frac{\partial F}{\partial x}\frac{dx}{dx} + \frac{\partial F}{\partial y}\frac{dy}{dx} = \frac{\partial F}{\partial x} + \frac{\partial F}{\partial y}\varphi'(x)$$

where $\dfrac{\partial F}{\partial x}$ and $\dfrac{\partial F}{\partial y}$ are also functions of x and y. Again by the Chain Rule and by equality of mixed partial derivatives, we get

$$\frac{d^2 z}{dx^2} = \frac{d}{dx}\left(\frac{\partial F}{\partial x} + \frac{\partial F}{\partial y}\varphi'(x)\right)$$

$$= \left(\frac{\partial^2 F}{\partial x^2}\frac{dx}{dx} + \frac{\partial^2 F}{\partial x \partial y}\frac{dy}{dx}\right) + \left(\frac{\partial^2 F}{\partial y \partial x}\frac{dx}{dx} + \frac{\partial^2 F}{\partial y^2}\frac{dy}{dx}\right)\varphi'(x) + \frac{\partial F}{\partial y}\varphi''(x)$$

$$= \left(\frac{\partial^2 F}{\partial x^2} + \frac{\partial^2 F}{\partial x \partial y}\varphi'(x)\right) + \left(\frac{\partial^2 F}{\partial y \partial x} + \frac{\partial^2 F}{\partial y^2}\varphi'(x)\right)\varphi'(x) + \frac{\partial F}{\partial y}\varphi''(x)$$

Chapter 8 Differential Calculus of Multivariable Functions

$$= \frac{\partial^2 F}{\partial x^2} + 2\frac{\partial^2 F}{\partial x \partial y}\varphi'(x) + \frac{\partial^2 F}{\partial y^2}(\varphi'(x))^2 + \frac{\partial F}{\partial y}\varphi''(x)$$

Example 6 If $z = f(x, \frac{x}{y})$, where f has continuous second partial derivatives, find $\frac{\partial^2 z}{\partial x \partial y}$.

Solution The Chain Rule gives

$$\frac{\partial z}{\partial x} = f'_1 \frac{\partial x}{\partial x} + f'_2 \frac{\partial}{\partial x}\left(\frac{x}{y}\right) = f'_1 + f'_2 \cdot \frac{1}{y}$$

Again by the Chain Rule and by equality of mixed partial derivatives, we get

$$\frac{\partial^2 z}{\partial x \partial y} = \frac{\partial}{\partial y}\left(f'_1 + f'_2 \cdot \frac{1}{y}\right) = \left(f''_{12}\frac{\partial}{\partial y}\left(\frac{x}{y}\right)\right) + \left(f''_{22}\frac{\partial}{\partial y}\left(\frac{x}{y}\right)\right)\frac{1}{y} + f'_2\left(-\frac{1}{y^2}\right)$$

$$= f''_{12}\left(\frac{-x}{y^2}\right) + f''_{22}\left(\frac{-x}{y^2}\right)\frac{1}{y} + f'_2\left(-\frac{1}{y^2}\right) = -\frac{1}{y^3}(xyf''_{12} + xf''_{22} + yf'_2)$$

Example 7 If $z = f(x,y)$ has continuous second partial derivatives, find the expression of $\left(\frac{\partial z}{\partial x}\right)^2 + \left(\frac{\partial z}{\partial y}\right)^2$ in terms of polar coordinates.

Solution Recall that rectangular and polar coordinates are related through the transformation equations $x = r\cos\theta$, $y = r\sin\theta$ or $r = \sqrt{x^2 + y^2}$, $\theta = \arctan\frac{y}{x}$. Then z is a function of r and θ

$$z = f(x,y) = f(r\cos\theta, r\sin\theta) = g(r,\theta)$$

By the Chain Rule, we have

$$\frac{\partial z}{\partial x} = \frac{\partial z}{\partial r}\frac{\partial r}{\partial x} + \frac{\partial z}{\partial \theta}\frac{\partial \theta}{\partial x}, \frac{\partial z}{\partial y} = \frac{\partial z}{\partial r}\frac{\partial r}{\partial y} + \frac{\partial z}{\partial \theta}\frac{\partial \theta}{\partial y}$$

where

$$\frac{\partial r}{\partial x} = \frac{x}{\sqrt{x^2+y^2}} = \frac{x}{r} = \cos\theta, \frac{\partial r}{\partial y} = \frac{y}{\sqrt{x^2+y^2}} = \frac{y}{r} = \sin\theta$$

$$\frac{\partial \theta}{\partial x} = -\frac{y}{x^2+y^2} = -\frac{\sin\theta}{r}, \frac{\partial \theta}{\partial y} = \frac{x}{x^2+y^2} = \frac{\cos\theta}{r}$$

Then

$$\frac{\partial z}{\partial x} = \frac{\partial z}{\partial r}\cos\theta - \frac{\partial z}{\partial \theta}\frac{\sin\theta}{r}, \frac{\partial z}{\partial y} = \frac{\partial z}{\partial r}\sin\theta + \frac{\partial z}{\partial \theta}\frac{\cos\theta}{r}$$

and therefore

Calculus(II)

$$\left(\frac{\partial z}{\partial x}\right)^2 + \left(\frac{\partial z}{\partial y}\right)^2 = \left(\frac{\partial z}{\partial r}\right)^2 \cos^2\theta - 2\frac{\partial z}{\partial r}\frac{\partial z}{\partial \theta}\frac{\sin\theta\cos\theta}{r} + \left(\frac{\partial z}{\partial \theta}\right)^2 \frac{\sin^2\theta}{r^2} +$$
$$\left(\frac{\partial z}{\partial r}\right)^2 \sin^2\theta + 2\frac{\partial z}{\partial r}\frac{\partial z}{\partial \theta}\frac{\sin\theta\cos\theta}{r} + \left(\frac{\partial z}{\partial \theta}\right)^2 \frac{\cos^2\theta}{r^2}$$
$$= \left(\frac{\partial z}{\partial r}\right)^2 + \frac{1}{r^2}\left(\frac{\partial z}{\partial \theta}\right)^2$$

The Invariance of the Total Differential Form

If $z = f(u,v)$, $u = u(x,y)$ and $v = v(x,y)$ are differentiable, then

$$dz = \frac{\partial z}{\partial x}dx + \frac{\partial z}{\partial y}dy = \left(\frac{\partial z}{\partial u}\frac{\partial u}{\partial x} + \frac{\partial z}{\partial v}\frac{\partial v}{\partial x}\right)dx + \left(\frac{\partial z}{\partial u}\frac{\partial u}{\partial y} + \frac{\partial z}{\partial v}\frac{\partial v}{\partial y}\right)dy$$
$$= \frac{\partial z}{\partial u}\left(\frac{\partial u}{\partial x}dx + \frac{\partial u}{\partial y}dy\right) + \frac{\partial z}{\partial v}\left(\frac{\partial v}{\partial x}dx + \frac{\partial v}{\partial y}dy\right) = \frac{\partial z}{\partial u}du + \frac{\partial z}{\partial v}dv$$

This shows that whether u and v are independent variables or functions of variables x and y, the total differential of $z = f(u,v)$ has the same form $dz = \frac{\partial z}{\partial u}du + \frac{\partial z}{\partial v}dv$. This property is called the invariance of the total differential form.

Example 8 If $u = f(x,y,z)$, $y = \varphi(x,t)$, $t = \psi(x,z)$, where f, φ, ψ are differentiable, find the partial derivatives of u.

Solution By the invariance of the total differential form, we have

$$du = \frac{\partial f}{\partial x}dx + \frac{\partial f}{\partial y}dy + \frac{\partial f}{\partial z}dz$$
$$= \frac{\partial f}{\partial x}dx + \frac{\partial f}{\partial y}\left(\frac{\partial \varphi}{\partial x}dx + \frac{\partial \varphi}{\partial t}dt\right) + \frac{\partial f}{\partial z}dz$$
$$= \frac{\partial f}{\partial x}dx + \frac{\partial f}{\partial y}\left[\frac{\partial \varphi}{\partial x}dx + \frac{\partial \varphi}{\partial t}\left(\frac{\partial \psi}{\partial x}dx + \frac{\partial \psi}{\partial z}dz\right)\right] + \frac{\partial f}{\partial z}dz$$
$$= \left(\frac{\partial f}{\partial x} + \frac{\partial f}{\partial y}\frac{\partial \varphi}{\partial x} + \frac{\partial f}{\partial y}\frac{\partial \varphi}{\partial t}\frac{\partial \psi}{\partial x}\right)dx + \left(\frac{\partial f}{\partial y}\frac{\partial \varphi}{\partial t}\frac{\partial \psi}{\partial z} + \frac{\partial f}{\partial z}\right)dz$$

Thus, we obtain

$$\frac{\partial u}{\partial x} = \frac{\partial f}{\partial x} + \frac{\partial f}{\partial y}\frac{\partial \varphi}{\partial x} + \frac{\partial f}{\partial y}\frac{\partial \varphi}{\partial t}\frac{\partial \psi}{\partial x}, \frac{\partial u}{\partial z} = \frac{\partial f}{\partial y}\frac{\partial \varphi}{\partial t}\frac{\partial \psi}{\partial z} + \frac{\partial f}{\partial z}$$

8.5 Implicit Differentiation

Using the Chain Rule for partial derivatives, we can give a more complete descrip-

Chapter 8 Differential Calculus of Multivariable Functions

tion of the process of implicit differentiation that was introduced in Section 3.2.

Implicit Differentiation for Single Equation

We suppose that F is a differntiable function of two variables x and y and the equation $F(x,y) = 0$ defines y implicitly as a differentiable function of one variable x, say $y = f(x)$. This means that $F(x,f(x)) = 0$ for all x in the domain of f. To find $\dfrac{dy}{dx}$, we can use the Chain Rule to differentiate both sides of the equation $F(x,y) = 0$ with respect to x

$$\frac{\partial F}{\partial x}\frac{dx}{dx} + \frac{\partial F}{\partial y}\frac{dy}{dx} = 0$$

But $\dfrac{dx}{dx} = 1$, so if $\dfrac{\partial F}{\partial y} \neq 0$ we solve for $\dfrac{dy}{dx}$ and obtain

$$\frac{dy}{dx} = -\frac{\dfrac{\partial F}{\partial x}}{\dfrac{\partial F}{\partial y}} \tag{1}$$

To derive equation (1), we assumed that $F(x,y) = 0$ defines y implicitly as a differentiable function of x. The following theorem, proved in advanced calculus, gives conditions under which this assumption is valid.

Theorem 8.5 (Implicit Function Theorem) Suppose that F is defined on a neighborhood of the point (x_0, y_0). If $F(x_0, y_0) = 0$, F'_x and F'_y are continuous on the neighborhood, and $F'_y(x_0, y_0) \neq 0$, then the equation $F(x,y) = 0$ defines y as a differentiable function of x near the point (x_0, y_0) and the derivative of this function is given by the formula in (1).

Now we suppose that F is a differntiable function of three variables x, y and z and the equation $F(x,y,z) = 0$ defines z implicitly as a differentiable function of two variables x and y, say $z = f(x,y)$. This means that $F(x,y,f(x,y)) = 0$ for all (x,y) in the domain of f. To find $\dfrac{\partial z}{\partial x}$, we can use the Chain Rule to differentiate both sides of the equation $F(x,y,z) = 0$ with respect to x

$$\frac{\partial F}{\partial x}\frac{\partial x}{\partial x} + \frac{\partial F}{\partial y}\frac{\partial y}{\partial x} + \frac{\partial F}{\partial z}\frac{\partial z}{\partial x} = 0$$

But $\dfrac{\partial x}{\partial x} = 1$ and $\dfrac{\partial y}{\partial x} = 0$, so this equation becomes

Calculus(II)

$$\frac{\partial F}{\partial x} + \frac{\partial F}{\partial z}\frac{\partial z}{\partial x} = 0$$

If $\frac{\partial F}{\partial z} \neq 0$, we solve for $\frac{\partial z}{\partial x}$ and obtain the first formula in (2). The formula for $\frac{\partial z}{\partial y}$, given by the second formula in (2), is obtaind in a similar manner

$$\frac{\partial z}{\partial x} = -\frac{\frac{\partial F}{\partial x}}{\frac{\partial F}{\partial z}}, \frac{\partial z}{\partial y} = -\frac{\frac{\partial F}{\partial y}}{\frac{\partial F}{\partial z}} \qquad (2)$$

To derive equations (2), we assumed that $F(x,y,z) = 0$ defines z implicitly as a differentiable function of two variables x and y. A version of the Implicit Function Theorem gives conditions under which this assumption is valid. Suppose that F is defined on a neighborhood of the point (x_0, y_0, z_0). If $F(x_0, y_0, z_0) = 0$, F'_x, F'_y and F'_z are continuous on the neighborhood, and $F'_z(x_0, y_0, z_0) \neq 0$, then the equation $F(x,y,z) = 0$ defines z as a differentiable function of x and y near the point (x_0, y_0, z_0) and the partial derivatives of this function is given by the formulas in (2).

Example 1 Verify that the equation $xy - e^x + e^y = 0$ defines y as a function of x near the point $(0,0)$ and find $\frac{dy}{dx}$.

Solution Let $F(x,y) = xy - e^x + e^y$. Since $F(0,0) = 0$, $F'_x(x,y) = y - e^x$ and $F'_y(x,y) = x - e^y$ are continuous near $(0,0)$, and $F'_y(0,0) \neq 0$, the Implicit Function Theorem says that the equation $F(x,y) = xy - e^x + e^y = 0$ defines y as a function of x near the point $(0,0)$ and the derivative of this function is

$$\frac{dy}{dx} = -\frac{F'_x}{F'_y} = -\frac{y - e^x}{x - e^y}$$

Example 2 Find $\frac{\partial z}{\partial x}, \frac{\partial z}{\partial y}$ and $\frac{\partial^2 z}{\partial x \partial y}$ if $\frac{x^2}{a^2} + \frac{y^2}{b^2} + \frac{z^2}{c^2} = 1$.

Solution Let $F(x,y,z) = \frac{x^2}{a^2} + \frac{y^2}{b^2} + \frac{z^2}{c^2} - 1$. Then

$$\frac{\partial F}{\partial x} = \frac{2x}{a^2}, \frac{\partial F}{\partial y} = \frac{2y}{b^2}, \frac{\partial F}{\partial z} = \frac{2z}{c^2}$$

By the equations (2), we obtain

Chapter 8 Differential Calculus of Multivariable Functions

$$\frac{\partial z}{\partial x} = -\frac{\dfrac{\partial F}{\partial x}}{\dfrac{\partial F}{\partial z}} = -\frac{c^2 x}{a^2 z}, \frac{\partial z}{\partial y} = -\frac{\dfrac{\partial F}{\partial y}}{\dfrac{\partial F}{\partial z}} = -\frac{c^2 y}{b^2 z} \quad (z \neq 0)$$

To find $\dfrac{\partial^2 z}{\partial x \partial y}$, we differentiate $\dfrac{\partial z}{\partial x} = -\dfrac{c^2 x}{a^2 z}$ with respect to y and obtain

$$\frac{\partial^2 z}{\partial x \partial y} = \frac{\partial}{\partial y}\left(-\frac{c^2 x}{a^2 z}\right) = -\frac{c^2 x}{a^2}\left(-\frac{1}{z^2}\frac{\partial z}{\partial y}\right) = \frac{c^2 x}{a^2 z^2}\left(-\frac{c^2 y}{b^2 z}\right) = -\frac{c^4 xy}{a^2 b^2 z^3}$$

Example 3 If $F\left(\dfrac{x}{z}, \dfrac{y}{z}\right) = 0$, where F has continuous partial derivatives and $xF'_1 + yF'_2 \neq 0$, find $\dfrac{\partial z}{\partial x}$ and $\dfrac{\partial z}{\partial y}$.

Solution Method 1 (Using Formulas). Let $G(x,y,z) = F\left(\dfrac{x}{z}, \dfrac{y}{z}\right)$. Then

$$\frac{\partial G}{\partial x} = F'_1 \cdot \frac{1}{z}, \quad \frac{\partial G}{\partial y} = F'_2 \cdot \frac{1}{z}, \quad \frac{\partial G}{\partial z} = F'_1 \cdot \left(-\frac{x}{z^2}\right) + F'_2 \cdot \left(-\frac{y}{z^2}\right)$$

By the formulas in (2), we get

$$\frac{\partial z}{\partial x} = -\frac{\dfrac{\partial G}{\partial x}}{\dfrac{\partial G}{\partial z}} = -\frac{F'_1 \cdot \dfrac{1}{z}}{F'_1 \cdot \left(-\dfrac{x}{z^2}\right) + F'_2 \cdot \left(-\dfrac{y}{z^2}\right)} = \frac{zF'_1}{xF'_1 + yF'_2}$$

$$\frac{\partial z}{\partial y} = -\frac{\dfrac{\partial G}{\partial y}}{\dfrac{\partial G}{\partial z}} = -\frac{F'_2 \cdot \dfrac{1}{z}}{F'_1 \cdot \left(-\dfrac{x}{z^2}\right) + F'_2 \cdot \left(-\dfrac{y}{z^2}\right)} = \frac{zF'_2}{xF'_1 + yF'_2}$$

Method 2 (Using the Chain Rule). We differentiate both sides of the equation with respect to x and y respectively

$$F'_1 \cdot \left(\frac{z - x\dfrac{\partial z}{\partial x}}{z^2}\right) + F'_2 \cdot \left(-\frac{y}{z^2}\frac{\partial z}{\partial x}\right) = 0$$

$$F'_1 \cdot \left(-\frac{x}{z^2}\frac{\partial z}{\partial y}\right) + F'_2 \cdot \left(\frac{z - y\dfrac{\partial z}{\partial y}}{z^2}\right) = 0$$

Solving for $\dfrac{\partial z}{\partial x}$ and $\dfrac{\partial z}{\partial y}$, we obtain

$$\frac{\partial z}{\partial x} = \frac{zF'_1}{xF'_1 + yF'_2}, \frac{\partial z}{\partial y} = \frac{zF'_2}{xF'_1 + yF'_2}$$

Method 3 (Using Total Differentials). We take differentials of both sides of the equation

$$F'_1 d\left(\frac{x}{z}\right) + F'_2 d\left(\frac{y}{z}\right) = 0$$

$$F'_1 \cdot \frac{zdx - xdz}{z^2} + F'_2 \cdot \frac{zdy - ydz}{z^2} = 0$$

Solving for dz, we obtain

$$dz = \frac{zF'_1}{xF'_1 + yF'_2} dx + \frac{zF'_2}{xF'_1 + yF'_2} dz$$

This gives

$$\frac{\partial z}{\partial x} = \frac{zF'_1}{xF'_1 + yF'_2}, \frac{\partial z}{\partial y} = \frac{zF'_2}{xF'_1 + yF'_2}$$

Implicit Differentiation for Two or More Equations

We suppose that F and G are differntiable functions of three variables x, y and z and the equations $F(x,y,z) = 0$ and $G(x,y,z) = 0$ define y and z implicitly as differentiable functions of one variable x, say $y = f(x)$ and $z = g(x)$. This means that $F(x,f(x), g(x)) = 0$ and $G(x,f(x),g(x)) = 0$ for all x in the domains of f and g. To find $\frac{dy}{dx}$ and $\frac{dz}{dx}$, we can use the Chain Rule to differentiate of both sides of the equations $F(x,y,z) = 0$ and $G(x,y,z) = 0$ with respect to x

$$\frac{\partial F}{\partial x}\frac{dx}{dx} + \frac{\partial F}{\partial y}\frac{dy}{dx} + \frac{\partial F}{\partial z}\frac{dz}{dx} = 0$$

$$\frac{\partial G}{\partial x}\frac{dx}{dx} + \frac{\partial G}{\partial y}\frac{dy}{dx} + \frac{\partial G}{\partial z}\frac{dz}{dx} = 0$$

But $\frac{dx}{dx} = 1$, so these two equations become

$$\frac{\partial F}{\partial x} + \frac{\partial F}{\partial y}\frac{dy}{dx} + \frac{\partial F}{\partial z}\frac{dz}{dx} = 0$$

$$\frac{\partial G}{\partial x} + \frac{\partial G}{\partial y}\frac{dy}{dx} + \frac{\partial G}{\partial z}\frac{dz}{dx} = 0$$

If

Chapter 8 Differential Calculus of Multivariable Functions

$$\begin{vmatrix} \dfrac{\partial F}{\partial y} & \dfrac{\partial F}{\partial z} \\ \dfrac{\partial G}{\partial y} & \dfrac{\partial G}{\partial z} \end{vmatrix} \neq 0$$

we solve for $\dfrac{dy}{dx}$ and $\dfrac{dz}{dx}$ and obtain

$$\dfrac{dy}{dx} = - \dfrac{\begin{vmatrix} \dfrac{\partial F}{\partial x} & \dfrac{\partial F}{\partial z} \\ \dfrac{\partial G}{\partial x} & \dfrac{\partial G}{\partial z} \end{vmatrix}}{\begin{vmatrix} \dfrac{\partial F}{\partial y} & \dfrac{\partial F}{\partial z} \\ \dfrac{\partial G}{\partial y} & \dfrac{\partial G}{\partial z} \end{vmatrix}}, \dfrac{dz}{dx} = - \dfrac{\begin{vmatrix} \dfrac{\partial F}{\partial y} & \dfrac{\partial F}{\partial x} \\ \dfrac{\partial G}{\partial y} & \dfrac{\partial G}{\partial x} \end{vmatrix}}{\begin{vmatrix} \dfrac{\partial F}{\partial y} & \dfrac{\partial F}{\partial z} \\ \dfrac{\partial G}{\partial y} & \dfrac{\partial G}{\partial z} \end{vmatrix}}$$

The above determinants involving partial derivatives are called Jacobian determinants. For convenience, we write

$$\dfrac{\partial(F,G)}{\partial(y,z)} = \begin{vmatrix} \dfrac{\partial F}{\partial y} & \dfrac{\partial F}{\partial z} \\ \dfrac{\partial G}{\partial y} & \dfrac{\partial G}{\partial z} \end{vmatrix}$$

Then

$$\dfrac{dy}{dx} = - \dfrac{\dfrac{\partial(F,G)}{\partial(x,z)}}{\dfrac{\partial(F,G)}{\partial(y,z)}}, \dfrac{dz}{dx} = - \dfrac{\dfrac{\partial(F,G)}{\partial(y,x)}}{\dfrac{\partial(F,G)}{\partial(y,z)}} \qquad (3)$$

Now we suppose that F and G are differntiable functions of four variables x, y, u and v and the equations $F(x,y,u,v) = 0$ and $G(x,y,u,v) = 0$ define u and v implicitly as differentiable functions of two variables x and y, say $u = f(x,y)$ and $v = g(x,y)$. This means that $F(x,y,f(x,y),g(x,y)) = 0$ and $G(x,y,f(x,y),g(x,y)) = 0$ for all (x,y) in the domains of f and g. To find $\dfrac{\partial u}{\partial x}$ and $\dfrac{\partial v}{\partial x}$, we can use the Chain Rule to differentiate both sides of the equations $F(x,y,u,v) = 0$ and $G(x,y,u,v) = 0$ with respect to x

$$\dfrac{\partial F}{\partial x}\dfrac{\partial x}{\partial x} + \dfrac{\partial F}{\partial y}\dfrac{\partial y}{\partial x} + \dfrac{\partial F}{\partial u}\dfrac{\partial u}{\partial x} + \dfrac{\partial F}{\partial v}\dfrac{\partial v}{\partial x} = 0$$

$$\dfrac{\partial G}{\partial x}\dfrac{\partial x}{\partial x} + \dfrac{\partial G}{\partial y}\dfrac{\partial y}{\partial x} + \dfrac{\partial G}{\partial u}\dfrac{\partial u}{\partial x} + \dfrac{\partial G}{\partial v}\dfrac{\partial v}{\partial x} = 0$$

Calculus(Ⅱ)

But $\dfrac{\partial x}{\partial x} = 1$ and $\dfrac{\partial y}{\partial x} = 0$, so these equations become

$$\dfrac{\partial F}{\partial x} + \dfrac{\partial F}{\partial u}\dfrac{\partial u}{\partial x} + \dfrac{\partial F}{\partial v}\dfrac{\partial v}{\partial x} = 0$$

$$\dfrac{\partial G}{\partial x} + \dfrac{\partial G}{\partial u}\dfrac{\partial u}{\partial x} + \dfrac{\partial G}{\partial v}\dfrac{\partial v}{\partial x} = 0$$

If $\dfrac{\partial(F,G)}{\partial(u,v)} \neq 0$, we solve for $\dfrac{\partial u}{\partial x}$ and $\dfrac{\partial v}{\partial x}$ and obtain

$$\dfrac{\partial u}{\partial x} = -\dfrac{\dfrac{\partial(F,G)}{\partial(x,v)}}{\dfrac{\partial(F,G)}{\partial(u,v)}}, \quad \dfrac{\partial v}{\partial x} = -\dfrac{\dfrac{\partial(F,G)}{\partial(u,x)}}{\dfrac{\partial(F,G)}{\partial(u,v)}} \tag{4}$$

Similarly, we can obtain the formulas for $\dfrac{\partial u}{\partial y}$ and $\dfrac{\partial v}{\partial y}$ as follows

$$\dfrac{\partial u}{\partial y} = -\dfrac{\dfrac{\partial(F,G)}{\partial(y,v)}}{\dfrac{\partial(F,G)}{\partial(u,v)}}, \quad \dfrac{\partial v}{\partial y} = -\dfrac{\dfrac{\partial(F,G)}{\partial(u,y)}}{\dfrac{\partial(F,G)}{\partial(u,v)}} \tag{5}$$

Two equations implicit differentiation can also be extended to cases involving three or more equations.

Example 4 Find $\dfrac{dy}{dx}$ and $\dfrac{dz}{dx}$ if $x^2 + y^2 + z^2 = 50$ and $x + 2y + 3z = 4$.

Solution Let $F(x,y,z) = x^2 + y^2 + z^2 - 50$ and $G(x,y,z) = x + 2y + 3z - 4$. Then

$$\dfrac{\partial(F,G)}{\partial(y,z)} = \begin{vmatrix} \dfrac{\partial F}{\partial y} & \dfrac{\partial F}{\partial z} \\ \dfrac{\partial G}{\partial y} & \dfrac{\partial G}{\partial z} \end{vmatrix} = \begin{vmatrix} 2y & 2z \\ 2 & 3 \end{vmatrix} = 2(3y - 2z)$$

$$\dfrac{\partial(F,G)}{\partial(x,z)} = \begin{vmatrix} \dfrac{\partial F}{\partial x} & \dfrac{\partial F}{\partial z} \\ \dfrac{\partial G}{\partial x} & \dfrac{\partial G}{\partial z} \end{vmatrix} = \begin{vmatrix} 2x & 2z \\ 1 & 3 \end{vmatrix} = 2(3x - z)$$

$$\dfrac{\partial(F,G)}{\partial(y,x)} = \begin{vmatrix} \dfrac{\partial F}{\partial y} & \dfrac{\partial F}{\partial x} \\ \dfrac{\partial G}{\partial y} & \dfrac{\partial G}{\partial x} \end{vmatrix} = \begin{vmatrix} 2y & 2x \\ 2 & 1 \end{vmatrix} = 2(y - 2x)$$

By the formulas in (3), we obtain

Chapter 8 Differential Calculus of Multivariable Functions

$$\frac{dy}{dx} = -\frac{\frac{\partial(F,G)}{\partial(x,z)}}{\frac{\partial(F,G)}{\partial(y,z)}} = -\frac{2(3x-z)}{2(3y-2z)} = \frac{z-3x}{3y-2z}$$

$$\frac{dz}{dx} = -\frac{\frac{\partial(F,G)}{\partial(y,x)}}{\frac{\partial(F,G)}{\partial(y,z)}} = -\frac{2(y-2x)}{2(3y-2z)} = \frac{2x-y}{3y-2z}$$

provided that $3y - 2z \neq 0$.

Example 5 If the functions $u = e^x + x\sin y$, $v = e^x + x\cos y$ define x and y implicitly as differentiable functions of two variables u and v, find $\frac{\partial x}{\partial u}$ and $\frac{\partial y}{\partial u}$.

Solution Let $F(x,y,u,v) = e^x + x\sin y - u$, $G(x,y,u,v) = e^x + x\cos y - v$. Since

$$\frac{\partial(F,G)}{\partial(x,y)} = \begin{vmatrix} \frac{\partial F}{\partial x} & \frac{\partial F}{\partial y} \\ \frac{\partial G}{\partial x} & \frac{\partial G}{\partial y} \end{vmatrix} = \begin{vmatrix} e^x + \sin y & x\cos y \\ e^x + \cos y & -x\sin y \end{vmatrix} = -xe^x\sin y - xe^x\cos y - x$$

$$\frac{\partial(F,G)}{\partial(u,y)} = \begin{vmatrix} \frac{\partial F}{\partial u} & \frac{\partial F}{\partial y} \\ \frac{\partial G}{\partial u} & \frac{\partial G}{\partial y} \end{vmatrix} = \begin{vmatrix} -1 & x\cos y \\ 0 & -x\sin y \end{vmatrix} = x\sin y$$

$$\frac{\partial(F,G)}{\partial(x,u)} = \begin{vmatrix} \frac{\partial F}{\partial x} & \frac{\partial F}{\partial u} \\ \frac{\partial G}{\partial x} & \frac{\partial G}{\partial u} \end{vmatrix} = \begin{vmatrix} e^x + \sin y & -1 \\ e^x + \cos y & 0 \end{vmatrix} = e^x + \cos y$$

by the formulas in (4) we have

$$\frac{\partial x}{\partial u} = -\frac{\frac{\partial(F,G)}{\partial(u,y)}}{\frac{\partial(F,G)}{\partial(x,y)}} = -\frac{x\sin y}{-xe^x\sin y - xe^x\cos y - x} = \frac{\sin y}{e^x\sin y + e^x\cos y + 1}$$

$$\frac{\partial y}{\partial u} = -\frac{\frac{\partial(F,G)}{\partial(x,u)}}{\frac{\partial(F,G)}{\partial(x,y)}} = -\frac{e^x + \cos y}{-xe^x\sin y - xe^x\cos y - x} = \frac{e^x + \cos y}{xe^x\sin y + xe^x\cos y + x}$$

Example 6 If the equations $u = f(x,y,z)$, $g(x,y,z) = 0$ and $h(x,z) = 0$ define u as a differentiable function of x, where f, g and h have continuous partial derivatives and $g'_y \neq 0, h'_x \neq 0$, find $\dfrac{du}{dx}$.

Solution Method 1 (Finding a Derivative with Constrained Variables). The variables y and z in $u = f(x,y,z)$ are constrained variables, which are differeniable functions of x defined implicitly by the equations $g(x,y,z) = 0$ and $h(x,z) = 0$. Since

$$\frac{\partial(g,h)}{\partial(y,z)} = \begin{vmatrix} g'_y & g'_z \\ 0 & h'_z \end{vmatrix} = g'_y h'_z, \quad \frac{\partial(g,h)}{\partial(x,z)} = \begin{vmatrix} g'_x & g'_z \\ h'_x & h'_z \end{vmatrix} = g'_x h'_z - g'_z h'_x$$

$$\frac{\partial(g,h)}{\partial(y,x)} = \begin{vmatrix} g'_y & g'_x \\ 0 & h'_x \end{vmatrix} = g'_y h'_x$$

by the formulas in (3) we obtain

$$\frac{dy}{dx} = -\frac{\dfrac{\partial(g,h)}{\partial(x,z)}}{\dfrac{\partial(g,h)}{\partial(y,z)}} = -\frac{g'_x h'_z - g'_z h'_x}{g'_y h'_z} = \frac{g'_z h'_x - g'_x h'_z}{g'_y h'_z}$$

$$\frac{dz}{dx} = -\frac{\dfrac{\partial(g,h)}{\partial(y,x)}}{\dfrac{\partial(g,h)}{\partial(y,z)}} = -\frac{g'_y h'_x}{g'_y h'_z} = -\frac{h'_x}{h'_z}$$

Then by the Chain Rule we get

$$\frac{du}{dx} = f'_x + f'_y \frac{dy}{dx} + f'_z \frac{dz}{dx} = f'_x + f'_y \frac{g'_z h'_x - g'_x h'_z}{g'_y h'_z} - f'_z \frac{h'_x}{h'_z}$$

Method 2 (Using Formulas). Let $F(x,y,z,u) = f(x,y,z) - u$. Since

$$\frac{\partial(F,g,h)}{\partial(y,z,u)} = \begin{vmatrix} f'_y & f'_z & -1 \\ g'_y & g'_z & 0 \\ 0 & h'_z & 0 \end{vmatrix} = -g'_y h'_z$$

$$\frac{\partial(F,g,h)}{\partial(y,z,x)} = \begin{vmatrix} f'_y & f'_z & f'_x \\ g'_y & g'_z & g'_x \\ 0 & h'_z & h'_x \end{vmatrix} = f'_x g'_y h'_z + f'_y(g'_z h'_x - g'_x h'_z) - f'_z y h'_x$$

we have

$$\frac{du}{dx} = -\frac{\dfrac{\partial(F,g,h)}{\partial(y,z,x)}}{\dfrac{\partial(F,g,h)}{\partial(y,z,u)}} = f'_x + f'_y \frac{g'_z h'_x - g'_x h'_z}{g'_y h'_z} - f'_z \frac{h'_x}{h'_z}$$

Chapter 8 Differential Calculus of Multivariable Functions

Method 3 (Using Total Differentials). We take differentials of both sides of each equation

$$du = f'_x dx + f'_y dy + f'_z dz$$
$$g'_x dx + g'_y dy + g'_z dz = 0$$
$$h'_x dx + h'_z dz = 0$$

Solving the last two equations for dy and dz and substituting them into the first equation, we get

$$du = \left(f'_x + f'_y \frac{g'_z h'_x - g'_x h'_z}{g'_y h'_z} - f'_z \frac{h'_x}{h'_z} \right) dx$$

Therefore

$$\frac{du}{dx} = f'_x + f'_y \frac{g'_z h'_x - g'_x h'_z}{g'_y h'_z} - f'_z \frac{h'_x}{h'_z}$$

8.6 Applications of Partial Derivatives to Analytic Geometry

8.6.1 Tangent Lines and Normal Planes for Curves in Space

Suppose that a curve C in space is given by parametric equations

$$x = x(t), y = y(t), z = z(t), t \in I$$

Let $P_0(x_0, y_0, z_0)$ and $P(x_0 + \Delta x, y_0 + \Delta y, z_0 + \Delta z)$ be two points on the curve, which correspond to parameters t_0 and $t_0 + \Delta t$. Then an equation of the secant line $P_0 P$ is

$$\frac{x - x_0}{\Delta x} = \frac{y - y_0}{\Delta y} = \frac{z - z_0}{\Delta z}$$

If $\Delta t \neq 0$, then this equation can be written as

$$\frac{x - x_0}{\frac{\Delta x}{\Delta t}} = \frac{y - y_0}{\frac{\Delta y}{\Delta t}} = \frac{z - z_0}{\frac{\Delta z}{\Delta t}}$$

If all $x'(t_0), y'(t_0)$ and $z'(t_0)$ exist and at least one of them is not zero, then by letting $\Delta t \to 0$ we obtian

$$\frac{x - x_0}{x'(t_0)} = \frac{y - y_0}{y'(t_0)} = \frac{z - z_0}{z'(t_0)} \tag{1}$$

Calculus(II)

We define
$$t = \{x'(t_0), y'(t_0), z'(t_0)\}$$
when different from 0, to be the vector tangent to the curve C at the point P_0, called a tangent vector of the curve C at P_0. The tangent line to the curve C at a point $P_0(x_0, y_0, z_0)$ is defined to be the line through the point parallel to t, that is, the line given by the equation (1).

The plane that passes through the point $P_0(x_0, y_0, z_0)$ orthogonal to the vector t is called the normal plane of the curve C at the point P_0. An equation of the normal plane is
$$x'(t_0)(x - x_0) + y'(t_0)(y - y_0) + z'(t_0)(z - z_0) = 0 \qquad (2)$$

We say a curve $C: x = x(t), y = y(t), z = z(t)$ is smooth on an interval if $x(t), y(t)$ and $z(t)$ are differentiable and $t = \{x'(t), y'(t), z'(t)\} \neq 0$ on that interval. Smooth curves have no sharp corners or cusps.

Example 1 Find the equations of the tangent line and normal plane to the curve $x = t, y = t^2, z = t^3$ at the point $P_0(1,1,1)$.

Solution Since $x' = 1, y' = 2t, z' = 3t^2$, and $P_0(1,1,1)$ corresponds to parameter $t = 1$, the tangent vector of the curve at the point P_0 is $t = \{1,2,3\}$. Thus, an equation of the tangent line is
$$\frac{x-1}{1} = \frac{y-1}{2} = \frac{z-1}{3}$$
and an equation of the normal plane is
$$1(x-1) + 2(y-1) + 3(z-1) = 0$$
that is
$$x + 2y + 3z - 6 = 0$$

Example 2 Suppose that the normal plane to a curve $x = x(t), y = y(t), z = z(t)$ at any point on the curve passes through the origin. Show that the curve must lie on a certain sphere centered at the origin.

Solution By equation (2), an equation of the normal plane of the curve at any point $(x(t), y(t), z(t))$ is
$$x'(t)(X - x(t)) + y'(t)(Y - y(t)) + z'(t)(Z - z(t)) = 0$$
Since $O(0,0,0)$ is on the normal plane, we have
$$x'(t)x(t) + y'(t)y(t) + z'(t)z(t) = 0$$

Chapter 8 Differential Calculus of Multivariable Functions

that is
$$[x^2(t) + y^2(t) + z^2(t)]' = 0$$
Thus
$$x^2(t) + y^2(t) + z^2(t) = C$$
This shows that the curve must lie on a certain sphere centered at the origin.

Now suppose that a curve C in space is given by equations
$$F(x,y,z) = 0, G(x,y,z) = 0 \tag{3}$$
Let $P_0(x_0, y_0, z_0)$ be a point on the curve. Suppose that F and G have continuous partial derivatives near P_0 and at least one of the Jacobian determinants $\dfrac{\partial(F,G)}{\partial(y,z)}\bigg|_{P_0}$, $\dfrac{\partial(F,G)}{\partial(z,x)}\bigg|_{P_0}$ and $\dfrac{\partial(F,G)}{\partial(x,y)}\bigg|_{P_0}$ is not zero. Without loss of generality, we assume that $\dfrac{\partial(F,G)}{\partial(y,z)}\bigg|_{P_0} \ne 0$. Then the equations (3) define y and z implicitly as differentiable functions of x near the point P_0, say $y = y(x)$ and $z = z(x)$, and
$$\frac{dy}{dx} = -\frac{\dfrac{\partial(F,G)}{\partial(x,z)}}{\dfrac{\partial(F,G)}{\partial(y,z)}} = \frac{\dfrac{\partial(F,G)}{\partial(z,x)}}{\dfrac{\partial(F,G)}{\partial(y,z)}}, \quad \frac{dz}{dx} = -\frac{\dfrac{\partial(F,G)}{\partial(y,x)}}{\dfrac{\partial(F,G)}{\partial(y,z)}} = \frac{\dfrac{\partial(F,G)}{\partial(x,y)}}{\dfrac{\partial(F,G)}{\partial(y,z)}}$$
Since the curve C can be written as parametric equations
$$x = x, y = y(x), z = z(x)$$
near P_0, an equation of the tangent line of the curve at P_0 is
$$\frac{x - x_0}{1} = \frac{y - y_0}{\dfrac{dy}{dx}\bigg|_{P_0}} = \frac{z - z_0}{\dfrac{dz}{dx}\bigg|_{P_0}}$$
that is
$$\frac{x - x_0}{\dfrac{\partial(F,G)}{\partial(y,z)}\bigg|_{P_0}} = \frac{y - y_0}{\dfrac{\partial(F,G)}{\partial(z,x)}\bigg|_{P_0}} = \frac{z - z_0}{\dfrac{\partial(F,G)}{\partial(x,y)}\bigg|_{P_0}} \tag{4}$$
This tells us that the tangent vector of the curve at the point P_0 is
$$t = \left\{\frac{\partial(F,G)}{\partial(y,z)}\bigg|_{P_0}, \frac{\partial(F,G)}{\partial(z,x)}\bigg|_{P_0}, \frac{\partial(F,G)}{\partial(x,y)}\bigg|_{P_0}\right\}$$
Then an equation of the normal plane of the curve at P_0 is
$$\frac{\partial(F,G)}{\partial(y,z)}\bigg|_{P_0}(x - x_0) + \frac{\partial(F,G)}{\partial(z,x)}\bigg|_{P_0}(y - y_0) + \frac{\partial(F,G)}{\partial(x,y)}\bigg|_{P_0}(z - z_0) = 0 \tag{5}$$

Calculus(II)

Example 3 Find the equations of the tangent line and normal plane at the point $(1, -1, 2)$ to the curve $2x^2 + 3y^2 + z^2 = 9$, $z^2 = 3x^2 + y^2$.

Solution Let $F(x,y,z) = 2x^2 + 3y^2 + z^2 - 9$, $G(x,y,z) = 3x^2 + y^2 - z^2$. Since

$$\frac{\partial(F,G)}{\partial(y,z)} = \begin{vmatrix} \frac{\partial F}{\partial y} & \frac{\partial F}{\partial z} \\ \frac{\partial G}{\partial y} & \frac{\partial G}{\partial z} \end{vmatrix} = \begin{vmatrix} 6y & 2z \\ 2y & -2z \end{vmatrix} = -16yz$$

$$\frac{\partial(F,G)}{\partial(z,x)} = \begin{vmatrix} \frac{\partial F}{\partial z} & \frac{\partial F}{\partial x} \\ \frac{\partial G}{\partial z} & \frac{\partial G}{\partial x} \end{vmatrix} = \begin{vmatrix} 2z & 4x \\ -2z & 6x \end{vmatrix} = 20xz$$

$$\frac{\partial(F,G)}{\partial(x,y)} = \begin{vmatrix} \frac{\partial F}{\partial x} & \frac{\partial F}{\partial y} \\ \frac{\partial G}{\partial x} & \frac{\partial G}{\partial y} \end{vmatrix} = \begin{vmatrix} 4x & 6y \\ 6x & 2y \end{vmatrix} = -28xy$$

the tangent vector of the curve at $(1, -1, 2)$ is

$$t = \{-16yz, 20xz, -28xy\}\big|_{(1,-1,2)} = \{32, 40, 28\}$$

By equation (4), an equation of the tangent line at $(1, -1, 2)$ is

$$\frac{x-1}{32} = \frac{y+1}{40} = \frac{z-2}{28}$$

or

$$\frac{x-1}{8} = \frac{y+1}{10} = \frac{z-2}{7}$$

By equation (5), an equation of the normal plane at $(1, -1, 2)$ is

$$32(x-1) + 40(y+1) + 28(z-2) = 0$$

or

$$8x + 10y + 7z - 12 = 0$$

8.6.2 Tangent Planes and Normal Lines

Suppose that a surface S in space is defined implicitly by the equation

$$F(x,y,z) = 0$$

Let $P_0(x_0, y_0, z_0)$ be a point on the surface. Suppose that F is differentiable at P_0 and at least one of the partial derivatives $F'_x\big|_{P_0}$, $F'_y\big|_{P_0}$ and $F'_z\big|_{P_0}$ is not zero. To find an e-

Chapter 8 Differential Calculus of Multivariable Functions

quation of the tangent plane to the surface S at the point P_0, we consider any smooth curve $C: x = x(t), y = y(t), z = z(t)$ on the surface S passing through P_0. Because the points of C lie on the surface, we have

$$F(x(t), y(t), z(t)) = 0$$

Differentiating both sides of this equation with respect to t gives

$$F'_x(x(t), y(t), z(t))x'(t) + F'_y(x(t), y(t), z(t))y'(t) + F'_z(x(t), y(t), z(t))z'(t) = 0$$

Let $P_0(x_0, y_0, z_0)$ correspond to the parameter $t = t_0$. Then $x(t_0) = x_0, y(t_0) = y_0, z(t_0) = z_0$. If we put $t = t_0$ in the above equation, we obtain

$$F'_x(x_0, y_0, z_0)x'(t_0) + F'_y(x_0, y_0, z_0)y'(t_0) + F'_z(x_0, y_0, z_0)z'(t_0) = 0$$

This shows that the vector

$$\boldsymbol{n} = \{F'_x(x_0, y_0, z_0), F'_y(x_0, y_0, z_0), F'_z(x_0, y_0, z_0)\}$$

is orthogonal to the tangent vector $\boldsymbol{t} = \{x'(t_0), y'(t_0), z'(t_0)\}$. Because this argument applies to all smooth curves on the surface passing through P_0, the tangent vectors for all these curves are orthogonal to $\boldsymbol{n}$, and thus they all lie in the same plane. This plane is called the tangent palne to the surface S at the point P_0, and the vector $\boldsymbol{n}$ is called a normal vector of the surface S at P_0. We can easily find an equation of the tangent plane since we know a point on the plane $P_0(x_0, y_0, z_0)$ and a normal vector $\boldsymbol{n}$. An equation of the tangent plane is

$$F'_x(x_0, y_0, z_0)(x - x_0) + F'_y(x_0, y_0, z_0)(y - y_0) + F'_z(x_0, y_0, z_0)(z - z_0) = 0 \tag{6}$$

The line passing through the point $P_0(x_0, y_0, z_0)$ parallel to the normal vector $\boldsymbol{n}$ is called a normal line of the surface S at the point P_0. An equation of the normal line is

$$\frac{x - x_0}{F'_x(x_0, y_0, z_0)} = \frac{y - y_0}{F'_y(x_0, y_0, z_0)} = \frac{z - z_0}{F'_z(x_0, y_0, z_0)} \tag{7}$$

A surface S in space is often defined explicitly in the form

$$z = f(x, y)$$

In this situation, the equation of the tangent plane is a special case of the general equation just derived. The equation $z = f(x, y)$ is written as

$$F(x, y, z) = f(x, y) - z = 0$$

and the normal vector of the surface at the point $P_0(x_0, y_0, f(x_0, y_0))$ is

Calculus(II)

$$n = \{f'_x(x_0,y_0), f'_y(x_0,y_0), -1\}$$

Proceeding as before, an equation of the tangent plane to the surface S at the point P_0 is

$$f'_x(x_0,y_0)(x-x_0) + f'_y(x_0,y_0)(y-y_0) - (z-f(x_0,y_0)) = 0 \qquad (8)$$

and an equation of the normal line to the surface S at the point P_0 is

$$\frac{x-x_0}{f'_x(x_0,y_0)} = \frac{y-y_0}{f'_y(x_0,y_0)} = \frac{z-f(x_0,y_0)}{-1} \qquad (9)$$

After some rearranging, we obtain an equation of the tangent plane

$$z = f(x_0,y_0) + f'_x(x_0,y_0)(x-x_0) + f'_y(x_0,y_0)(y-y_0)$$

This agrees with the result obtained in Section 8.3.

Example 4 Find the equations of the tangent plane and the normal line at the point $(1,2,3)$ to the ellipsoid $\frac{x^2}{3} + \frac{y^2}{12} + \frac{z^2}{27} = 1$.

Solution Let $F(x,y,z) = \frac{x^2}{3} + \frac{y^2}{12} + \frac{z^2}{27} - 1$. Since

$$F'_x(x,y,z) = \frac{2x}{3}, F'_y(x,y,z) = \frac{y}{6}, F'_z(x,y,z) = \frac{2z}{27}$$

the normal vector of the ellipsoid at the point $(1,2,3)$ is

$$n = \left\{\frac{2x}{3}, \frac{y}{6}, \frac{2z}{27}\right\}\bigg|_{(1,2,3)} = \left\{\frac{2}{3}, \frac{1}{3}, \frac{2}{9}\right\}$$

Then equation (6) gives the equation of the tangent plane at $(1,2,3)$ as

$$\frac{2}{3}(x-1) + \frac{1}{3}(y-2) + \frac{2}{9}(z-3) = 0$$

which simplifies to

$$6x + 3y + 2z - 18 = 0$$

By equation (7), the equation of the normal line at $(1,2,3)$ is

$$\frac{x-1}{\frac{2}{3}} = \frac{y-2}{\frac{1}{3}} = \frac{z-3}{\frac{2}{9}}$$

or

$$\frac{x-1}{6} = \frac{y-2}{3} = \frac{z-3}{2}$$

Example 5 Find the equations of the tangent plane and the normal line to the surface $z = x\cos y - ye^x$ at the point $(0,0,0)$.

Solution We calculate the partial derivatives of $f(x,y) = x\cos y - ye^x$

Chapter 8 Differential Calculus of Multivariable Functions

$$f'_x(x,y) = \cos y - ye^x, f'_y(x,y) = - x\sin y - e^x$$

Evaluated at $(0,0)$, we have

$$f'_x(0,0) = 1, f'_y(0,0) = -1$$

By equation (8), an equation of the tangent plane at the point $(0,0,0)$ is

$$1 \cdot (x - 0) - 1 \cdot (y - 0) - (z - 0) = 0$$

or

$$x - y - z = 0$$

Using equation (9), an equation of the normal line at $(0,0,0)$ is

$$\frac{x}{1} = \frac{y}{-1} = \frac{z}{-1}$$

8.7 Extreme Values of Functions of Several Variables

8.7.1 Taylor's Formula for Functions of Two Variables

The Taylor polynomial approximation to functions of one variable that we discussed in Chapter 4 can be extended to functions of two or more variables. In this subsection, we investigate polynomial approximations to functions of two variables.

Theorem 8.6 (Taylor's Formula for $f(x,y)$ at the Point (x_0, y_0)) Suppose that $f(x,y)$ and its partial derivatives through order $n + 1$ are continuous on a neighborhood $U(P_0)$ of a point $P_0(x_0, y_0)$. Then, throughout $U(P_0)$

$$f(x_0 + h, y_0 + k) = f(x_0, y_0) + \left(h\frac{\partial}{\partial x} + k\frac{\partial}{\partial y}\right)f(x_0, y_0) +$$

$$\frac{1}{2!}\left(h\frac{\partial}{\partial x} + k\frac{\partial}{\partial y}\right)^2 f(x_0, y_0) + \cdots + \frac{1}{n!}\left(h\frac{\partial}{\partial x} + k\frac{\partial}{\partial y}\right)^n f(x_0, y_0) +$$

$$\frac{1}{(n+1)!}\left(h\frac{\partial}{\partial x} + k\frac{\partial}{\partial y}\right)^{n+1} f(x_0 + \theta h, y_0 + \theta k) \qquad (1)$$

where $0 < \theta < 1$ and $\left(h\frac{\partial}{\partial x} + k\frac{\partial}{\partial y}\right)^m f(x_0, y_0) = \sum_{i=0}^{m} C_m^i \frac{\partial^m f}{\partial x^i \partial y^{m-i}}\bigg|_{(x_0, y_0)} h^i k^{m-i}$.

Proof Let h and k be increments small enough to put the point $P(x_0 + h, y_0 + k)$ inside $U(P_0)$. We parametrize the segment $P_0 P$ as $x = x_0 + th, y = y_0 + tk, 0 \leq t \leq 1$ and let $F(t) = f(x_0 + th, y_0 + tk)$. Then $F(t)$ and its derivatives through order $n + 1$ are continuous on $[0,1]$, and these derivatives are

Calculus(II)

$$F^{(m)}(t) = \sum_{i=0}^{m} C_m^i \frac{\partial^m f}{\partial x^i \partial y^{m-i}} h^i k^{m-i} \left(h\frac{\partial}{\partial x} + k\frac{\partial}{\partial y} \right)^m f(x,y), m = 1,2,\cdots,n+1 \quad (2)$$

Thus, we can apply the Maclaurin's formula to obtain

$$F(t) = F(0) + F'(0)t + \frac{F''(0)}{2!}t^2 + \cdots + \frac{F^{(n)}(0)}{n!}t^n + \frac{F^{(n+1)}(\theta t)}{(n+1)!}t^{n+1}$$

and take $t = 1$ to obtain

$$F(1) = F(0) + F'(0) + \frac{F''(0)}{2!} + \cdots + \frac{F^{(n)}(0)}{n!} + \frac{F^{(n+1)}(\theta)}{(n+1)!}$$

for some θ between 0 and 1. When we replace the first n derivatives on the right of this equation by their equivalent expressions from equation (2) evaluated at $t = 0$ and replace the $(n+1)$-order derivative in the remainder term by its equivalent expression from equation (2) evaluated at $t = \theta$, we arrive at the Taylor's formula (1).

If $(x_0, y_0) = (0,0)$ and we treat h and k as independent variables and denote them by x and y, then the Taylor's formula (1) becomes the following simpler form

$$f(x,y) = f(0,0) + \left(x\frac{\partial}{\partial x} + y\frac{\partial}{\partial y}\right)f(0,0) + \frac{1}{2!}\left(x\frac{\partial}{\partial x} + y\frac{\partial}{\partial y}\right)^2 f(0,0) + \cdots +$$

$$\frac{1}{n!}\left(x\frac{\partial}{\partial x} + y\frac{\partial}{\partial y}\right)^n f(0,0) + \frac{1}{(n+1)!}\left(x\frac{\partial}{\partial x} + y\frac{\partial}{\partial y}\right)^{n+1} f(\theta x, \theta y) \quad (3)$$

Taylor's formula (1) provides polynomial approximations of functions of two variables. The first n derivative terms give the polynomial; the last term gives the approximation error.

Example 1 Find a quadratic approximation of the function $f(x,y) = x^y$ near the point $(1,4)$ and use it to calculate $(1.08)^{3.96}$.

Solution We take $n = 2$, $(x_0, y_0) = (1,4)$ and $(x,y) = (x_0 + h, y_0 + k)$ in equation (1)

$$f(x,y) = f(1,4) + f'_x(1,4)(x-1) + f'_y(1,4)(y-4) +$$

$$\frac{1}{2}f''_{xx}(1,4)(x-1)^2 + f''_{xy}(1,4)(x-1)(y-4) +$$

$$\frac{1}{2}f''_{yy}(1,4)(y-4)^2 + \text{remainder}$$

with

$$f(1,4) = x^y\big|_{(1,4)} = 1, \ f'_x(1,4) = yx^{y-1}\big|_{(1,4)} = 4$$

$$f'_y(1,4) = x^y \ln x\big|_{(1,4)} = 0, \ f''_{xx}(1,4) = y(y-1)x^{y-2}\big|_{(1,4)} = 12$$

Chapter 8 Differential Calculus of Multivariable Functions

$$f''_{xy}(1,4) = (x^{y-1} + yx^{y-1}\ln x)\big|_{(1,4)} = 1, f''_{yy}(1,4) = x^y (\ln x)^2 \big|_{(1,4)} = 0$$

Then the corresponding quadratic approximation to f at $(1,4)$ is

$$x^y \approx 1 + 4(x - 1) + 6(x - 1)^2 + (x - 1)(y - 4)$$

Letting $x = 1.08$ and $y = 3.96$, we obtain

$$(1.08)^{3.96} \approx 1 + 4 \times 0.08 + 6 \times 0.08^2 + 0.08 \times (-0.04) = 1.3552$$

8.7.2 Maximum and Minimum Values

In this subsection, we show how to use partial derivatives to find maximum and minimum values of functions of several variables.

Local Maximum and Minimum Values

The concepts of local maximum and minimum values encounters in Chapter 4 extend readily to functions of two variables of the form $z = f(x,y)$.

Definition 8.8 A function f of two variables has a local maximum at (x_0, y_0) if $f(x,y) \leq f(x_0, y_0)$ for all (x,y) in some open region containing (x_0, y_0). The number $f(x_0, y_0)$ is called a local maximum value of f. Similarly, f has a local minimum at (x_0, y_0) if $f(x,y) \geq f(x_0, y_0)$ for all (x,y) in some open region containing (x_0, y_0) and the number $f(x_0, y_0)$ is called a local minimum value of f. Local maximum and minimum values of f are also called local extreme values of f.

For example, the function $f(x,y) = 3x^2 + 4y^2$ has a local minimum value $f(0,0) = 0$ at $(0,0)$ and the function $g(x,y) = 1 - \sqrt{x^2 + (y-1)^2}$ has a local maximum value $g(0,1) = 1$ at $(0,1)$.

Theorem 8.7 If f has a local maximum or minimum at (x_0, y_0) and the partial derivatives of f exist there, then $f'_x(x_0, y_0) = 0$ and $f'_y(x_0, y_0) = 0$.

Proof Suppose that f has a local maximum at (x_0, y_0). The function of one variable $g(x) = f(x, y_0)$, obtained by holding $y = y_0$ fixed, has a local maximum at x_0. By Fermat's Theorem (see Section 4.4), $f'_x(x_0, y_0) = g'(x_0) = 0$. Similarly, the function $h(y) = f(x_0, y)$, obtained by holding $x = x_0$ fixed, has a local maximum at y_0, which implies that $f'_y(x_0, y_0) = h'(y_0) = 0$. An analogous argument is used for the local minimum case.

Recall that for a function of one variable the condition $f'(x_0) = 0$ does not guarantee a local extreme value at x_0. A similar precaution must be taken with Theorem 8.7. The conditions $f'_x(x_0,y_0) = 0$ and $f'_y(x_0,y_0) = 0$ do not imply that f has a local extremum at (x_0,y_0). For example, the function $f(x,y) = x^2 - y^2$ has zero partial derivatives at $(0, 0)$, but f has no local maximum or minimum at $(0,0)$. A function f may have a local extremum at a point where one or both of the partial derivatives of f do not exist. For example, the function $f(x,y) = \sqrt{x^2 + y^2}$ has a local minimum value $f(0,0) = 0$ at $(0,0)$ where both partial derivatives of f do not exist.

Definition 8.9 An interior point (x_0,y_0) in the domain of f is a critical point of f if either $f'_x(x_0,y_0) = 0$ and $f'_y(x_0,y_0) = 0$, or one or both of the partial derivatives of f at (x_0,y_0) do not exist.

Critical points are candidates for local maximum and minimum values. Therefore, the procedure for locating local maximum and minimum values is to find the critical points and then determine whether these candidates correspond to genuine local maximum and minimum values.

Theorem 8.8 (Second Derivative Test for Local Extrema) Suppose that the second partial derivatives of f are continuous on some open region containing (x_0,y_0), and suppose that $f'_x(x_0,y_0) = 0$ and $f'_y(x_0,y_0) = 0$. Let $D(x,y) = f''_{xx}(x,y)f''_{yy}(x,y) - (f''_{xy}(x,y))^2$. Then

(a) f has a local maximum at (x_0,y_0) if $D(x_0,y_0) > 0$ and $f''_{xx}(x_0,y_0) < 0$.
(b) f has a local minimum at (x_0,y_0) if $D(x_0,y_0) > 0$ and $f''_{xx}(x_0,y_0) > 0$.
(c) f has no local extremum at (x_0,y_0) if $D(x_0,y_0) < 0$.

Proof By Taylor formula for $f(x,y)$ at (x_0,y_0) with $n = 2$, we have
$$f(x_0 + h,y_0 + k) = f(x_0,y_0) + f'_x(x_0,y_0)h + f'_y(x_0,y_0)k +$$
$$\frac{1}{2}(f''_{xx}(x_0 + \theta h, y_0 + \theta k)h^2 + 2f''_{xy}(x_0 + \theta h, y_0 + \theta k)hk +$$
$$f''_{yy}(x_0 + \theta h, y_0 + \theta k)k^2)$$

Since $f'_x(x_0,y_0) = 0$ and $f'_y(x_0,y_0) = 0$, this equation reduces to
$$f(x_0 + h,y_0 + k) - f(x_0,y_0)$$
$$= \frac{1}{2}(f''_{xx}(x_0 + \theta h, y_0 + \theta k)h^2 + 2f''_{xy}(x_0 + \theta h, y_0 + \theta k)hk +$$
$$f''_{yy}(x_0 + \theta h, y_0 + \theta k)k^2)$$

The presence of an extremum of f at (x_0,y_0) is determined by the sign of $f(x_0 + h, y_0 + k) - f(x_0,y_0)$. By the above equation, this is the same as the sign of

Chapter 8 Differential Calculus of Multivariable Functions

$$F(\theta) = f''_{xx}(x_0 + \theta h, y_0 + \theta k)h^2 + 2f''_{xy}(x_0 + \theta h, y_0 + \theta k)hk + f''_{yy}(x_0 + \theta h, y_0 + \theta k)k^2$$

Now, if $F(0) \neq 0$, the sign of $F(\theta)$ will be the same as the sign of $F(0)$ for sufficiently small values of h and k because the second partial derivatives of f at (x_0, y_0) are continuous. Since

$$f''_{xx}(x_0, y_0) F(0) = f''_{xx}(x_0, y_0) [f''_{xx}(x_0, y_0)h^2 + 2f''_{xy}(x_0, y_0)hk + f''_{yy}(x_0, y_0)]$$
$$= [f''_{xx}(x_0, y_0)h + f''_{xy}(x_0, y_0)k]^2 + D(x_0, y_0)k^2$$

we know that

(a) If $D(x_0, y_0) > 0$ and $f''_{xx}(x_0, y_0) < 0$, then $F(0) < 0$ for sufficiently small values of h and k, and f has a local maximum at (x_0, y_0).

(b) If $D(x_0, y_0) > 0$ and $f''_{xx}(x_0, y_0) > 0$, then $F(0) > 0$ for sufficiently small values of h and k, and f has a local minimum at (x_0, y_0).

(c) If $D(x_0, y_0) < 0$, there are combinations of arbitrarily small nonzero values of h and k for which $F(0) < 0$, and other values for which $F(0) > 0$. So f does not have a local extremum at (x_0, y_0).

Example 2 Find the local extreme values of the function $f(x, y) = x^3 - y^2 + 3x^2 + 4y - 9x$.

Solution The function f is differentiable everywhere, so the critical points occur only at the points where

$$f'_x(x, y) = 3x^2 + 6x - 9 = 0$$
$$f'_y(x, y) = -2y + 4 = 0$$

Solving the first equation, we get $x = -3$ and $x = 1$; solving the second equation, we get $y = 2$. Therefore, the critical points are $(-3, 2)$ and $(1, 2)$. The second partial derivatives and $D(x, y)$ are

$$f''_{xx}(x, y) = 6x + 6, \ f''_{xy}(x, y) = 0, \ f''_{yy}(x, y) = -2$$
$$D(x, y) = f''_{xx}(x, y)f''_{yy}(x, y) - (f''_{xy}(x, y))^2 = -12(x + 1)$$

Since $D(-3, 2) = 24 > 0$ and $f''_{xx}(-3, 2) = -12 < 0$, $f(-3, 2) = 31$ is a local maximum value by the Second Derivative Test. But since $D(1, 2) = -24 < 0$, the Second Derivative Test tells us that f does not have a local extremum at $(1, 2)$.

Example 3 Find the shortest distance from the point $(1, 0, -2)$ to the plane $x + 2y + z - 4 = 0$.

Solution The distance from any point (x,y,z) to the point $(1,0,-2)$ is
$$d = \sqrt{(x-1)^2 + y^2 + (z+2)^2}$$
but if (x,y,z) lies on the plane $x + 2y + z - 4 = 0$, then $z = 4 - x - 2y$ and so we have
$$d = \sqrt{(x-1)^2 + y^2 + (6 - x - 2y)^2}$$
It is easier to minimize
$$f(x,y) = d^2 = (x-1)^2 + y^2 + (6 - x - 2y)^2$$
which has the same critical points as d. The critical points of f satisfy the equations
$$f'_x(x,y) = 4x + 4y - 14 = 0$$
$$f'_y(x,y) = 4x + 10y - 24 = 0$$
By solving these equations, we find the only critical point is $\left(\dfrac{4}{3}, -\dfrac{4}{3}\right)$. We calculate the second partial derivatives and $D(x,y)$
$$f''_{xx}(x,y) = 4, f''_{xy}(x,y) = 4, f''_{yy}(x,y) = 10$$
$$D(x,y) = f''_{xx}(x,y)f''_{yy}(x,y) - (f''_{xy}(x,y))^2 = 24$$
Since $D\left(\dfrac{4}{3}, -\dfrac{4}{3}\right) = 24 > 0$ and $f''_{xx}\left(\dfrac{4}{3}, -\dfrac{4}{3}\right) = 4 > 0$, by the Second Derivative Test f has a local minimum at $\left(\dfrac{4}{3}, -\dfrac{4}{3}\right)$. This local minimum is actually an absolute minimum because there must be a point on the given plane that is closest to $(1,0,-2)$. If $x = \dfrac{4}{3}$ and $y = -\dfrac{4}{3}$, then
$$d = \sqrt{\left(\dfrac{5}{6}\right)^2 + \left(\dfrac{5}{3}\right)^2 + \left(\dfrac{5}{6}\right)^2} = \dfrac{5\sqrt{6}}{6}$$
The shortest distance from $(1,0,-2)$ to the plane $x + 2y + z - 4 = 0$ is $\dfrac{5\sqrt{6}}{6}$.

Absolute Extreme Values on Closed Bounded Set

We know that a continuous function on a closed bounded set attains its absolute maximum and absolute minimum values on that set (Section 8.1). Absolute maximum and minimum values on a closed bounded set D occur in two ways: (i) They may be local maximum or minimum values at interior points of D, where they are associated with critical points. (ii) They may occur on the boundary of D. Therefore, the search for absolute maximum and minimum values of a continuous function f on a closed bounded

Chapter 8 Differential Calculus of Multivariable Functions

set D is accomplished in the following steps:
1. Find the values of f at all critical points of f in D.
2. Find the maximum and minimum values of f on the boundary of D.
3. The largest of the values from Steps 1 and 2 is the absolute maximum value; the smallest of these values is the absolute minimum value.

Example 4 Find the absolute maximum and minimum values of the function $f(x,y) = 2 + 2x + 2y - x^2 - y^2$ on the triangular plate $D = \{(x,y) \mid 0 \leq x \leq 9, 0 \leq y \leq 9 - x\}$.

Solution Since f is continuous on the closed bounded triangular plate D, it has both an absolute maximum and an absolute minimum on D. We first find the critical points and the values of these points. The critical points satisfy the equations

$$f'_x(x,y) = 2 - 2x = 0$$
$$f'_y(x,y) = 2 - 2y = 0$$

so the only critical point inside D is $(1,1)$, and the value of f there is

$$f(1,1) = 4$$

Next we determine the maximum and minimum values of f on the boundary of D. On the line segment $y = 0, 0 \leq x \leq 9$, we have

$$f(x,0) = 2 + 2x - x^2, 0 \leq x \leq 9$$

Its extreme values may occur at the endpoints $x = 0$ and $x = 9$ and at the interior points where $\frac{d}{dx}f(x,0) = 2 - 2x = 0$. The only interior point where $\frac{d}{dx}f(x,0) = 0$ is $x = 1$. At these values of x, we have

$$f(0,0) = 2, f(1,0) = 3, f(9,0) = -61$$

On the line segment $x = 0, 0 \leq y \leq 9$, we have

$$f(0,y) = 2 + 2y - y^2, 0 \leq y \leq 9$$

We know from the symmetry of f in x and y and from the results we just obtained that the candidates on this line segment are

$$f(0,0) = 2, f(0,1) = 3, f(0,9) = -61$$

On the line segment $y = 9 - x, 0 \leq x \leq 9$, we have

$$f(x, 9 - x) = -61 + 18x - 2x^2, 0 \leq x \leq 9$$

Since we have already accounted for the values of f at the endpoints of this line segment, we need only look at the interior points of this line segment where $\frac{d}{dx}f(x,9-x) = 18 - 4x = 0$. The only interior point is $x = \frac{9}{2}$, where

$$f\left(\frac{9}{2},\frac{9}{2}\right) = -\frac{41}{2}$$

Thus, on the boundary, the absolute maximum value of f is $f(0,1) = f(1,0) = 3$ and the absolute minimum value is $f(0,9) = f(9,0) = -61$.

Having completed the first two steps of this procedure, we have three function values to consider

$$f(1,1) = 4, f(0,1) = f(1,0) = 3, f(0,9) = f(9,0) = -61$$

We compare these values and conclude that the absolute maximum value of f on D is $f(1,1) = 4$ and the absolute minimum value is $f(0,9) = f(9,0) = -61$.

8.7.3 Lagrange Multipliers

We sometimes need to find the extreme values of a function subject to one or more constraints. Such a constrained extremum problem may be solved by substitution. For example, in Example 3 we minimized the distance function $d = \sqrt{(x-1)^2 + y^2 + (z+2)^2}$ subject to the constraint $x + 2y + z - 4 = 0$ by minimizing $d = \sqrt{(x-1)^2 + y^2 + (6-x-2y)^2}$. However, attempts to solve a constrained extremum problem by substitution do not always go smoothly. In this section, we explore a powerful method for finding extreme values of constrained functions: the method of Lagrange multipliers.

Suppose that we want to find the maximum and minimum values of a function $z = f(x,y)$ subject to a constraint $g(x,y) = 0$. We suppose that f has such an extreme value at a point (x_0, y_0) and suppose also that f and g have continuous partial derivatives and $g'_y(x_0, y_0) \neq 0$. By the Implicit Function Theorem, we know that the equation $g(x,y) = 0$ defines $y = y(x)$ as a differentiable function of x near the point (x_0, y_0) ($y_0 = y(x_0)$) and

$$\frac{dy}{dx} = -\frac{g'_x(x,y)}{g'_y(x,y)}$$

Then the one variable function

Chapter 8 Differential Calculus of Multivariable Functions

$$z = f(x, y(x))$$

attains its local extremum at x_0. Therefore, this function has zero derivative at x_0

$$\frac{dz}{dx}\bigg|_{x=x_0} = f'_x(x_0, y_0) + f'_y(x_0, y_0)\frac{dy}{dx}\bigg|_{x=x_0} = 0$$

that is

$$f'_x(x_0, y_0) - \frac{f'_y(x_0, y_0)}{g'_y(x_0, y_0)} g'_x(x_0, y_0) = 0$$

Let

$$\lambda_0 = -\frac{f'_y(x_0, y_0)}{g'_y(x_0, y_0)}$$

Then x_0, y_0 and λ_0 must satisfy

$$f'_x(x_0, y_0) + \lambda_0 g'_x(x_0, y_0) = 0$$
$$f'_y(x_0, y_0) + \lambda_0 g'_y(x_0, y_0) = 0$$
$$g(x_0, y_0) = 0$$

For easy to remember, we introduce a new function

$$F(x, y, \lambda) = f(x, y) + \lambda g(x, y)$$

which is called the Lagrange function. It is obvious that (x_0, y_0, λ_0) is a critical point of the Lagrange function F.

The procedure above is as follows.

The Method of Lagrange Multipliers with One Constraint

Suppose that $f(x, y)$ and $g(x, y)$ have continuous partial derivatives. To find the absolute maximum and minimum values of $f(x, y)$ subject to the constraint $g(x, y) = 0$ (assuming that these absolute extreme values exist):

1. Find all values of x, y and λ (λ is called a Lagrange multiplier) that simultaneously satisfy the equations

$$\frac{\partial F}{\partial x} = f'_x(x, y) + \lambda g'_x(x, y) = 0$$

$$\frac{\partial F}{\partial y} = f'_y(x, y) + \lambda g'_y(x, y) = 0$$

$$\frac{\partial F}{\partial \lambda} = g(x, y) = 0$$

2. Evaluate f at all the points (x, y) found in Step 1. The largest of these values is the absolute maximum value of f subject to the constraint; the smallest is the absolute

minimum value of f.

Example 5 A rectangular box without a lid is to be made from 12 m² of cardboard. Find the maximum volume of such a box.

Solution Let x, y and z be the length, width and height, respectively, of the box in meters. Then we wish to maximize
$$V = f(x,y,z) = xyz$$
subject to the constraint
$$g(x,y,z) = 2xz + 2yz + xy - 12$$
Let
$$F(x,y,z,\lambda) = f(x,y,z) + \lambda g(x,y,z) = xyz + \lambda(2xz + 2yz + xy - 12)$$
We calculate all the partial derivatives of F and set these partial derivatives equal to zero
$$\frac{\partial F}{\partial x} = yz + \lambda(2z + y) = 0$$
$$\frac{\partial F}{\partial y} = xz + \lambda(2z + x) = 0$$
$$\frac{\partial F}{\partial z} = xy + \lambda(2x + 2y) = 0$$
$$\frac{\partial F}{\partial \lambda} = 2xz + 2yz + xy - 12 = 0$$

Solving this system of equations, we get $x = y = 2, z = 1, \lambda = -\frac{1}{2}$ (noticing that x, y and z are all positve). We argue from the physical nature of this problem that there must be an absolute maximum volume, which must occur at the point $(2,2,1)$ by the method of Lagrange multipliers. Thus, the maximum volume of the box is $V = 2 \times 2 \times 1 = 4$ m³.

Suppose now that we want to find the maximum and minimum values of a function $f(x,y,z)$ subject to two constraints $g(x,y,z) = 0$ and $h(x,y,z) = 0$. To solve the problem, we introduce the following Lagrange function
$$F(x,y,z,\lambda,\mu) = f(x,y,z) + \lambda g(x,y,z) + \mu h(x,y,z)$$
Then the procedure for solving the problem is as follows.

The Method of Lagrange Multipliers with Two Constraints

Suppose that $f(x,y,z)$, $g(x,y,z)$ and $h(x,y,z)$ have continuous partial derivatives. To find the absolute maximum and minimum values of $f(x,y,z)$ subject to the two

Chapter 8 Differential Calculus of Multivariable Functions

constraints $g(x,y,z) = 0$ and $h(x,y,z) = 0$ (assuming that these absolute extreme values exist):

1. Find all values of x, y, z, λ and μ (λ and μ are called Lagrange multipliers) that simultaneously satisfy the equations

$$\frac{\partial F}{\partial x} = f'_x(x,y,z) + \lambda g'_x(x,y,z) + \mu h'_x(x,y,z) = 0$$

$$\frac{\partial F}{\partial y} = f'_y(x,y,z) + \lambda g'_y(x,y,z) + \mu h'_y(x,y,z) = 0$$

$$\frac{\partial F}{\partial z} = f'_z(x,y,z) + \lambda g'_z(x,y,z) + \mu h'_z(x,y,z) = 0$$

$$\frac{\partial F}{\partial \lambda} = g(x,y,z) = 0$$

$$\frac{\partial F}{\partial \mu} = h(x,y,z) = 0$$

2. Evaluate f at all the points (x,y,z) found in Step 1. The largest of these values is the absolute maximum value of f subject to the constraints; the smallest is the absolute minimum value of f.

Example 6 The paraboloid $z = x^2 + y^2$ and the plane $x + y + z = 1$ intersect in a curve. Find the points on the curve that lie closest to and farthest from the origin.

Solution The distance from any point (x,y,z) to the origin is

$$d = \sqrt{x^2 + y^2 + z^2}$$

but it is easier to maximize and minimize the square of the distance

$$f(x,y,z) = d^2 = x^2 + y^2 + y^2$$

subject to the constraints

$$g(x,y,z) = x^2 + y^2 - z = 0$$
$$h(x,y,z) = x + y + z - 1 = 0$$

Let

$$F(x,y,z,\lambda,\mu) = f(x,y,z) + \lambda g(x,y,z) + \mu h(x,y,z)$$
$$= x^2 + y^2 + z^2 + \lambda(x^2 + y^2 - z) + \mu(x + y + z - 1)$$

We compute the partial derivatives of F and set these partial derivatives equal zero

$$\frac{\partial F}{\partial x} = 2x + 2\lambda x + \mu = 0$$

$$\frac{\partial F}{\partial y} = 2y + 2\lambda y + \mu = 0$$

$$\frac{\partial F}{\partial z} = 2z - \lambda + \mu = 0$$

$$\frac{\partial F}{\partial \lambda} = x^2 + y^2 - z = 0$$

$$\frac{\partial F}{\partial \mu} = x + y + z - 1 = 0$$

Solving this system of equations, we obtain

$$x = y = \frac{-1 \pm \sqrt{3}}{2}, z = 2 \mp \sqrt{3}, \lambda = -3 \pm \frac{5}{3}\sqrt{3}, \mu = -7 \pm \frac{11}{3}\sqrt{3}$$

We argue from the geometric nature of this problem that there must be both absolute maximum and minimum distances. Therefore, the distance function d must have absolute extreme values at the points

$$P_1\left(\frac{\sqrt{3}-1}{2}, \frac{\sqrt{3}-1}{2}, 2-\sqrt{3}\right), P_2\left(-\frac{\sqrt{3}+1}{2}, -\frac{\sqrt{3}+1}{2}, 2+\sqrt{3}\right)$$

Evaluating d at these two points, we find that

$$d|_{P_1} = \sqrt{\left(\frac{\sqrt{3}-1}{2}\right)^2 + \left(\frac{\sqrt{3}-1}{2}\right)^2 + (2-\sqrt{3})^2} = \sqrt{9 - 5\sqrt{3}}$$

$$d|_{P_2} = \sqrt{\left(-\frac{\sqrt{3}+1}{2}\right)^2 + \left(-\frac{\sqrt{3}+1}{2}\right)^2 + (2+\sqrt{3})^2} = \sqrt{9 + 5\sqrt{3}}$$

Thus, the closest point is P_1 and The farthest is P_2.

The method of Lagrange multipliers can be extended to the general case. If we want to find the maximum and minimum values of a function $f(x_1, x_2, \cdots, x_n)$ subject to m constraints $g_i(x_1, x_2, \cdots, x_n) = 0, i = 1, 2, \cdots, m$ ($m < n$), then we can solve this problem by introducing the following Lagrange function

$$F(x_1, \cdots, x_n, \lambda_1, \cdots, \lambda_m) = f(x_1, \cdots, x_n) + \sum_{i=1}^{m} \lambda_i g(x_1, \cdots, x_n)$$

and finding out all critical points of the function F.

Chapter 8 Differential Calculus of Multivariable Functions

8.8 Directional Derivatives and The Gradient Vector

Partial derivatives tell us the rate of change of a function in coordinate directions. However, they do not directly tell us the rate of change of the function in directions other than the coordinate directions. For example, suppose that you are standing at a point $(x_0, y_0, f(x_0, y_0))$ on the surface $z = f(x,y)$. The partial derivatives $f'_x(x_0, y_0)$ and $f'_y(x_0, y_0)$ tell you the rate of change of the surface at that point in the directions parallel to the x-axis and y-axis, respectively. But you could walk in an infinite number of directions from that point and find a different rate of change in every direction. In this section we introduce a type of derivative, called the directional derivative, that enables us to find the rate of change of a function of two or more variables in any direction.

Suppose that we wish to find the rate of change of a function f at a point $P_0(x_0, y_0, z_0)$ in the direction of an arbitrary vector l. The derivative we seek must be computed along the line L through P_0 in the direction of l. The points on L satisfy the parametric equations

$$x = x_0 + t\cos\alpha, y = y_0 + t\cos\beta, z = z_0 + t\cos\gamma, t \in \mathbf{R}$$

where $\cos\alpha, \cos\beta, \cos\gamma$ are the directional cosines of l.

Definition 8.10 The directional derivative of a function f at a point $P_0(x_0, y_0, z_0)$ in the direction of a vector l, denoted by $\left.\dfrac{\partial f}{\partial l}\right|_{P_0}$ (or $\left.\dfrac{\partial u}{\partial l}\right|_{P_0}$ if we write $u = f(x,y,z)$), is

$$\left.\frac{\partial f}{\partial l}\right|_{P_0} = \lim_{t \to 0} \frac{f(x_0 + t\cos\alpha, y_0 + t\cos\beta, z_0 + t\cos\gamma) - f(x_0, y_0, z_0)}{t}$$

if this limit exists, where $\cos\alpha, \cos\beta, \cos\gamma$ are the directional cosines of l.

The directional derivatives in coordinate directions are just partial derivatives. For example, the directional derivative of f at $P_0(x_0, y_0, z_0)$ in the x-direction, that is, in the direction of the vector i, is $f'_x(x_0, y_0, z_0)$.

Theorem 8.9 If the function $f(x,y,z)$ is differentiable at $P_0(x_0, y_0, z_0)$, then f has a directional derivative at P_0 in the direction of any vector l and

$$\left.\frac{\partial f}{\partial l}\right|_{P_0} = \left.\frac{\partial f}{\partial x}\right|_{P_0}\cos\alpha + \left.\frac{\partial f}{\partial y}\right|_{P_0}\cos\beta + \left.\frac{\partial f}{\partial z}\right|_{P_0}\cos\gamma \qquad (1)$$

where $\cos\alpha, \cos\beta, \cos\gamma$ are the directional cosines of l.

Proof If we define a function g of the single variable t by

$$g(t) = f(x_0 + t\cos\alpha, y_0 + t\cos\beta, z_0 + t\cos\gamma)$$

then

$$g'(0) = \frac{\partial f}{\partial l}\Big|_{P_0}$$

On the other hand, we can write $z = f(x,y,z)$, where $x = x_0 + t\cos\alpha$, $y = y_0 + t\cos\beta$, $z = z_0 + t\cos\gamma$. By the Chain Rule, we obtain

$$\frac{\partial f}{\partial l}\Big|_{P_0} = g'(0) = \frac{\partial f}{\partial x}\Big|_{P_0}\cos\alpha + \frac{\partial f}{\partial y}\Big|_{P_0}\cos\beta + \frac{\partial f}{\partial z}\Big|_{P_0}\cos\gamma$$

For a function f of two variables x and y, the directional derivative of f at the point $P_0(x_0, y_0)$ in the direction of a vector l is

$$\frac{\partial f}{\partial l}\Big|_{P_0} = \frac{\partial f}{\partial x}\Big|_{P_0}\cos\alpha + \frac{\partial f}{\partial y}\Big|_{P_0}\cos\beta = \frac{\partial f}{\partial x}\Big|_{P_0}\cos\alpha + \frac{\partial f}{\partial y}\Big|_{P_0}\sin\alpha \qquad (2)$$

Example 1 Find the directional derivative of the function $f(x,y) = xe^y + \cos(xy)$ at the point $(2,0)$ in the direction of $l = \{3, -4\}$.

Solution The unit vector in the direction of $l = \{3, -4\}$ is

$$l^0 = \frac{l}{|l|} = \frac{1}{5}\{3, -4\} = \left\{\frac{3}{5}, -\frac{4}{5}\right\}$$

Then we obtain

$$\cos\alpha = \frac{3}{5}, \sin\alpha = -\frac{4}{5}$$

Computing the partial derivatives of f with respect to x and y, we find that

$$f'_x(x,y) = e^y - y\sin(xy)$$
$$f'_y(x,y) = xe^y - x\sin(xy)$$

Evaluated at $(2,0)$, we have

$$f'_x(2,0) = 1, f'_y(2,0) = 2$$

By formula (2), the directional derivative of f at $(2,0)$ in the direction of l is

$$\frac{\partial f}{\partial l}\Big|_{(2,0)} = f'_x(2,0)\cos\alpha + f'_y(2,0)\sin\alpha = 1\times\frac{3}{5} + 2\times\left(-\frac{4}{5}\right) = -1$$

Example 2 Find the directional derivative of the function $u = x^2y + y^2z + z^2x$ at the point $(1,1,1)$ in the direction of $l = \{1, -2, 1\}$.

Solution The unit vector in the direction of $l = \{1, -2, 1\}$ is

Chapter 8 Differential Calculus of Multivariable Functions

$$l^0 = \frac{l}{|l|} = \frac{1}{\sqrt{6}}\{1, -2, 1\} = \left\{\frac{1}{\sqrt{6}}, -\frac{2}{\sqrt{6}}, \frac{1}{\sqrt{6}}\right\}$$

Then we have

$$\cos\alpha = \frac{1}{\sqrt{6}}, \cos\beta = -\frac{2}{\sqrt{6}}, \cos\gamma = \frac{1}{\sqrt{6}}$$

We compute the partial derivatives of u

$$\frac{\partial u}{\partial x} = 2xy + z^2, \frac{\partial u}{\partial y} = 2yz + x^2, \frac{\partial u}{\partial z} = 2zx + y^2$$

At the given point $(1,1,1)$, we have

$$\frac{\partial u}{\partial x}\Big|_{(1,1,1)} = 3, \frac{\partial u}{\partial y}\Big|_{(1,1,1)} = 3, \frac{\partial u}{\partial z}\Big|_{(1,1,1)} = 3$$

By formula (1), the directional derivative of u at $(1,1,1)$ in the direction of l is

$$\frac{\partial u}{\partial l}\Big|_{(1,1,1)} = \frac{\partial u}{\partial x}\Big|_{(1,1,1)}\cos\alpha + \frac{\partial f}{\partial y}\Big|_{(1,1,1)}\cos\beta + \frac{\partial f}{\partial z}\Big|_{(1,1,1)}\cos\gamma$$

$$= 3 \times \frac{1}{\sqrt{6}} + 3 \times \left(-\frac{2}{\sqrt{6}}\right) + 3 \times \frac{1}{\sqrt{6}} = 0$$

Notice from Theorem 8.9 that the directional derivative of f at the point (x,y,z) can be written as the dot product of two vectors

$$\frac{\partial f}{\partial l} = \frac{\partial f}{\partial x}\cos\alpha + \frac{\partial f}{\partial y}\cos\beta + \frac{\partial f}{\partial z}\cos\gamma = \left\{\frac{\partial f}{\partial x}, \frac{\partial f}{\partial y}, \frac{\partial f}{\partial z}\right\} \cdot \{\cos\alpha, \cos\beta, \cos\gamma\}$$

The first vector in this dot product is important not only in calculating directinal derivatives; it plays many other roles in multivariable calculus. We call it the gradient of f.

Definition 8.11 If f is a function of three variables x, y and z, then the gradient of f, denoted by grad f or ∇f, is the vector function

$$\text{grad } f = \nabla f = \frac{\partial f}{\partial x}i + \frac{\partial f}{\partial y}j + \frac{\partial f}{\partial z}k = \left\{\frac{\partial f}{\partial x}, \frac{\partial f}{\partial y}, \frac{\partial f}{\partial z}\right\}$$

For a function f of two variables x and y, the gradient vector is

$$\text{grad } f = \nabla f = \frac{\partial f}{\partial x}i + \frac{\partial f}{\partial y}j = \left\{\frac{\partial f}{\partial x}, \frac{\partial f}{\partial y}\right\}$$

The gradient satisfies sum, difference, product and quotient rules analogous to those for ordinary derivatives.

Algebraic Rules for Gradients

If f and g are differentiable functions, then
(a) $\operatorname{grad}(f + g) = \operatorname{grad} f + \operatorname{grad} g$;
(b) $\operatorname{grad}(f - g) = \operatorname{grad} f - \operatorname{grad} g$;
(c) $\operatorname{grad}(fg) = f\operatorname{grad} g + g\operatorname{grad} f$;
(d) $\operatorname{grad}\left(\dfrac{f}{g}\right) = \dfrac{g\operatorname{grad} f - f\operatorname{grad} g}{g^2} \; (g \neq 0)$.

With the definition of the gradient, the directional derivative of f at each point in its domain in the direction of the vector l can be written as

$$\frac{\partial f}{\partial l} = \operatorname{grad} f \cdot l^0$$

where l^0 is the unit vector in the direction of l. Using properties of the dot product, we have

$$\frac{\partial f}{\partial l} = \operatorname{grad} f \cdot l^0 = |\operatorname{grad} f| \cos \theta$$

where θ is the angle between $\operatorname{grad} f$ and l. It follows that $\dfrac{\partial f}{\partial l}$ has its maximum value when $\cos \theta = 1$, which corresponds to $\theta = 0$. Therefore, $\dfrac{\partial f}{\partial l}$ has its maximum value and f has its greatest rate of increase when $\operatorname{grad} f$ and l point in the same direction. Notice that when $\cos \theta = 1$, the actual rate of increase is $\dfrac{\partial f}{\partial l} = |\operatorname{grad} f|$.

Similarly, when $\theta = \pi$, we have $\cos \theta = -1$, and f has its greatest rate of decrease when $\operatorname{grad} f$ and l point in opposite directions. The actual rate of decrease is $\dfrac{\partial f}{\partial l} = -|\operatorname{grad} f|$. Notice that $\dfrac{\partial f}{\partial l} = 0$ when $\theta = \dfrac{\pi}{2}$, which means $\operatorname{grad} f$ and l are orthogonal.

Example 3 Suppose that the temperature at a point (x,y,z) in space is given by $f(x,y,z) = x^2 + 2y^2 + 4z^2$, where f is measured in degrees Celsius and x, y and z in meters. In which direction does the temperature increase fastest at the point $(1,1,1)$? What is the maximum rate of increase?

Solution The gradient of f is

$$\operatorname{grad} f = \left\{\frac{\partial f}{\partial x}, \frac{\partial f}{\partial y}, \frac{\partial f}{\partial z}\right\} = \{2x, 4y, 8z\}$$

Chapter 8 Differential Calculus of Multivariable Functions

At the point $(1,1,1)$ the gradient vector is
$$\text{grad } f|_{(1,1,1)} = \{2,4,8\}$$
The temperature increases fastest in the direction of the gradient vector $\text{grad } f|_{(1,1,1)} = \{2,4,8\}$ and the maximum rate of increase is the length of the gradient vector
$$|\text{grad } f|_{(1,1,1)} = \sqrt{2^2 + 4^2 + 8^2} = 2\sqrt{21}$$
Therefore, the maximum rate of increase of temperature is $2\sqrt{21} \approx 9.17\,°C/m$.

The set of points (x,y,z) in space where a function of three independent variables has a constant value $f(x,y,z) = c$ is called a level surface of f. At every point $P(x,y,z)$ in the domain of $f(x,y,z)$, the gradient $\text{grad } f$ is normal to the level surface through P (Fig. 8.7).

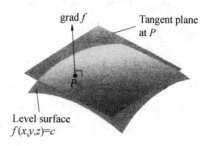

Fig. 8.7

8.9 Examples

Example 1 Let $u = f(t), t = \ln\sqrt{x^2 + y^2 + z^2}$ with
$$\frac{\partial^2 u}{\partial x^2} + \frac{\partial^2 u}{\partial y^2} + \frac{\partial^2 u}{\partial z^2} = (x^2 + y^2 + z^2)^{-\frac{3}{2}} \tag{1}$$
Find $f(t)$.

Solution
$$\frac{\partial u}{\partial x} = f'(t) \frac{x}{x^2 + y^2 + z^2}$$
$$\frac{\partial^2 u}{\partial x^2} = f''(t) \frac{x^2}{(x^2 + y^2 + z^2)^2} + f'(t) \frac{y^2 + z^2 - x^2}{(x^2 + y^2 + z^2)^2}$$

Because x, y, z appear symmetrically in the expression for u, we can directly get $\frac{\partial^2 u}{\partial y^2}$ and $\frac{\partial^2 u}{\partial z^2}$ by interchanging x with y and z respectively in the expression for $\frac{\partial^2 u}{\partial x^2}$, and substituting them into equation (1), we have
$$f''(t) \frac{1}{x^2 + y^2 + z^2} + f'(t) \frac{1}{x^2 + y^2 + z^2} = \frac{1}{(x^2 + y^2 + z^2)^{3/2}}$$

Note that $\sqrt{x^2+y^2+z^2} = e^t$, we conclude
$$f''(t) + f'(t) = e^{-t}$$
It is a second order non-homogeneous linear differential equation with constant cofficients, and its general solution is
$$f(t) = C_1 + C_2 e^{-t} - te^{-t}$$

Example 2 Suppose that $z = f(x,u,v)$, $u = g(x,y)$, $v = h(x,y,u)$, are all differentiable. Find $\dfrac{\partial z}{\partial x}$ and $\dfrac{\partial z}{\partial y}$.

Solution For multivariable composite functions, if a variable is both the intermediate variable and the independent variable, then sometimes it is very convenient to find partial derivatives by using the invariance of the total differential form. Taking the total differential on both sides of the three formulas respectively we have
$$dz = f'_x dx + f'_u du + f'_v dv$$
$$du = g'_x dx + g'_y dy$$
$$dv = h'_x dx + h'_y dy + h'_u du$$
Substitute du, dv into the first equality, we get
$$dz = f'_x dx + f'_u [g'_x dx + g'_y dy] + f'_v [h'_x dx + h'_y dy + h'_u (g'_x dx + g'_y dy)]$$
$$= [f'_x + f'_u g'_x + f'_v h'_x + f'_v h'_u g'_x] dx + [f'_u g'_y + f'_v h'_y + f'_v h'_u g'_y] dy$$
Thus
$$\frac{\partial z}{\partial x} = f'_x + f'_u g'_x + f'_v h'_x + f'_v h'_u g'_x$$
$$\frac{\partial z}{\partial y} = f'_u g'_y + f'_v h'_y + f'_v h'_u g'_y$$

It is also very convenient to use the implicit differentiation. Let
$$F_1 = z - f(x,u,v), F_2 = u - g(x,y), F_3 = v - h(x,y,u)$$
Then
$$\frac{\partial(F_1,F_2,F_3)}{\partial(z,u,v)} = \begin{vmatrix} 1 & -f'_u & -f'_v \\ 0 & 1 & 0 \\ 0 & -h'_u & 1 \end{vmatrix} = 1$$

$$\frac{\partial(F_1,F_2,F_3)}{\partial(x,u,v)} = \begin{vmatrix} -f'_x & -f'_u & -f'_v \\ -g'_x & 1 & 0 \\ -h'_x & -h'_u & 1 \end{vmatrix}$$

Chapter 8 Differential Calculus of Multivariable Functions

$$\frac{\partial(F_1,F_2,F_3)}{\partial(y,u,v)} = \begin{vmatrix} =-f'_x - f'_v h'_u g'_x - f'_v h'_x - f'_u g'_x \\ 0 & -f'_u & -f'_v \\ -g'_y & 1 & 0 \\ -h'_y & -h'_u & 1 \end{vmatrix}$$
$$= -f'_v h'_u g'_y - f'_v h'_y - f'_u g'_y$$

Hence
$$\frac{\partial z}{\partial x} = f'_x + f'_v h'_u g'_x + f'_v h'_x + f'_u g'_x$$

$$\frac{\partial z}{\partial y} = f'_v h'_u g'_y + f'_v h'_y + f'_u g'_y$$

Example 3 Find the extreme values of the implicit function $z = z(x,y)$ defined by the following equation
$$x^2 + y^2 + 4z^2 - 4xz + 2z - 1 = 0$$

Solution Let
$$\frac{\partial z}{\partial x} = -\frac{x - 2z}{4z - 2x + 1} = 0, \quad \frac{\partial z}{\partial y} = -\frac{y}{4z - 2x + 1} = 0$$

Solving the equations yields $x = 2z$, $y = 0$. Substitute them into the initial equation to get $z = \frac{1}{2}$, so the critical point is $(1,0)$. Since

$$A = \frac{\partial^2 z}{\partial x^2}\bigg|_{(1,0)} = -1, \quad B = \frac{\partial^2 z}{\partial x \partial y}\bigg|_{(1,0)} = 0, \quad C = \frac{\partial^2 z}{\partial y^2}\bigg|_{(1,0)} = -1$$

$A < 0$, $AC - B^2 = 1 > 0$, as $x = 1$, $y = 0$, z has a local maximum $\frac{1}{2}$.

If we regard $f = z$ as an objective function, then this question is changed into the conditional extreme value question of f under the constraint condition
$$x^2 + y^2 + 4z^2 - 4xz + 2z - 1 = 0$$

Example 4 Find the maximum and the minimum values of the real quadratic form
$$f(x,y,z) = Ax^2 + By^2 + Cz^2 + 2Dyz + 2Ezx + 2Fxy$$
in the unit sphere $x^2 + y^2 + z^2 = 1$.

Solution We set up the Lagrange function
$$F = f(x,y,z) + \lambda(1 - x^2 - y^2 - z^2)$$
According to the last equation in the equation system

$$\begin{cases} F'_x = 2[(A-\lambda)x + Fy + Ez] = 0 \\ F'_y = 2[Fx + (B-\lambda)y + Dz] = 0 \\ F'_z = 2[Ex + Dy + (C-\lambda)z] = 0 \\ x^2 + y^2 + z^2 = 1 \end{cases} \quad (2)$$

we get x,y,z are not all zero. Therefore, the homogeneous linear equation system consists of the first three equations have non-zero solutions. So, the coefficient determinant is zero, i. e.

$$\begin{vmatrix} A-\lambda & F & E \\ F & B-\lambda & D \\ E & D & C-\lambda \end{vmatrix} = 0$$

And the Lagrange multiplier λ_0 is the eigenvalue of the quadratic form matrix

$$G = \begin{pmatrix} A & F & E \\ F & B & D \\ E & D & C \end{pmatrix}$$

The solution $(x_0, y_0, z_0)^T$ of the system (2) is a unit eigenvector belonging to the eigenvalue λ_0 of the matrix G. Then the value of the quadratic form is as follows

$$f(x_0, y_0, z_0) = (x_0, y_0, z_0) G \begin{pmatrix} x_0 \\ y_0 \\ z_0 \end{pmatrix} = \lambda_0(x_0^2 + y_0^2 + z_0^2) = \lambda_0$$

Thus, the maximum value of the conditional extreme values problem is the maximum eigenvalue $\lambda_{\max}$ of the matrix G, and the minimum value of that is the minimum eigenvalue $\lambda_{\min}$ of the matrix G.

Example 5 An item E buried underground is emitting a distinctive scent in the atmosphere, and the distribution of the concentration of the scent on the ground is $v = e^{-k(x^2+2y^2)}$ (k is a positive constant). A police dog always searchs in the direction of strongest scent. Find the police dog's search route starting from the point (x_0, y_0).

Solution 1 Suppose that the police dog's search route is the curve $y = y(x)$, and that it's forward direction at the point (x,y) is the direction of the tangent vector $\{1, \frac{dy}{dx}\}$ of the curve $y = y(x)$. Note that the direction of strongest scent is the direction of the gradient of v

Chapter 8 Differential Calculus of Multivariable Functions

$$\text{grad } v = e^{-k(x^2+2y^2)}(-k)(2xi + 4yj)$$

Hence, $y = y(x)$ satisfies the following initial condition

$$\begin{cases} x\dfrac{dy}{dx} = 2y \\ y\vert_{x_0} = y_0 \end{cases}$$

Solving the equation by the method of separation of variables, we get:

as $x_0 \neq 0$, the search curve is $y = \dfrac{y_0}{x_0^2}x^2$;

as $x_0 = 0$, the search curve is $x = 0$.

Solution 2   The equation of the isoline of the scent is

$$x^2 + 2y^2 = C$$

Differentiating implicitly both sides with respect to x gives a equation that the isoline of the scent satisfies

$$x + 2yy' = 0$$

Since the police dog searchs in the direction of the gradient, the search curve is orthogonal to the isoline of the scent. Therefore, we obtain the equation for the search curve with initial condition

$$\begin{cases} xy' - 2y = 0 \\ y\vert_{x_0} = y_0 \end{cases}$$

The next steps are the same as shown in solution 1.

Exercises 8

8.1

1. Suppose that there is a circle. Let L denote the chord length corresponding to the arc, r and θ ($\theta < \pi$) denote the radius and central angle respectively. (1) Express L as a function of r and θ; (2) Express L as a function of r and d, where d denotes the distance from the center to the chord.

2. Suppose that the cross section of a rain gutter is an isosceles trapezoid (as shown in Fig. 8.8). Let $AB = x$, $BC = y$, z be the depth, and A denote the area of the cross section of the rain gutter. Express A as a function of x, y, z.

3. A particle with mass M is located in a fixed point (a,b,c), and another particle with mass m is located in the point (x,y,z). Let F denote the gravitational force between them. Express the projections F_x, F_y, F_z of F on three coordinate axes as a function of x, y, z respectively.

4. Find the domain of each of the given functions and sketch the graph of the domain. Point out which of the domains are open or closed, which are connected or disconnected, and which are bounded or unbounded

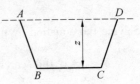

Fig. 8.8

(1) $z = \sqrt{x} - \sqrt{y}$;

(2) $z = \sqrt{2 - x^2 - y^2} + \dfrac{1}{\sqrt{x^2 + y^2 - 1}}$;

(3) $z = \ln[x\ln(y - x)]$;

(4) $u = \dfrac{1}{\arccos(x^2 + y^2 + z^2)}$.

5. Let $z = x + y + f(x - y)$, and $z = x^2$ as $y = 0$. Find the functions f and z.

6. Prove that $f(tx, ty) = t^2 f(x,y)$, given $f(x,y) = \sqrt{x^4 + y^4} - 2xy$.

7. Find $f(x,y)$ given $f(x + y, \dfrac{y}{x}) = x^2 - y^2$.

8. Find the following limits.

(1) $\lim\limits_{\substack{x \to 0 \\ y \to \pi}} [1 + \sin(xy)]^{\frac{y}{x}}$;

(2) $\lim\limits_{(x,y) \to (0,0)} \dfrac{xy}{\sqrt{xy + 1} - 1}$.

9. Find discontinuities of each of the following functions.

(1) $z = \dfrac{1}{x^2 + y^2}$;

(2) $z = \ln|4 - x^2 - y^2|$;

(3) $u = e^{\frac{1}{z}}/(x - y^2)$.

10. Discuss the continuity of the function

$$f(x,y) = \begin{cases} \dfrac{\sin(x^2 + y^2)}{2(x^2 + y^2)}, & x^2 + y^2 \ne 0 \\ \dfrac{1}{2}, & x^2 + y^2 = 0 \end{cases}$$

11. Let the functin $f(x,y)$ be continuous in the closed domain: $|x| \le a$, $|y| \le b$ and be positive definite (i.e., as $(x,y) \ne (0,0)$, $f(x,y) > 0, f(0,0) = 0$). Verify

Chapter 8 Differential Calculus of Multivariable Functions

that for the appropriate small positive number C, the graph of the equation $f(x,y) = C$ contains a closed curve enclosing the origin $(0,0)$.

8.2

1. Let $f(x,y) = x + (y-1)\arcsin\sqrt{\dfrac{x}{y}}$, find $f'_x(x,1)$.

2. Let $f(x,y) = \begin{cases} \dfrac{1}{2xy}\sin(x^2 y), & xy \neq 0 \\ 0, & xy = 0 \end{cases}$, find $f'_x(0,1), f'_y(0,1)$.

3. Find the partial derivatives of the given functions.
   (1) $z = (1 + xy)^y$;
   (2) $z = e^{-x}\sin(x + 2y)$;
   (3) $z = \arctan\dfrac{y}{x}$;
   (4) $z = \arcsin(y\sqrt{x})$;
   (5) $u = xe^{\pi xyz}$;
   (6) $u = z\ln\dfrac{x}{y}$.

4. Determine the second order partial derivatives of the given functions.
   (1) $z = \cos(xy)$;
   (2) $z = x^{2y}$;
   (3) $z = e^x \cos y$;
   (4) $z = \ln(e^x + e^y)$.

5. Verify that the following given function satisfies the specified equation.
   (1) $z = \dfrac{xy}{x+y}$ satisfies $x\dfrac{\partial z}{\partial x} + y\dfrac{\partial z}{\partial y} = z$;
   (2) $z = e^{x/y^2}$ satisfies $2x\dfrac{\partial z}{\partial x} + y\dfrac{\partial z}{\partial y} = 0$;
   (3) $z = \ln\sqrt{x^2 + y^2}$ satisfies $\dfrac{\partial^2 z}{\partial x^2} + \dfrac{\partial^2 z}{\partial y^2} = 0$;
   (4) $z = 2\cos^2\left(x - \dfrac{t}{2}\right)$ satisfies $2\dfrac{\partial^2 z}{\partial t^2} + \dfrac{\partial^2 z}{\partial x \partial t} = 0$.

6. Suppose $u = \varphi\left(\dfrac{y}{x}\right)$ is the solution of the equation $\dfrac{\partial^2 u}{\partial x^2} + \dfrac{\partial^2 u}{\partial y^2} = 0$. Find $u = \varphi\left(\dfrac{y}{x}\right)$.

7. Let

$$f(x,y) = \begin{cases} \dfrac{x^3 y}{x^6 + y^6}, & x^2 + y^2 \neq 0 \\ 0, & x^2 + y^2 = 0 \end{cases}$$

Prove that $f(x,y)$ is discontinuous at the point $(0,0)$, but the two partial derivatives of $f(x,y)$ exist and are discontinuous at $(0,0)$.

8. Suppose that $f(x,y) = \dfrac{2xy}{x^2+y^2}$ as $x^2+y^2 \neq 0$, and that $f(0,0) = 0$. Discuss the existence of $f''_{xy}(0,0)$.

9. If $f'_x(x,y) > 0$ in the domain D, then what kind of geometry information of the function $z = f(x,y)$ can you get?

10. Let the partial derivatives f'_x and f'_y of f which is a function of two variables be bounded in some neighbourhood $U(P_0)$ of the point P_0. Verify that f is continuous in the neighbourhood $U(P_0)$.

8.3

1. Find the total differentials of the following function at the indicated point M_0 and arbitrary point M, respectively.

(1) $z = x^2 y^3$, $M_0(2,1)$;

(2) $z = e^{xy}$, $M_0(0,0)$;

(3) $z = x\ln(xy)$, $M_0(-1,-1)$;

(4) $u = \cos(xy + xz)$, $M_0\left(1, \dfrac{\pi}{6}, \dfrac{\pi}{6}\right)$.

2. Use the definition of the total differential to determine the total differential of the function $z = 4 - \dfrac{1}{4}(x^2 + y^2)$ at the point $\left(\dfrac{3}{2}, \dfrac{3}{2}\right)$.

3. Verify that the partial derivatives of the function
$$f(x,y) = \begin{cases} \dfrac{x^3 - y^3}{x^2 + y^2}, & (x,y) \neq (0,0) \\ 0, & (x,y) = (0,0) \end{cases}$$
at the origin $(0,0)$ exist, but $f(x,y)$ is not differentiable at the orign.

4. Calculate the approximate value of $(10.1)^{2.03}$.

5. There is a right angle triangle whose two right-angle sides are measured as 7 cm and 24 cm respectively, and the measurement accuracy is ± 0.1 cm. Find the error of the hypotenuse length produced from the errors of the measured values.

6. Let the function $z = f(x,y)$ be well defined on the convex domain D (A convex domain is the region such that, for every pair of points within the region, every point on

Chapter 8　Differential Calculus of Multivariable Functions

the straight line segment that joins the pair of points is also within the region). What is the necessary and sufficient condition for $\dfrac{\partial z}{\partial x} \equiv 0$? What is the necessary and sufficient condition for $\dfrac{\partial^2 z}{\partial x \partial y} \equiv 0$? And what is the necessary and sufficient condition for $dz \equiv 0$?

7. Let $f'_x(x_0, y_0)$ exist and $f'_y(x, y)$ be continuous at the point (x_0, y_0). Prove that the function $f(x, y)$ is differentiable at the point (x_0, y_0).

8. Suppose that the function $z = f(x, y)$ of two variables is differentiable with $f(0, 0) = 1$, and that two partial increments are
$$\Delta_x z = (2 + 3x^2 y^2) \Delta x + 3xy^2 (\Delta x)^2 + y^2 (\Delta x)^3$$
$$\Delta_y z = 2x^3 y \Delta y + x^3 (\Delta y)^2$$
Determine $f(x, y)$.

8.4

1. Find partial derivatives of the given functions by using the Chain Rule.

(1) $z = (x^2 + y^2) \exp\left(\dfrac{x^2 + y^2}{xy}\right)$;　(2) $z = \dfrac{xy}{x + y} \arctan(x + y + xy)$.

2. Determine the total derivatives of the following functions.

(1) $u = \tan(3t + 2x^2 - y), x = \dfrac{1}{t}, y = \sqrt{t}$;

(2) $u = e^{x - 2y} + \dfrac{1}{t}, x = \sin t, y = t^3$.

3. Let $z = e^u \sin v, u = xy, v = x - y$, find $\dfrac{\partial z}{\partial x}, \dfrac{\partial z}{\partial y}$.

4. Let f and g be differentiable functions. Find the first order partial derivatives of the given composite functions:

(1) $z = f(x + y, x^2 + y^2)$;　　(2) $z = f\left(\dfrac{x}{y}, \dfrac{y}{x}\right)$;

(3) $u = f(xy) g(yz)$;　　(4) $u = f(x - y^2, y - x^2, xy)$.

5. Suppose that f has continuous second order partial derivatives. Find the indicated partial derivatives of the following functions.

(1) $z = f(u, x, y), u = xe^y$, find $\dfrac{\partial^2 z}{\partial x \partial y}$;

(2) $z = x^3 f(xy, \frac{y}{x})$, find $\frac{\partial z}{\partial y}, \frac{\partial^2 z}{\partial y^2}$ and $\frac{\partial^2 z}{\partial x \partial y}$.

6. Verify that the following function u satisfies the given equation.

(1) Let $u = \varphi(x + at) + \psi(x - at)$, where φ, ψ have continuous second order derivatives. Prove that u satisfies the equation

$$\frac{\partial^2 u}{\partial t^2} = a^2 \frac{\partial^2 u}{\partial x^2}$$

(2) Suppose that $z = f[x + \varphi(y)]$, where φ is differentiable, and f has continuous second order derivative. Verify that

$$\frac{\partial z}{\partial x} \frac{\partial^2 z}{\partial x \partial y} = \frac{\partial z \partial^2 z}{\partial y \partial x^2}$$

(3) If the function $u = f(x,y,z)$ satisfies

$$f(tx, ty, tz) = t^k f(x, y, z)$$

then u is called a homogeneous function with degree k. Prove that if f is differentiable then the homogeneous function with degree k satisfies the following equation

$$x \frac{\partial f}{\partial x} + y \frac{\partial f}{\partial y} + z \frac{\partial f}{\partial z} = kf(x, y, z)$$

Conversely, if a function satisfies the above equation, it must be a homogeneous function with degree k.

7. Given that $z = f(x,y)$ have continuous second order partial derivatives and satisfies $a^2 \frac{\partial^2 z}{\partial x^2} - \frac{\partial^2 z}{\partial y^2} = 0$. Make a substitution: $u = x + ay, v = x - ay$, then find the equation which z as a function of u, v satisfies.

8. The equation $6 \frac{\partial^2 z}{\partial x^2} + \frac{\partial^2 z}{\partial x \partial y} - \frac{\partial^2 z}{\partial y^2} = 0$ can be simplified as $\frac{\partial^2 z}{\partial u \partial v} = 0$ by making a substitution: $u = x - 2y, v = x + ay$. Find the constant a.

9. Let $u = f(x, y, z)$ be differentiable and $\frac{u'_x}{x} = \frac{u'_y}{y} = \frac{u'_z}{z}$. Verify that u is a function of one variable ρ by making a substitution: $x = \rho \sin \varphi \cos \theta$, $y = \rho \sin \varphi \sin \theta$, $z = \rho \cos \varphi$.

10. Find the total differentials and partial derivatives of the following functions by using the invariance of the total differential form and differential algorithm.

Chapter 8 Differential Calculus of Multivariable Functions

(1) $u = f(x - y, x + y)$;　　　(2) $u = f(xy, \dfrac{x}{y})$;

(3) $u = f(\sin x + \sin y, \cos x - \cos z)$.

8.5

1. Find the first and second order partial derivatives of the implicit function z determined by the following equation.

(1) $\dfrac{x}{z} = \ln \dfrac{z}{y}$;　　　(2) $x^2 - 2y^2 + z^2 - 4x + 2z - 5 = 0$.

2. Find the total differential and partial derivatives of the implicit function z by using the invariance of the total differential form.

(1) $xyz + \sqrt{x^2 + y^2 + z^2} = \sqrt{2}$;　　(2) $z - y - x + xe^{z-y-x} = 0$.

3. Let $z = z(x, y)$ be determined by the equation $ax + by + cz = \Phi(x^2 + y^2 + z^2)$, where Φ is differentiable. Verify that

$$(cy - bz)\dfrac{\partial z}{\partial x} + (az - cx)\dfrac{\partial z}{\partial y} = bx - ay$$

4. Let $z = z(x, y)$ be defined by the equation $F(x + zy^{-1}, y + zx^{-1}) = 0$. Prove that

$$x\dfrac{\partial z}{\partial x} + y\dfrac{\partial z}{\partial y} = z - xy$$

5. Let $z = z(x, y)$ be determined by the equation $\dfrac{x}{z} = \varphi(\dfrac{y}{z})$, where φ has continuous second order derivative. Verify that

$$\dfrac{\partial^2 z}{\partial x^2} \cdot \dfrac{\partial^2 z}{\partial y^2} = \left(\dfrac{\partial^2 z}{\partial x \partial y}\right)^2$$

6. Let $z = z(x, y)$ be determined by $F(x + y + z, x^2 + y^2 + z^2) = 0$, where F has continuous second order partial derivatives. Find $\dfrac{\partial^2 z}{\partial x \partial y}$.

7. Suppose that the fuction $z = z(x, y)$ is differentiable with $\dfrac{\partial z}{\partial x} \neq 0$, and it satisfies the equation $(x - z)\dfrac{\partial z}{\partial x} + y\dfrac{\partial z}{\partial y} = 0$. If x is a function of y and z, then what kind of equation should x satisfy?

8. Find the curvature for the curve $F(x, y) = 0$, where F has continuous second partial derivatives.

9. Find derivatives or partial derivatives of the implicit function determined by the following equations.

(1) $\begin{cases} z = x^2 + y^2 \\ x^2 + 2y^2 + 3z^2 = 20 \end{cases}$, find $\dfrac{dy}{dx}, \dfrac{dz}{dx}$;

(2) $\begin{cases} u = f(ux, v + y) \\ v = g(u - x, v^2 y) \end{cases}$, where f have continuous first partial derivatives. Find $\dfrac{\partial u}{\partial x}$, $\dfrac{\partial v}{\partial x}$;

(3) $\begin{cases} x = e^u + u\sin v \\ y = e^u - u\cos v \end{cases}$, find $\dfrac{\partial u}{\partial x}$ and $\dfrac{\partial v}{\partial y}$.

10. Suppose that $y = f(x, t)$, and t determined by the equation $F(x, y, t) = 0$ is a function of x, where f and F have continuous first partial derivatives. Find $\dfrac{dy}{dx}$.

11. Let $u = f(x, y, z)$, $\varphi(x^2, e^y, z) = 0$, $y = \sin x$, where f, φ have continuous first partial derivatives, and $\dfrac{\partial \varphi}{\partial z} \neq 0$. Find $\dfrac{du}{dx}$.

12. Assume that the function $z = f(x, y)$ have continuous second partial derivatives with $\dfrac{\partial z}{\partial y} \neq 0$. Prove that for any constant C which belongs to the range of the function z, the necessary and sufficient condition such that $f(x, y) = C$ denotes a straight line is
$$(z'_y)^2 z''_{xx} - 2z'_x z'_y z''_{xy} + (z'_x)^2 z''_{yy} = 0$$

8.6

1. Find the tangent line and the normal plane to each of the following curves at the specified points.

(1) $x = at, y = bt^2, z = ct^3$, at the point where $t = 1$;

(2) $x = \cos t + \sin^2 t, y = \sin t(1 - \cos t), z = \cos t$, at the point where $t = \dfrac{\pi}{2}$;

(3) $x = y^2, z = x^2$, at the point $(1, 1, 1)$;

(4) $2x^2 + y^2 + z^2 = 45, x^2 + 2y^2 = z$, at the point $(-2, 1, 6)$.

2. Find a point of the curve $x = t, y = t^2, z = t^3$ such that the tangent line to the curve at that point is parallel to the plane $x + 2y + z = 4$.

3. Prove that the included angle between the tangent vector to the spiral line $x =$

Chapter 8 Differential Calculus of Multivariable Functions

$a\cos\theta, y = a\sin\theta, z = k\theta$ at any point and the positive direction of the z- axis is a constant.

4. Find the equations of the tangent plane and the normal line to the given surface at the indicated points.

(1) $z = \sqrt{x^2 + y^2}$, at the point $(3,4,5)$;

(2) $x^3 + y^3 + z^3 + xyz - 6 = 0$, at the point $(1,2,-1)$;

(3) $x = u + v, y = u^2 + v^2, z = u^3 + v^3$, at the point where $(u_0, v_0) = (2,1)$.

5. Find a point of the surface $z = xy$ such that the normal line at that point is perpendicular to the plane $x + 3y + z + 9 = 0$, and write out the equation of the normal line.

6. Given that $f(u,v)$ is differentiable. Show that the tangent plane to the surface $f(ax - bz, ay - cz) = 0$ at any point is parallel to a fixed straight line, where a, b and c are constants, and not all zero.

7. Let $f(u,v)$ be differentiable. Show that the tangent plane to the surface $f\left(\dfrac{y-b}{x-a}, \dfrac{z-c}{x-a}\right) = 0$ at any point must pass through a fixed point.

8. Show that the volume of a tetrahedron enclosed by three coordinate planes and the tangent plane to the surface $xyz = a^3 (a > 0)$ at any point is a constant.

9. Given $f'(x) \neq 0$. Prove that the normal line to the rotating surface $z = f(\sqrt{x^2 + y^2})$ at any point always intersects the axis of rotation, the z-axis.

10. Find the included angle θ between the normal line to the helical surface $x = u\cos v, y = u\sin v, z = av$ and the z- axis.

11. Show that the surface $e^{2x-z} = f(\pi y - \sqrt{2}z)$ is a cylinder, where f is differentiable.

8.7

1. Find the Taylor formula of degree 1 for $f(x,y) = \sin x \sin y$ at the point $\left(\dfrac{\pi}{4}, \dfrac{\pi}{4}\right)$.

2. Find extreme values of the given functions.

(1) $z = 3axy - x^3 - y^3, a > 0$;   (2) $z = e^{2x}(x + 2y + y^2)$.

3. Find maximum and minimum values of the function $f(x,y) = 2x^3 - 4x^2 + 2xy - y^2$ in the rectangular closed region: $-2 \leq x \leq 2, -1 \leq y \leq 1$. What does the result of this problem indicate?

4. Find a point in the xOy plane such that the sum of the squares of the distances

Calculus(II)

from this point to three straight lines $x = 0, y = 0$ and $x + 2y - 16 = 0$ is minimized.

5. Suppose the function $z = z(x,y)$ in the region D satisfies the equation $\dfrac{\partial^2 z}{\partial x^2} + \dfrac{\partial^2 z}{\partial y^2} + a\dfrac{\partial z}{\partial x} + b\dfrac{\partial z}{\partial y} + c = 0$ (where the constant $c > 0$). Show that the function $z = z(x,y)$ has no extreme value in the region D.

6. Show that among all triangles with a given perimeter $2p$, the one with the largest area is the equilateral triangle.

7. Find the extreme points of the following functions with given constraint conditions.

(1) $u = x - 2y + 2z$, with the constraint condition $x^2 + y^2 + z^2 = 1$.

(2) $u = xyz$, with the constraint condition $x^2 + y^2 + z^2 = 1$, $x + y + z = 0$.

8. A company plan to do some kind of commodity sales advertising through the radio and newspapers. Let R, x and y denote sales revenue, radio advertising costs and newspaper advertising costs (unit: million), respectively. The experience relationship between R, x and y is as follows

$$R = 15 + 14x + 32y - 8xy - 2x^2 - 10y^2$$

(1) Find the best advertising strategy if the advertising fee is not limited;

(2) If 1.5 (million yuans) is provided for the cost of advertising fee, find the corresponding optimal advertising strategy.

9. Suppose that the production of a product must be put into two productive factors, that x_1 and x_2 are the input amount of two productive factors respectively, and that Q is the output amount of the product. If the production function is $Q = 2x_1^\alpha x_2^\beta$, where α, β are positive constants with $\alpha + \beta = 1$, and the price of the two productive factors are P_1 and P_2 respectively, then what are the values of x_1 and x_2 in order to minimize total cost of input as $Q = 12$?

10. Find a point on the surface $z = \sqrt{2 + x^2 + 4y^2}$ with the minimum distance to the plane $x - 2y + 3z = 1$.

11. Find the semi-major axis and the semi-minor axis of an ellipse formed by a plane $x + y + z = 0$ intersecting an ellipsoid surface $\dfrac{x^2}{3} + \dfrac{y^2}{2} + z^2 = 1$.

12. Find the value of the positive number a such that the ellipsoid $x^2 + \dfrac{y^2}{4} + \dfrac{z^2}{9} = a^2$

Chapter 8 Differential Calculus of Multivariable Functions

and the plane $3x - 2y + z = 34$ are tangent.

13. An uncovered cuboid pool with volume V is to be made. Known the ratio of the base "unit area cost to the side" is $3:2$. How to design the length, width, height of the pool to make the total construction cost the lowest?

14. A line segment with length l is cut into three pieces. One piece is bent in the shape of a circle, the other is bent in the shape of a square, and the third is bent in the shape of a regular triangle. How should this to be done to minimize the total area sum of the three figures? What is the minimum?

15. Divide the positive real number a into the sum of n real numbers such that the product of these n real numbers is the largest, and then obtain the result that the geometric mean value does not exceed the arithmetic mean value of n.

16. Three vertexes of a triangle lie respectively in three curves $f(x,y) = 0$, $\varphi(x,y) = 0$ and $\psi(x,y) = 0$ which do not intersect each other, where f, φ, ψ are all differentiable with $f'_y, \varphi'_y, \psi'_y \neq 0$. If the area of the triangle can attain the extreme value, then prove that the normal line to the curve at each vertex of the triangle must pass through the orthocenter of the triangle as the triangle area attain the extreme value.

17. Show that the normal line to the smooth surface $G(x,y,z) = 0$ at the point with the nearest distance to the origin must pass through the origin.

8.8

1. Let $u = x^2 + y^2 - 2z^2 + 3xy + xyz - 2z - 3y$ be a scalar field. Find its gradient at the point $(1,2,3)$ and its directional derivative in direction of $\boldsymbol{l} = \{1, -1, 0\}$.

2. Given a scalar field $u = x^2 + 2y^2 + 3z^2 + xy + 3x - 2y - 6z$.

(1) Find a point at which the gradient of u is a zero vector;

(2) In what direction at $(2,0,1)$ the rate of change of u is the largest? And determine the maximum rate of change;

(3) Find a point at which the gradient of u is perpendicular to the z-axis.

3. Let $u = u(x,y,z)$ be a scalar field. Discuss the relation between the gradient, the directional derivative, the isosurface and the total differential at the point (x_0, y_0, z_0).

4. Find the directional derivative of the function $u = xyz$ at the point $M(3,4,5)$ in the direction of the normal line to the conical surface $z = \sqrt{x^2 + y^2}$.

5. Determine the directional derivative of the function $u = 1 - \dfrac{x^2}{a^2} - \dfrac{y^2}{b^2}$ at the point $\left(\dfrac{a}{\sqrt{2}}, \dfrac{b}{\sqrt{2}}\right)$ in the direction of the inner normal line to the curve $\dfrac{x^2}{a^2} + \dfrac{y^2}{b^2} = 1$.

6. Determine the directional derivative of the function $w = e^{-2y}\ln(x + z^2)$ at the point $(e^2, 1, e)$ in direction of the normal vector to the surface $x = e^{u+v}, y = e^{u-v}, z = e^{uv}$.

7. Calculate $\operatorname{grad}\left[\boldsymbol{c} \cdot \boldsymbol{r} + \dfrac{1}{2}\ln(\boldsymbol{c} \cdot \boldsymbol{r})\right]$, where $\boldsymbol{c}$ is a constant vector, $\boldsymbol{r}$ is a radius vector and $\boldsymbol{c} \cdot \boldsymbol{r} > 0$.

8. Verify that ∇u is a constant vector if and only if u is a linear function, i. e. $u = ax + by + cz + d$.

9. Suppose that the blood concentration in the ocean at the point (x, y) in parts per million units of water is given by $C = \exp((x^2 + 2y^2)/10^4)$. A shark always swims in the direction of the strongest scent of blood. Find the shark's approach path starting from the point (x_0, y_0).

8.9

1. Let $f(x, y, z)$ be continuous at the origin and differentiable at other points, and $x\dfrac{\partial f}{\partial x} + y\dfrac{\partial f}{\partial y} + z\dfrac{\partial f}{\partial z} > a\sqrt{x^2 + y^2 + z^2} > 0$. Then $f(0, 0, 0)$ (   ) of $f(x, y, z)$.

(A) is the maximum

(B) is the minimum

(C) is a local maximum, but may not be the maximum

(D) is a local minimum, but may not be the minimum

2. Suppose the directional derivatives of the function $f(x, y)$ at the point (x_0, y_0) in any directions exist and are equal. Determine if the partial derivatives of $f(x, y)$ at the point (x_0, y_0) exist, and determine whether $f(x, y)$ is differentiable at the point (x_0, y_0).

3. Given $z = \sin(xy)$, find $\dfrac{\partial^3 z}{\partial x \partial y^2}, \dfrac{\partial^3 z}{\partial y \partial x \partial y}, \dfrac{\partial^3 z}{\partial y^2 \partial x}$.

4. Suppose that the functions u, v, w of three independent variables x, y, z are determined by the equations: $x = f(u, v, w), y = g(u, v, w), z = h(u, v, w)$. Find $\dfrac{\partial u}{\partial x}$.

Chapter 8 Differential Calculus of Multivariable Functions

5. Let the equations $x = \varphi(u,v)$, $y = \psi(u,v)$, $z = f(u,v)$ define a function z of two independent variables x and y. Find computational formulas of partial derivatives $\dfrac{\partial z}{\partial x}$ and $\dfrac{\partial z}{\partial y}$.

6. Let $z = f(x,y)$ be differentalble at the point P_0, $l_1 = \{2, -2\}$, $l_2 = \{-2, 0\}$, and $\left.\dfrac{\partial u}{\partial l_1}\right|_{P_0} = 1$, $\left.\dfrac{\partial u}{\partial l_2}\right|_{P_0} = -3$. Determine the gradient, the total differential and the directional derivative in the direction of $l = \{3,2\}$ at the point P_0, respectively.

7. Suppose that the function $u = F(x,y,z)$ attains the extreme value m at the point (x_0, y_0, z_0) with constraint conditions $\varphi(x,y,z) = 0$ and $\psi(x,y,z) = 0$. Show that three normal lines to three surfaces $F(x,y,z) = m$, $\varphi(x,y,z) = 0$ and $\psi(x,y,z) = 0$ at the point (x_0, y_0, z_0) lie in the same plane, where F, φ, ψ all have continuous first partial derivatives, and show that the three partial derivatives of each function are not all zero.

8. Applying the method of conditional extremum, verify that for any positive numbers a, b and c, the following inequality always holds
$$abc^3 \leqslant 27 \left(\dfrac{a+b+c}{5}\right)^5$$

9. Given a quadrilateral with four sides a, b, c, d. When does the quadrilateral have the largest area?

10. The roof of the National Grand Theatre planned to be built is an elliptical shell type (called the century egg). Let the equation of the roof be $\dfrac{x^2}{4} + \dfrac{y^2}{3} + \dfrac{z^2}{2} = 1$. When the rain falls at the point (x_0, y_0, z_0) of the top of the roof, it will slide downward under the action of gravity. Find the curve equation of the rain's path.

Chapter 9 Multiple Integrals

9.1 Double Integrals

9.1.1 Concepts of Double Integrals

Practical Example 1 Suppose $z = f(x,y)$ is a non-negative continuous function on a bounded closed region σ on the xOy plane. We study the volume of the curved roof cylinder (Fig. 9.1) with the surface $z = f(x,y)$ as the roof, the region σ as the base, and the cylindrical surface with the boundary of σ as the directrix and generatrix parallel to the z-axis as the side surface.

As we know, the height of a flat roof cylinder is fixed and its volume can be calculated by the following formula

   Volume = Height × Base Area

For the curved roof cylinder, when the point (x,y) varies in the region σ, the height $f(x,y)$ is a variable, so the volume cannot calculated by the above formula. To settle this problem, we recall the concept of definite integral introduced in Chapter 6.

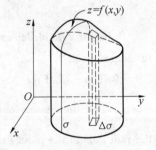

Fig. 9.1

First, we use a family of curves to divide the region σ into n sub-regions

$$\Delta\sigma_1, \Delta\sigma_2, \cdots, \Delta\sigma_n$$

and using the boundary of each sub-region as the directrix we make a thin cylinder with generatrix parallel to the z-axis, then these cylinders divide the geometric body into n small curved roof cylinders. Since $f(x,y)$ is continuous, then within the same sub-region $f(x,y)$ varies very little, then the curved roof cylinders can be thought of as flat roof cylinders approximately. We take an arbitrary point (ξ_i, η_i) in each sub-region $\Delta\sigma_i$ (whose area is also denoted by $\Delta\sigma_i$), then the volume of the flat roof cylinder with

Chapter 9 Multiple Integrals

base $\Delta\sigma_i$ and height $f(\xi_i, \eta_i)$ is
$$f(\xi_i, \eta_i)\Delta\sigma_i, \quad i = 1, 2, \cdots, n$$
The sum of the volumes of these flat roof cylinders is
$$\sum_{i=1}^{n} f(\xi_i, \eta_i)\Delta\sigma_i$$
which can be thought of as the approximation of the volume of the big curved roof cylinder. Let λ be the maximum diameter of the sub-regions, then as λ approaches zero the limit of the above sum is naturally defined as the volume of the big curved roof cylinder.

Practical Example 2 Given the mass surface density $\rho = \rho(x, y)$ of a plate σ, find the mass of the plate σ.

We apply the similar idea as for Practical Example 1. We first divide the plate σ into n sub-regions using a family of curves as following
$$\Delta\sigma_1, \Delta\sigma_2, \cdots, \Delta\sigma_n$$
Since $\rho(x, y)$ is continuous, then within the same sub-region $f(x, y)$ varies very little, and the density of this sub-region can be thought of as uniform. We take an arbitrary point (ξ_i, η_i) in each sub-region $\Delta\sigma_i$ (whose area is also denoted by $\Delta\sigma_i$ (Fig. 9.2)) and use $\rho(\xi_i, \eta_i)$ as the density of $\Delta\sigma_i$, then the mass of $\Delta\sigma_i$ is
$$\Delta m_i \approx \rho(\xi_i, \eta_i)\Delta\sigma_i, \quad i = 1, 2, \cdots, n$$
The sum of the masses of these sub-regions is approximately

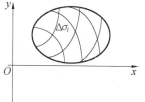

$$\sum_{i=1}^{n} \rho(\xi_i, \eta_i)\Delta\sigma_i$$
which can be thought of as an approximation of the mass of the plate σ. Let λ be the maximum diameter of the sub-regions, then as λ approaches zero the limit of the above sum is naturally defined as the mass of the plate σ.

Fig. 9.2

Although the above two examples have the different practical significance, then can both be boiled down to a limit of a product of a binary function and a small area. We can abstract the following definition.

Definition 9.1 Suppose that $f(x, y)$ is a bounded function on a closed bounded region σ. Divide the region σ into n closed sub-regions
$$\Delta\sigma_1, \Delta\sigma_2, \cdots, \Delta\sigma_n$$

Calculus(II)

and use these to denote their areas. Let $d_i = \sup\limits_{P_1,P_2 \in \Delta\sigma_i} d(P_1,P_2)$ be the diameter of $\Delta\sigma_i$, denote

$$\lambda = \max_{1 \leqslant i \leqslant n}(d_i)$$

Take an arbitrary point $P_i(\xi_i,\eta_i) \in \Delta\sigma_i (i = 1,2,\cdots,n)$, and take the sum of the products

$$\sum_{i=1}^{n} f(\xi_i,\eta_i)\Delta\sigma_i$$

If the limit $\lim\limits_{\lambda \to 0}\sum\limits_{i=1}^{n} f(\xi_i,\eta_i)\Delta\sigma_i$ always exists and gives the same value no matter how we divide the region σ and how we pick the points (ξ_i,η_i), then this limit is called the double integral of the function $f(x,y)$ on the closed bounded region σ, denoted by $\iint_\sigma f(x,y)\,d\sigma$, i.e.

$$\iint_\sigma f(x,y)\,d\sigma = \lim_{\lambda \to 0}\sum_{i=1}^{n} f(\xi_i,\eta_i)\Delta\sigma_i \tag{1}$$

At this point we say that the function $f(x,y)$ is integrable on σ, the function $f(x,y)$ is called the integrant, the expression $f(x,y)\,d\sigma$ is called the integral expression, the region σ is called the integral region, and the notation $d\sigma$ is called the area element of σ.

9.1.2 Properties of Double Integrals

From the definition of the double integral and properties of limits, we can obtain the following properties for double integral. For simplicity, we assume that the integrals involved in the following all exist.

1. When $f(x,y) \equiv 1$, the integral of $f(x)$ on σ equals to the area of σ, i.e.

$$\iint_\sigma 1\,d\sigma = \sigma$$

2. **Linear property**

$$\iint_\sigma [af(x,y) + bg(x,y)]\,d\sigma = a\iint_\sigma f(x,y)\,d\sigma + b\iint_\sigma g(x,y)\,d\sigma$$

where a,b are constants.

3. **Additivity property for integral regions** If σ is divided into two parts σ_1, σ_2, then

Chapter 9　Multiple Integrals

$$\iint_\sigma f(x,y)\,d\sigma = \iint_{\sigma_1} f(x,y)\,d\sigma + \iint_{\sigma_2} f(x,y)\,d\sigma$$

4. Comparison property

(i) If $f(x,y) \leq g(x,y)$, $\forall (x,y) \in \sigma$, then

$$\iint_\sigma f(x,y)\,d\sigma \leq \iint_\sigma g(x,y)\,d\sigma$$

(ii) $\iint_\sigma |f(x,y)|\,d\sigma \leq \left|\iint_\sigma f(x,y)\,d\sigma\right|$.

5. Estimation property　　If $m \leq f(x,y) \leq M$, $\forall (x,y) \in \sigma$, then

$$m\sigma \leq \iint_\sigma f(x,y)\,d\sigma \leq M\sigma$$

6. Mean value theorem of integrals　　If $f(x,y)$ is continuous on the closed bounded region σ, then there exists at least one point (ξ,η) on σ, such that

$$\iint_\sigma f(x,y)\,d\sigma = f(\xi,\eta)\sigma$$

Proof　Since $f(x,y) \in C(\sigma)$, then there exist a maximum value M and a minimum value m, such that

$$m \leq f(x,y) \leq M, \forall (x,y) \in \sigma$$

By the estimation property we get

$$m\sigma \leq \iint_\sigma f(x,y)\,d\sigma \leq M\sigma$$

then

$$m \leq \frac{1}{\sigma}\iint_\sigma f(x,y)\,d\sigma \leq M$$

then by the intermediate value theorem of continuous functions on closed regions, there exists a point $(\xi,\eta) \in \sigma$, such that

$$f(\xi,\eta) = \frac{1}{\sigma}\iint_\sigma f(x,y)\,d\sigma$$

7. Symmetric property　　Under the rectangular coordinate system Oxy, suppose that the integral region σ is symmetric with respect to the coordinate axis $x = 0$. If the integrant is an odd function of x (i.e. $f(-x,y) = -f(x,y)$), then

$$\iint_\sigma f(x,y)\,d\sigma = 0$$

If the integrant is an even function of x (i.e. $f(-x,y) = f(x,y)$), then

Calculus(II)

$$\iint_\sigma f(x,y)\,d\sigma = 2\iint_{\sigma^+} f(x,y)\,d\sigma$$

where $\sigma^+ = \{(x,y) \mid (x,y) \in \sigma, x \geq 0\}$.

By the additivity property for integral regions and the definition of double integral the previous property is not hard to be proved.

The following theorem is about the existence of double integrals.

Theorem 9.1 If $f(x,y)$ is continuous on the closed bounded region σ, then $f(x,y)$ is integrable on σ.

9.2 Calculating Double Integrals

In this section we will study the method for calculating double integrals by just following the geometric significance. By Practical Example 1 of 9.1, when $f(P) \geq 0$, the double integral $\iint_\sigma f(P)\,d\sigma$ can be thought of as the volume of a curved roof cylinder, when $f(P) < 0$, the integral $\iint_\sigma f(P)\,d\sigma$ equals to the negative value of the volume of a curved roof cylinder; when $f(P)$ takes on both positive and negative values on σ, the integral $\iint_\sigma f(P)\,d\sigma$ equals to the algebraic sum of the volumes of the cylinders above and below the plane region σ.

9.2.1 Calculating Double Integrals under Rectangular Coordinate System

Suppose σ is a bounded region on the xOy plane, $f(x,y) \in C(\sigma)$, then the double integral

$$\iint_\sigma f(x,y)\,d\sigma = \lim_{\lambda \to 0} \sum_{i=1}^{n} f(\xi_i, \eta_i)\,\Delta\sigma_i \qquad (1)$$

exists. Since the limit in equation (1) is irrelevant to the partition of σ, we use straight lines parallel to the coordinate axis to divide σ, a particular sub-region $\Delta\sigma$ is a rectangle with area $\Delta\sigma = \Delta x \Delta y$, so, under the rectangular coordinate system the area element $d\sigma = dxdy$ (Fig. 9.3). At this point, the double integral can be expressed as

$$\iint_\sigma f(x,y)\,dxdy$$

Chapter 9 Multiple Integrals

When σ is an x-type region, i.e. σ can be expressed as
$$a \leqslant x \leqslant b, y_1(x) \leqslant y \leqslant y_2(x)$$
where $y_1(x), y_2(x) \in C[a,b]$. In other words, the integral region is between the straight lines $x = a, x = b$, bounded below by $y = y_1(x)$, and bounded above by $y = y_2(x)$ (Fig. 9.4).

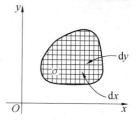

Fig. 9.3

For any number x in the interval $[a,b]$, we use a plane perpendicular to the x-axis to cut the curved roof cylinder, then the cross section is a trapezoid with curve side (Fig. 9.5) with area
$$S(x) = \int_{y_1(x)}^{y_2(x)} f(x,y) \, dy$$

By the volume formula of a solid with known areas of parallel cross sections, we get the volume of this curved roof cylinder
$$V = \int_a^b S(x) \, dx = \int_a^b \left[\int_{y_1(x)}^{y_2(x)} f(x,y) \, dy \right] dx$$

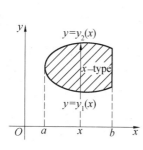

Fig. 9.4

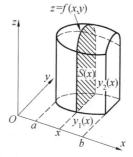

Fig. 9.5

Traditionally, we denote the twice definite integral on the right hand side of the previous equation as
$$\int_a^b dx \int_{y_1(x)}^{y_2(x)} f(x,y) \, dy$$
And we call such a multiple definite integral of a multivariate function an iterated integral. Then we obtain a formula for calculating double integrals under rectangular coordinate system
$$\iint_\sigma f(x,y) \, dx dy = \int_a^b dx \int_{y_1(x)}^{y_2(x)} f(x,y) \, dy \tag{2}$$

Formula (2) transforms a double integral into an iterated integral. When calculating, we first treat x as a constant, and treat $f(x,y)$ as a function of a single variable y, we evaluate the definite integral from $y_1(x)$ to $y_2(x)$; the result is a function of x and can be thought of as the integrant of the second integral which we integrate with respect to x from a to b.

When σ is an y-type region, i.e. σ can be expressed as $c \leqslant y \leqslant d, x_1(y) \leqslant x \leqslant x_2(y)$, where $x_1(y), x_2(y) \in C[c,d]$ (Fig. 9.6). Similar to the discussion in the previous paragraphs, we obtain another formula for calculating double integrals under rectangular coordinate system

$$\iint_\sigma f(x,y)\,dxdy = \int_c^d dy \int_{x_1(y)}^{x_2(y)} f(x,y)\,dx \tag{3}$$

Formula (3) transforms a double integral into another type of iterated integral. When calculating, we first treat y as a constant, and treat $f(x,y)$ as a function of a single variable x, we evaluate the definite integral from $x_1(y)$ to $x_2(y)$; the result is a function of y and then we integrate with respect to y from c to d.

If the function is not always positive on σ, the formulas (2) and (3) are still hold. If the region σ is neither x-type nor y-type, we can divide σ into several parts, such that each part is either x-type or y-type, then we calculate the integral according to the additivity property for integral regions. Formulas (2) and (3) transform double integrals into two iterated integrals with different orders. When we calculate double integrals, we need to determine which formula to use according to the integral region and the integrant.

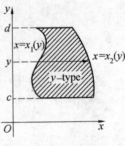

Fig. 9.6

Example 1 Evaluate $\iint_\sigma xy\,dxdy$, where σ is the bounded region enclosed by the curves $y = x^2, y^2 = x$.

Solution We sketch the region σ as shown in Fig. 9.7, from the system

$$\begin{cases} y = x^2 \\ y^2 = x \end{cases}$$

we obtain the coordinates of the vertices $O(0,0), B(1,1)$. Apparently, σ is both x-type and y-type; and from the integrant we can integrate either variable first. Here we apply formula (2), since

Chapter 9 Multiple Integrals

$\sigma: 0 \leqslant x \leqslant 1, x^2 \leqslant y \leqslant \sqrt{x}$

then

$$\iint_\sigma xy\,dx\,dy = \int_0^1 dx \int_{x^2}^{\sqrt{x}} xy\,dy = \int_0^1 \frac{1}{2}xy^2 \Big|_{x^2}^{\sqrt{x}}\,dx$$

$$= \frac{1}{2}\int_0^1 (x^2 - x^5)\,dx = \frac{1}{12}$$

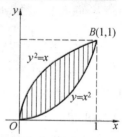

Fig. 9.7

Example 2 Evaluate $\iint_\sigma \dfrac{x}{y} dx\,dy$, where σ is the bounded region enclosed by curves $xy = 1$, $x = \sqrt{y}$ and $y = 2$.

Solution We sketch the region σ as shown in Fig. 9.8, find the coordinates of the vertices $A\left(\dfrac{1}{2},2\right)$, $B(\sqrt{2},2)$, $C(1,1)$. Here σ is an y-type region

$$\sigma: 1 \leqslant y \leqslant 2,\ \frac{1}{y} \leqslant x \leqslant \sqrt{y}$$

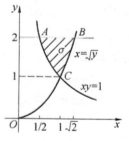

Fig. 9.8

According to the integral region and the integrant, it's better if we integrate with respect to x first, then by formula (3)

$$\iint_\sigma \frac{x}{y} dx\,dy = \int_1^2 dy \int_{1/y}^{\sqrt{y}} \frac{x}{y} dx = \int_1^2 \frac{x^2}{2y}\Big|_{1/y}^{\sqrt{y}} dy$$

$$= \frac{1}{2}\int_1^2 (1 - y^{-3})\,dy = \frac{7}{16}$$

If we use formula (2), and integrate for y first. Then we need to divide σ into two parts by using the straight line $x = 1$, and we need to apply integration by parts when we calculating, which is more complicate.

Example 3 Evaluate $\iint_\sigma e^{x^2} dx\,dy$, where σ is determined by the inequalities $0 \leqslant x \leqslant 1, 0 \leqslant y \leqslant x$.

Solution We sketch the integral region as shown in Fig. 9.9. If we use the integral formula (3) which we need to integrate x first then y, then

$$\iint_\sigma e^{x^2} dx\,dy = \int_0^1 dy \int_y^1 e^{x^2} dx$$

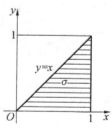

Fig. 9.9

we will face an integral which cannot be expressed as elementary functions, $\int e^{x^2} dx$. If we use formula (2) which we need to integrate y first then x, then

$$\iint_\sigma e^{x^2} dxdy = \int_0^1 dx \int_0^x e^{x^2} dy = \int_0^1 e^{x^2} x dx = \frac{1}{2} e^{x^2} \Big|_0^1 = \frac{1}{2}(e-1)$$

It's very important to choose the order of the iterated integral when calculating double integrals. It's not just about the simplicity of the calculation but is also about the calculability. When calculating double integral, we first need to identify the integral region (including sketching the graph, identifying the boundaries, intersections and vertices), then we need to determine the order of the iterated integral and the upper and lower limits of the definite integral according to the integrant and the integral region to transform the double integral into an iterated integral. At last, we evaluate the iterated integral.

Example 4 Find the volume of the solid enclosed by two orthogonal cylinders both with base radii R.

Solution Suppose the equations of these two cylinders are

$$x^2 + y^2 = R^2, x^2 + z^2 = R^2$$

According to the symmetry property of the solid with respect to the coordinate planes, we just need to calculate the volume of the part of the solid that lies in the first octant V_1, then multiply the result by 8. The top surface of the part of the solid is $z = \sqrt{R^2 - x^2}$ and the base surface is $x^2 + y^2 \leq R^2, x > 0, y > 0$, as shown in Fig. 9.10 and Fig. 9.11, so the volume is

$$V = 8V_1 = 8\iint_D \sqrt{R^2 - x^2}\, dxdy = 8\int_0^R dx \int_0^{\sqrt{R^2-x^2}} \sqrt{R^2 - x^2}\, dy$$

$$= 8\int_0^R (R^2 - x^2)\, dx = \frac{16}{3} R^3$$

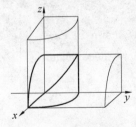

Fig. 9.10

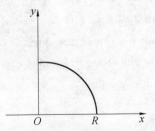

Fig. 9.11

Chapter 9 Multiple Integrals

When $f(x,y) \in C(\sigma)$, the double integral of $f(x,y)$ on σ exists and can be transformed into two iterated integrals with different orders. Because of the difference in calculation, we often need to transform an iterated integral into another iterated integral with different order, and we call it changing order of iterated integrals.

Example 5 Change the order of the iterated integral
$$\int_a^b dx \int_a^x f(x,y) dy$$

Solution We first identify the integral region of the double integral according to the given upper and lower limits
$$\sigma: a \leqslant x \leqslant b, a \leqslant y \leqslant x$$
as shown in Fig. 9.12, σ is an x-type region. Now we can express σ as an y-type region
$$\sigma: a \leqslant y \leqslant b, y \leqslant x \leqslant b$$

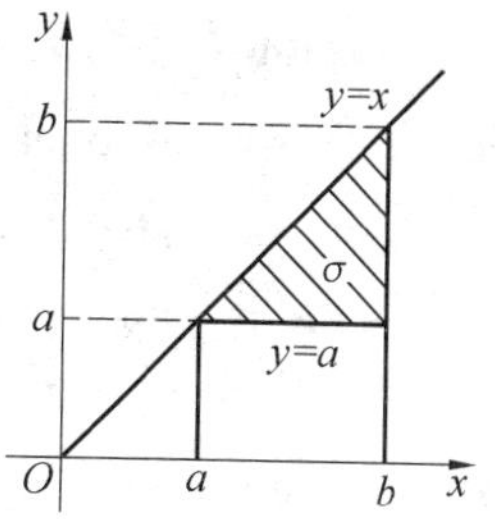

Fig. 9.12

Then
$$\int_a^b dx \int_a^x f(x,y) dy = \iint_\sigma f(x,y) d\sigma = \int_a^b dy \int_y^b f(x,y) dx$$

Example 6 Change the order of the following iterated integral
$$\int_0^{2a} dx \int_{\sqrt{2ax-x^2}}^{\sqrt{2ax}} f(x,y) dy \quad (a > 0)$$

Solution The integral region for the corresponding double integral is
$$\sigma: 0 \leqslant x \leqslant 2a, \sqrt{2ax - x^2} \leqslant y \leqslant \sqrt{2ax}$$
which is an x-type region, see Fig. 9.13. To express σ as an y-type region, we need to divide σ into three pieces $\sigma_1, \sigma_2, \sigma_3$

$\sigma_1: 0 \leqslant y \leqslant a, y^2/(2a) \leqslant x \leqslant a - \sqrt{a^2 - y^2}$

$\sigma_2: 0 \leqslant y \leqslant a, a + \sqrt{a^2 - y^2} \leqslant x \leqslant 2a$

$\sigma_3: a \leqslant y \leqslant 2a, y^2/(2a) \leqslant x \leqslant 2a$

then

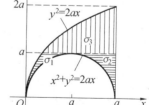

Fig. 9.13

$$\int_0^{2a} dx \int_{\sqrt{2ax-x^2}}^{\sqrt{2ax}} f(x,y) dy$$
$$= \int_0^a dy \int_{y^2/(2a)}^{a-\sqrt{a^2-y^2}} f(x,y) dx + \int_0^a dy \int_{a+\sqrt{a^2-y^2}}^{2a} f(x,y) dx +$$

83

$$\int_a^{2a} \int_{y^2/(2a)}^{2a} f(x,y)\,dx$$

Example 7 Prove that
$$\int_0^a dx \int_0^x f(y)\,dy = \int_0^a (a-x)f(x)\,dx \quad (a>0)$$

Solution Since in the iterated integral on the left hand side $f(y)$ is an abstract function of y, which cannot be calculated. Then we first change the order of the iterated integral, to express the integral region (Fig. 9.14)
$$\sigma: 0 \leqslant x \leqslant a, 0 \leqslant y \leqslant x$$
as an y-type region
$$\sigma: 0 \leqslant y \leqslant a, y \leqslant x \leqslant a$$

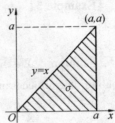

Fig. 9.14

then
$$\int_0^a dx \int_0^x f(y)\,dy = \int_0^a dy \int_y^a f(y)\,dx = \int_0^a f(y)(a-y)\,dy$$
$$= \int_0^a (a-x)f(x)\,dx$$

9.2.2 Calculating Double Integrals under Polar Coordinate System

There is another method to calculate double integral $\iint_\sigma f(x,y)\,d\sigma$, calculating double integral under polar coordinate system. First, we need to express the integral region σ under polar coordinate system
$$\sigma: \alpha \leqslant \theta \leqslant \beta, r_1(\theta) \leqslant r \leqslant r_2(\theta)$$
where $r_1(\theta), r_2(\theta)$ are single value continuous on the interval $[\alpha, \beta]$ (Fig. 9.15).

We use the curves r = constants and θ = constants to divide the region σ, a particular sub-region is a circular sector with area $\Delta\sigma \approx r\Delta r\Delta\theta$, then the area element under the polar coordinate system is
$$d\sigma = r\,dr\,d\theta$$

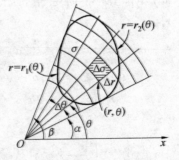

Fig. 9.15

Then we can transform the double integral in rectangular coordinate system to a double integral in polar coordinate system

Chapter 9 Multiple Integrals

$$\iint_\sigma f(x,y)\,dxdy = \iint_\sigma f(r\cos\theta, r\sin\theta)\,rdrd\theta \qquad (4)$$

If we treat $f(r\cos\theta, r\sin\theta)r$ as the integrant and r, θ as the independent variables equivalent to x, y, then analog to equation (3) of Section 9.2.1 we can transform the double integral under polar coordinate system into an iterated integral

$$\iint_\sigma f(r\cos\theta, r\sin\theta)\,rdrd\theta = \int_\alpha^\beta d\theta \int_{r_1(\theta)}^{r_2(\theta)} f(r\cos\theta, r\sin\theta)\,rdr \qquad (5)$$

What we should put our emphasis on is that how to express the integral region σ under polar coordinate system. First, we look for the interval $[\alpha, \beta]$ of polar angles that σ lies on, i.e. σ is between the two rays $\theta = \alpha, \theta = \beta$, then we look for the polar equations of the two boundary curves of σ that close to and far away from the pole, $r = r_1(\theta), r = r_2(\theta)$, then

$$\sigma: \alpha \leq \theta \leq \beta, r_1(\theta) \leq r \leq r_2(\theta)$$

Specially, for a sector area with the pole on its boundary (as shown in Fig. 9.16 (a)), then

$$\sigma: \alpha \leq \theta \leq \beta, 0 \leq r \leq r(\theta)$$

If the pole is inside the region σ, and the boundary is $r = r(\theta)$ (as shown in Fig. 9.16 (b)), then

$$\sigma: 0 \leq \theta \leq 2\pi, 0 \leq r \leq r(\theta)$$

When the integral region σ is a disk, a circular ring or a circular sector, or the integrant is $x^2 + y^2, x^2 - y^2, xy$ or y/x, or the composite function of them, it's more convenient to transform the integral into double integral under polar coordinate system.

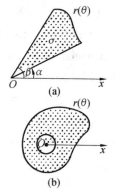

Fig. 9.16

Example 8 Evaluate the integral $\iint_{x^2+y^2\leq 1} (2x+y)^2\,dxdy$.

Solution By symmetric property

$$\iint_{x^2+y^2\leq 1} xy\,d\sigma = 0 \text{ and } \iint_{x^2+y^2\leq 1} x^2\,d\sigma = \iint_{x^2+y^2\leq 1} y^2\,d\sigma$$

Then the original integral becomes

$$\iint_{x^2+y^2\leq 1}(4x^2+y^2)\,d\sigma = \frac{5}{2}\iint_{x^2+y^2\leq 1}(x^2+y^2)\,d\sigma = \frac{5}{2}\int_0^{2\pi}d\theta\int_0^1 r^3\,dr = \frac{5\pi}{4}$$

Example 9 Evaluate $\iint_{x^2+y^2\leq x+y}(x+y)\,dxdy$.

Solution 1 We use the polar coordinate system (Fig. 9.17)

Calculus(II)

$$I = \int_{-\frac{\pi}{4}}^{\frac{3\pi}{4}} d\theta \int_{0}^{(\sin\theta+\cos\theta)} r^2(\sin\theta + \cos\theta) dr$$

$$= \frac{1}{3}\int_{-\frac{\pi}{4}}^{\frac{3\pi}{4}} (\sin\theta + \cos\theta)^4 d\theta$$

$$= \frac{1}{3}\int_{-\frac{\pi}{4}}^{\frac{3\pi}{4}} (1 + 2\sin 2\theta + \sin^2 2\theta) = \frac{\pi}{2}$$

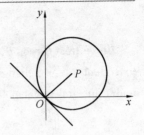

Fig. 9.17

Solution 2 Make transformation: $x = u + \frac{1}{2}, y = v + \frac{1}{2}$, then

$$I = \iint_{u^2+v^2 \leq \frac{1}{2}} (1 + u + v) du dv = \iint_{u^2+v^2 \leq \frac{1}{2}} du dv = \frac{\pi}{2}$$

Example 10 Calculate the area of the region enclosed by the lemniscates $(x^2 + y^2)^2 = 2a^2(x^2 - y^2)$ $(a > 0)$

Solution By the relation between the rectangular coordinate and polar coordinate, the polar equation of the lemniscates is

$$r^2 = 2a^2 \cos 2\theta$$

The graph of it is as shown in Fig. 9.18, and the area of the enclosed region is

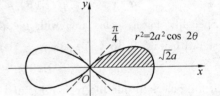

Fig. 9.18

$$S = \iint_{\sigma} d\sigma = 4\int_{0}^{\pi/4} d\theta \int_{0}^{a\sqrt{2\cos 2\theta}} r dr = 2a^2$$

Example 11 Prove that the probability integral

$$\int_{0}^{+\infty} e^{-x^2} dx = \frac{\sqrt{\pi}}{2}$$

Solution By the definition of improper integral

$$\int_{0}^{+\infty} e^{-x^2} dx = \lim_{b \to +\infty} \int_{0}^{b} e^{-x^2} dx$$

And

$$\left(\int_{0}^{b} e^{-x^2} dx\right)^2 = \int_{0}^{b} e^{-x^2} dx \int_{0}^{b} e^{-y^2} dy = \iint_{D} e^{-(x^2+y^2)} dx dy$$

where D is the square $0 \leq x \leq b, 0 \leq y \leq b$, since

$$\iint_{\sigma_1} e^{-(x^2+y^2)} dx dy \leq \iint_{D} e^{-(x^2+y^2)} dx dy \leq \iint_{\sigma_2} e^{-(x^2+y^2)} dx dy$$

where σ_1 (σ_2 respectively) is the portion in the first quadrant of the disk centered at or-

Chapter 9 Multiple Integrals

igin and with radius b ($\sqrt{2}b$ respectively) (Fig. 9.19), and

$$\iint_{\sigma_1} e^{-(x^2+y^2)}dxdy = \int_0^{\frac{\pi}{2}}d\theta \int_0^b e^{-r^2}rdr = \frac{\pi}{4}(1-e^{-b^2})$$

$$\iint_{\sigma_2} e^{-(x^2+y^2)}dxdy = \int_0^{\frac{\pi}{2}}d\theta \int_0^{\sqrt{2}b} e^{-r^2}rdr = \frac{\pi}{4}(1-e^{-2b^2})$$

then we have

$$\frac{\sqrt{\pi}}{2}\sqrt{1-e^{-b^2}} \leqslant \int_0^b e^{-x^2}dx \leqslant \frac{\sqrt{\pi}}{2}\sqrt{1-e^{-2b^2}}$$

Let $b \rightarrow +\infty$, by the Squeeze Rule

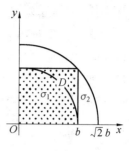

Fig. 9.19

$$\int_0^{+\infty} e^{-x^2}dx = \frac{\sqrt{\pi}}{2}$$

Example 12 Transform the following iterated integral under rectangular coordinate system into an iterated integral under polar coordinate system

$$\int_0^1 dx \int_{\sqrt{1-x^2}}^{\sqrt{4-x^2}} f(x,y)dy + \int_1^2 dx \int_0^{\sqrt{4-x^2}} f(x,y)dy$$

Solution The corresponding integral region is

$$\sigma_1: 0 \leqslant x \leqslant 1, \sqrt{1-x^2} \leqslant y \leqslant \sqrt{4-x^2}$$

$$\sigma_2: 1 \leqslant x \leqslant 2, 0 \leqslant y \leqslant \sqrt{4-x^2}$$

they together form the portion of a circular ring in the first quadrant (Fig. 9.20), its polar expression is

$$\sigma_1 + \sigma_2: 0 \leqslant \theta \leqslant \pi/2, 1 \leqslant r \leqslant 2$$

Then

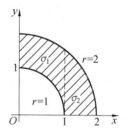

Fig. 9.20

$$\int_0^1 dx \int_{\sqrt{1-x^2}}^{\sqrt{4-x^2}} f(x,y)dy + \int_1^2 dx \int_0^{\sqrt{4-x^2}} f(x,y)dy$$

$$= \int_0^{\pi/2} d\theta \int_1^2 f(r\cos\theta, r\sin\theta)rdr$$

9.2.3 Calculating Surface Areas by Double Integrals

Suppose the surface S is given by a continuous single-valued function

$$z = f(x,y), (x,y) \in \sigma$$

and $f'_x(x,y), f'_y(x,y)$ are continuous on the closed bounded region σ (Fig. 9.21).

The main idea of calculating the area of the surface S is partition. We divide σ into many small pieces, and suppose $\Delta\sigma$ is a particular piece, take an arbitrary point $M(x, y) \in \Delta\sigma$ and make the tangent plane T of the surface S at the corresponding point $P(x, y, f(x,y))$. Then we make a cylinder using the boundary of $\Delta\sigma$ as the directrix and generatrix parallel to the z-axis. The cylinder cut off ΔS and ΔT from the surface S and the tangent plane T, respectively. If we also use ΔS and ΔT to denote the areas, then from Fig. 9.22 we have

$$\Delta S \approx \Delta T, \Delta T = \frac{\Delta\sigma}{\cos\gamma}$$

where γ is the dihedral angle formed by the tangent plane T and the Oxy plane, which equals to the included angle between the normal vector of the surface S at the point P $\boldsymbol{n} = \{-f'_x(x,y), -f'_y(x,y), 1\}$ and the positive direction of the Oz-axis, thus

$$\cos\gamma = \frac{1}{\sqrt{1 + f'^2_x(x,y) + f'^2_y(x,y)}}$$

Then

$$\Delta S \approx \Delta T = \sqrt{1 + f'^2_x(x,y) + f'^2_y(x,y)}\,\Delta\sigma$$

Then by the definition of the double integral, we get the formula for calculating surface area

$$S = \iint_\sigma \sqrt{1 + f'^2_x(x,y) + f'^2_y(x,y)}\,d\sigma = \iint_\sigma \sqrt{1 + \left(\frac{\partial z}{\partial x}\right)^2 + \left(\frac{\partial z}{\partial y}\right)^2}\,d\sigma \quad (6)$$

Traditionally, the integral expression of the above integral is called the surface area element, denoted by dS, i.e.

$$dS = \sqrt{1 + \left(\frac{\partial z}{\partial x}\right)^2 + \left(\frac{\partial z}{\partial y}\right)^2}\,d\sigma \quad (7)$$

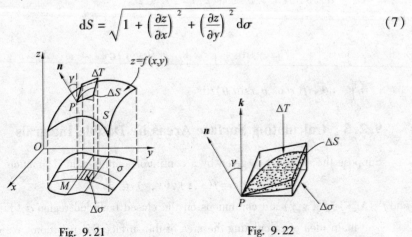

Fig. 9.21　　　　　　　　Fig. 9.22

Example 13 Find the area of the portion of the upper half sphere $z = \sqrt{R^2 - x^2 - y^2}$ that inside the cylinder $x^2 + y^2 = Rx$ (Fig. 9.23).

Solution Since

$$\frac{\partial z}{\partial x} = \frac{-x}{\sqrt{R^2 - x^2 - y^2}}, \quad \frac{\partial z}{\partial y} = \frac{-y}{\sqrt{R^2 - x^2 - y^2}}$$

$$\sqrt{1 + \left(\frac{\partial z}{\partial x}\right)^2 + \left(\frac{\partial z}{\partial y}\right)^2} = \frac{R}{\sqrt{R^2 - x^2 - y^2}}$$

The projection of the surface onto the Oxy plane is a disk

$$\sigma: x^2 + y^2 \leqslant Rx$$

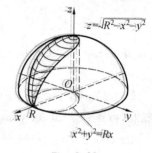

Fig. 9.23

Then

$$S = \iint_\sigma \frac{R}{\sqrt{R^2 - x^2 - y^2}} d\sigma = \int_0^{\frac{\pi}{2}} d\theta \int_0^{R\cos\theta} \frac{Rr}{\sqrt{R^2 - r^2}} dr = \pi R^2 - 2R^2$$

9.3 Calculating Triple Integrals

9.3.1 Concepts of Triple Integral

Definition 9.2 Suppose $f(x,y,z)$ is a bounded function on a closed bounded space region Ω. Divide Ω into n closed sub-regions

$$\Delta V_1, \Delta V_2, \cdots, \Delta V_n$$

and also use them to denote the volumes of the corresponding sub-regions. Let $d_i = \sup_{P_1, P_2 \in \Delta V_i}(P_1, P_2)$ be the diameter of ΔV_i, and let

$$\lambda = \max_{1 \leqslant i \leqslant n}(d_i)$$

Take an arbitrary point $P_i(\xi_i, \eta_i, \zeta_i) \in \Delta V_i (i = 1, 2, \cdots, n)$, and make the sum of the products

$$\sum_{i=1}^n f(\xi_i, \eta_i, \zeta_i) \Delta V_i$$

If no matter how to divide Ω and how to pick the sample points (ξ_i, η_i, ζ_i), the limit

$$\lim_{\lambda \to 0} \sum_{i=1}^n f(\xi_i, \eta_i, \zeta_i) \Delta V_i$$

always exists and gives the same value, then we call this limit value the triple integral of the function $f(x,y,z)$ on the closed bounded region Ω, denoted by $\iiint_\Omega f(x,y,z)\,\mathrm{d}V$, i. e.

$$\iiint_\Omega f(x,y,z)\,\mathrm{d}V = \lim_{\lambda \to 0} \sum_{i=1}^n f(P_i)\Delta V_i \tag{1}$$

In this case, we say the function $f(x,y,z)$ is integrable on Ω, where $\mathrm{d}V$ is the volume element.

Under the rectangular coordinate system, if we use three families of planes that parallel to the coordinate planes to divide the integral region V, then ΔV_i is a small rectangular cube, then the volume element under the rectangular coordinate system is

$$\mathrm{d}V = \mathrm{d}x\mathrm{d}y\mathrm{d}z$$

Then under the rectangular coordinate system, the triple integral can be expressed as

$$\iiint_V f(P)\,\mathrm{d}V = \iiint_V f(x,y,z)\,\mathrm{d}x\mathrm{d}y\mathrm{d}z$$

When the function $f(x,y,z)$ is continuous on the closed region Ω, the limit on the right hand side of equation (1) must exist, i. e. the triple integral of the function on the closed bounded region must exist. From now on we always assume that the function is continuous on the closed region. Terminologies for double integrals, such as integrant, integral region and so on, can also be applied to triple integrals, and the properties for triple integrals are also analogous to those of double integrals as showed in the first section. The physical significance of the triple integrals is as following:

Given the mass density $\mu = \mu(P)$ of the object V, then the mass of this object is

$$m = \iiint_V \mu(P)\,\mathrm{d}V$$

9.3.2 Calculating Triple Integrals under Rectangular Coordinate System

Similar as calculating double integrals, we need to transform triple integrals into iterated integrals. Here we introduce two methods.

1. Method of projection

Suppose $f(x,y,z) \in C(V)$, the projection of V on the Oxy plane is σ_{xy}, the lower and upper bound of V are

$$z = z_1(x,y), z = z_2(x,y), (x,y) \in \sigma_{xy}$$

Chapter 9 Multiple Integrals

where $z_1 \leq z_2$, and z_1, z_2 are single valued and continuous on σ_{xy} (Fig. 9.24). Then the integral region V can be expressed as

$$V:(x,y) \in \sigma_{xy}, z_1(x,y) \leq z \leq z_2(x,y)$$

Intuitively, we suppose $f(x,y,z)$ is the mass density, then the triple integral can be interpreted as the total mass distributed on the solid V. We can also think of this total mass as distributed on the projection σ_{xy}.

First, we use two families of planes $x = x_i$, $y = y_j$ to divide V and σ_{xy}. Suppose $\Delta\sigma_{xy}$ is a typical sub-region of σ_{xy} and ΔV is the corresponding small sub-solid of V, i.e. the projection of ΔV on the Oxy plane is $\Delta\sigma_{xy}$. Then we use the planes $z = z_k$ to divide ΔV into $\Delta V_1, \Delta V_2, \cdots, \Delta V_l$. Suppose $(x, y, z_k) \in \Delta V_k$, then the mass of ΔV (i.e. the mass on $\Delta\sigma_{xy}$) approximately equals to

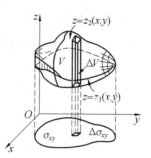

Fig. 9.24

$$\sum_{k=1}^{l} f(x,y,z_k) \Delta V_k = \left[\sum_{k=1}^{l} f(x,y,z_k) \Delta z_k\right] \Delta\sigma_{xy}$$

$$\approx \left[\int_{z_1(x,y)}^{z_2(x,y)} f(x,y,z) dz\right] \Delta\sigma_{xy}$$

(where Δz_k is the height of ΔV_k) this is the mass element of the solid ΔV over $\Delta\sigma_{xy}$. Apply double integral to this we will get the total mass of the solid V over σ, then we have the formula

$$\iiint_V f(x,y,z) dxdydz = \iint_{\sigma_{xy}} d\sigma \int_{z_1(x,y)}^{z_2(x,y)} f(x,y,z) dz \qquad (2)$$

This is the method of projection for calculating triple integrals. First, we treat x, y as constants and integrate z from the lower bound of V to the upper bound of V, then we take the double integral on the projective region σ_{xy}.

When σ_{xy} is an x-type closed region: $a \leq x \leq b, y_1(x) \leq y \leq y_2(x)$, i.e.

$$V: a \leq x \leq b, y_1(x) \leq y \leq y_2(x), z_1(x,y) \leq z \leq z_2(x,y)$$

By equation (2) and the formula for calculating double integrals, we get a formula for transforming triple integrals into iterated integrals

$$\iiint_V f(x,y,z) dxdydz = \int_a^b dx \int_{y_1(x)}^{y_2(x)} dy \int_{z_1(x,y)}^{z_2(x,y)} f(x,y,z) dz \qquad (3)$$

Analogously, we can get the formula for transforming triple integrals into iterated integrals when σ_{xy} is an y-type closed region. Similarly, we can also project the integral region V onto Oyz or Ozx. So, the triple integral can be transformed into six different iterated integrals.

Example 1 Evaluate $\iiint_V \dfrac{1}{(1+x+y+z)^3} dV$, where V is enclosed by the plane $x + y + z = 1$ and the coordinate planes.

Solution We first sketch the integral region as shown in Fig. 9.25, the projection of V on Oxy is the shaded part in the figure, obviously
$$V: 0 \leqslant x \leqslant 1, 0 \leqslant y \leqslant 1 - x, 0 \leqslant z \leqslant 1 - x - y$$
Then
$$\iiint_V \dfrac{1}{(1+x+y+z)^3} dV = \int_0^1 dx \int_0^{1-x} dy \int_0^{1-x-y} \dfrac{dz}{(1+x+y+z)^3} = \dfrac{1}{2}\left(\ln 2 - \dfrac{5}{8}\right)$$

Example 2 Evaluate $I = \iiint_V z dV$, where V is the upper half sphere enclosed by the surface $z = \sqrt{1 - x^2 - y^2}$ and the plane $z = 0$.

Solution The integral region V is as shown in Fig. 9.26, and the projection of it on the Oxy plane is the disk $x^2 + y^2 \leqslant 1$, then
$$V: -1 \leqslant x \leqslant 1, -\sqrt{1-x^2} \leqslant y \leqslant \sqrt{1-x^2}$$
$$0 \leqslant z \leqslant \sqrt{1-x^2-y^2}$$
$$I = \iint_{\sigma_{xy}} d\sigma \int_0^{\sqrt{1-x^2-y^2}} z dz = \dfrac{1}{2} \iint_{\sigma_{xy}} (1 - x^2 - y^2) d\sigma = \dfrac{\pi}{4}$$

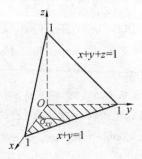

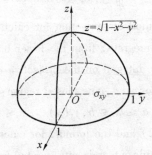

Fig. 9.25   Fig. 9.26

When calculating triple integrals, the symmetries of the integral region about the coordinate planes and the odd and even of the integrant with respect to the variables may simplify the calculation. For example, on the integral region V of example 2, we have

Chapter 9 Multiple Integrals

$$\iiint_V xz\,dV = 0, \iiint_V (y^3 + z)\,dV = \iiint_V z\,dV = \frac{\pi}{4}$$

2. Method of cross section

In equation (2), if we think of the iterated integral about z, y as a double integral and notice that in this case x is a constant, then the integral region of this double integral is the cross section σ_x of the plane perpendicular to the x-axis and V, then if the integral region is between the planes $x = a, x = b$, we take a point x arbitrarily on $[a,b]$ and use a plane perpendicular to the x-axis to intersect V, suppose the cross section is σ_x, then

$$\iiint_V f(x,y,z)\,dV = \int_a^b dx \iint_{\sigma_x} f(x,y,z)\,dydz \tag{4}$$

This is the method of cross section for calculating triple integrals. When the integrant is only related to the variable x, and the area of the cross section σ_x is easy to get, it is easier to apply equation (4). There are two other formulas for the method of cross section, please find them.

For example, if we use the method of cross section to calculate the triple integral in example 2, since σ_z is the disk $x^2 + y^2 = 1 - z^2, 0 \leqslant z \leqslant 1$, then

$$I = \int_0^1 z\,dz \iint_{\sigma_z} dxdy = \int_0^1 \pi(1 - z^2)z\,dz = \frac{\pi}{4}$$

Example 3 Evaluate $I = \iiint_V z^2\,dV$, where $V: \dfrac{x^2}{a^2} + \dfrac{y^2}{b^2} + \dfrac{z^2}{c^2} \leqslant 1$.

Solution As showed in Fig. 9.27, use a plane perpendicular to the z-axis at an arbitrary point on the interval $[-c, c]$ on the z-axis to intersect V to get

$$\sigma_z : \frac{x^2}{a^2} + \frac{y^2}{b^2} \leqslant 1 - \frac{z^2}{c^2}$$

then

$$\iiint_V z^2\,dV = \int_{-c}^c z^2\,dz \iint_{\sigma_z} dxdy$$

where $\iint_{\sigma_z} dxdy$ equals to the area of the ellipse $\dfrac{x^2}{a^2} + \dfrac{y^2}{b^2} \leqslant 1 - \dfrac{z^2}{c^2}$, that is

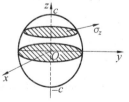

Fig. 9.27

$$\pi ab\sqrt{1-\frac{z^2}{c^2}}c\sqrt{1-\frac{z^2}{c^2}} = \pi ab\left(1-\frac{z^2}{c^2}\right)$$

then

$$\iiint_V z^2 dV = 2\pi ab \int_0^c z^2\left(1-\frac{z^2}{c^2}\right)dx = \frac{4}{15}\pi abc^3$$

9.3.3 Calculating Triple Integrals under Cylindrical Coordinate System

Suppose the polar coordinates of the projection point M of the point $P(x,y,z)$ on the Oxy plane is (r,θ), then the ordered tuple (r,θ,z) is the cylindrical coordinate of the point P. Use r to denote the distance from the point P to the z-axis, $0 \leqslant r < +\infty$; θ is the turning angle of the half plane $Ozx(x \geqslant 0)$ rotates about the positive direction of the z-axis counterclockwisely to the point P; where $-\infty < z < +\infty$.

The three families of coordinate planes of the cylindrical coordinate system are (Fig. 9.28):

$r = $ constant, which is the family of cylinders with axis the z-axis;

$\theta = $ constant, which is the family of half planes through the z-axis;

$z = $ constant, which is the family of planes parallel to the Oxy plane.

Obviously, the relation between the rectangular coordinate and the cylindrical coordinate of the point P is

$$x = r\cos\theta, y = r\sin\theta, z = z$$

Use these three families of coordinate planes to divide the integral region V, a typical sub-solid is a triangular cylinder (Fig. 9.29), the volume of the triangular cylinder obtained by increments $\Delta r, \Delta\theta, \Delta z$ on the variables r,θ,z is $\Delta V \approx r\Delta r\Delta\theta\Delta z$. Then the volume element under the cylindrical coordinate system is

$$dV = rdrd\theta dz$$

According to the above analysis, the relation between the triple integrals under the rectangular coordinate system and the cylindrical coordinate system is

Chapter 9　Multiple Integrals

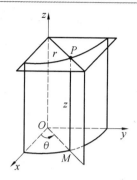
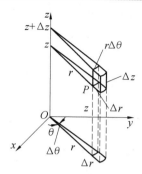

Fig. 9.28　　　　　　　　　Fig. 9.29

$$\iiint_V f(x,y,z)\,dxdydz = \iiint_V f(r\cos\theta, r\sin\theta, z)\,rdrd\theta dz \tag{5}$$

When calculating triple integrals under cylindrical coordinate system, we only need to treat $f(r\cos\theta, r\sin\theta, z)r$ as the integrant, and treat r,θ,z as independent variables equivalent to x,y,z, analogous to equation (2) we can get the formula for transforming triple integrals into iterated integrals under cylindrical coordinate system. For example, first we express the projection σ_{xy} of V on the Oxy plane: $\alpha \leq \theta \leq \beta$, $r_1(\theta) \leq r \leq r_2(\theta)$, then we determine the lower and upper bound of V, $z = z_1(r,\theta)$, $z = z_2(r,\theta)$. So

$$V: \alpha \leq \theta \leq \beta, r_1(\theta) \leq r \leq r_2(\theta), z_1(r,\theta) \leq z \leq z_2(r,\theta)$$

Then

$$\iiint_V f(r\cos\theta, r\sin\theta, z)\,rdrd\theta dz = \int_\alpha^\beta d\theta \int_{r_1(\theta)}^{r_2(\theta)} rdr \int_{z_1(r,\theta)}^{z_2(r,\theta)} f(r\cos\theta, r\sin\theta, z)\,dz$$

When the projection of the integral region V on the Oxy plane is a disk, a circular ring, a circular sector and the integrant is the composition of z with one of $x^2 + y^2, x^2 - y^2, xy, x/y$, then it is better to use cylindrical coordinate.

Example 4　Find the mass m of the solid enclosed by the surface $2z = x^2 + y^2$ and $z = 2$, given that the mass density μ of any point in the solid is proportional to the square of the distance from this point to the z-axis (Fig. 9.30).

Solution　By assumption, the mass density function $\mu = k(x^2 + y^2)$ ($k > 0$) then

$$m = \iiint_V k(x^2 + y^2)\,dV$$

Since the intersection of $2z = x^2 + y^2$ and $z = 2$ is the circle $x^2 + y^2 = 2^2$ on the plane $z = 2$, then the projection σ_{xy} of V on the Oxy plane is a disk with radius 2

Calculus(II)

$\sigma_{xy}: 0 \leq \theta \leq 2\pi, 0 \leq r \leq 2$

The lower bound of V is $z = \dfrac{1}{2}(x^2 + y^2)$, i.e. $z = \dfrac{1}{2}r^2$; upper bound is $z = 2$, then

$$m = \iiint_V kr^3 \, dr\,d\theta\,dz = \int_0^{2\pi} d\theta \int_0^2 kr^3 \, dr \int_{r^2/2}^2 dz = \dfrac{16}{3} k\pi$$

When transforming triple integrals into iterated integrals under cylindrical coordinate system, we should also pay attention to the order.

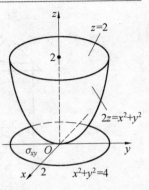

Fig. 9.30

Example 5 Suppose Ω is enclosed by $z = 16(x^2 + y^2)$, $z = 4(x^2 + y^2)$ and $z = 64$, evaluate $\iiint_\Omega (x^2 + y^2) \, dv$.

Solution $\iiint_\Omega (x^2 + y^2) \, dv = \int_0^{64} dz \int_0^{2\pi} d\theta \int_{\frac{\sqrt{z}}{4}}^{\frac{\sqrt{z}}{2}} r^3 \, dr = 2560\pi$.

Example 6 Evaluate $\iiint_V \dfrac{e^{z^2}}{\sqrt{x^2 + y^2}} \, dx\,dy\,dz$, where V is the part of the solid cone enclosed by the conical surface $z = \sqrt{x^2 + y^2}$ and the planes $z = 1, z = 2$.

Solution The integral region is as shown in Fig. 9.31. Notice the denominator of the integrant and the fact that the z-axis is passing through the integral region, so this is an improper triple integral, but can be transformed into an ordinary triple integral under cylindrical coordinate system

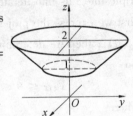

Fig. 9.31

$$\iiint_V \dfrac{e^{z^2}}{\sqrt{x^2 + y^2}} \, dx\,dy\,dz = \iiint_V e^{z^2} \, dr\,d\theta\,dz$$

If we apply formula (5) and integrate for z first, we will face the integral $\int e^{z^2} dz$. So we should integrate for r, θ first, then z

$$\iiint_V e^{z^2} \, dr\,d\theta\,dz = \int_1^2 e^{z^2} dz \int_0^{2\pi} d\theta \int_0^z dr = 2\pi \int_1^2 e^{z^2} z \, dz = \pi(e^4 - e)$$

For this problem, we could also integrate for r first, then z and at last for θ. Both of these two methods are actually methods of cross section, the former is for z-cross section, the later is for θ-cross section.

9.3.4 Calculating Triple Integrals under Spherical Coordinate System

Suppose $P(x,y,z)$ is an arbitrary point in space, the distance from P to the origin O is $\rho = |OP|, 0 \leq \rho < +\infty$; the included angle between the directed line segment $\overrightarrow{OP}$ and the positive direction of the z-axis is denoted by $\varphi, 0 \leq \varphi \leq \pi$; the turning angle of the half plane $Ozx(x \geq 0)$ rotates about the positive direction of the z-axis counter-clockwisely to the point P is denoted by $\theta, 0 \leq \theta \leq 2\pi$, then the ordered tuple (ρ, φ, θ) is the spherical coordinate of the point P.

The three families of coordinate planes of the spherical coordinate system are (Fig. 9.32):

ρ = constant, which is the family of spheres centered at the origin;

φ = constant, which is the family of conical surfaces with vertex the origin and axis the z-axis;

θ = constant, which is the family of half planes through the z-axis.

Obviously, the relation between the rectangular coordinate (x,y,z) and the spherical coordinate (ρ,φ,θ) of the point P is

$$x = \rho\sin\varphi\cos\theta, y = \rho\sin\varphi\sin\theta, z = \rho\cos\varphi$$

Use these three families of coordinate planes to divide the integral region V, a typical sub-solid is a right hexahedron (Fig. 9.33), the volume of the hexahedron obtained by increments $\Delta\rho, \Delta\varphi, \Delta\theta$ on the variables ρ, φ, θ is $\Delta V \approx \rho^2 \sin\varphi \Delta\rho\Delta\varphi\Delta\theta$. Then the volume element under the spherical coordinate system is

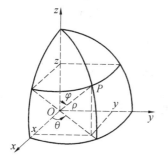

Fig. 9.32

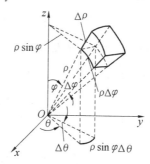

Fig. 9.33

$$dV = \rho^2 \sin\varphi \, d\rho d\varphi d\theta$$

According to the above analysis, the relation between the triple integrals under the rectangular coordinate system and the spherical coordinate system is

$$\iiint_V f(x,y,z) \, dxdydz = \iiint_V f(\rho\sin\varphi\cos\theta, \rho\sin\varphi\sin\theta, \rho\cos\varphi)\rho^2 \sin\varphi \, d\rho d\varphi d\theta \tag{6}$$

When calculating triple integrals under spherical coordinate system, we only need to treat $f(\rho\sin\varphi\cos\theta, \rho\sin\varphi\sin\theta, \rho\cos\varphi)\rho^2\sin\varphi$ as the integrant, and treat ρ, φ, θ as independent variables equivalent to x, y, z, analogous to equation (2) we can get the formula for transforming triple integrals into iterated integrals under spherical coordinate system.

First, we need to determine the two half-planes $\theta = \alpha, \theta = \beta$ which bounded the integral region V, i.e. $\alpha \leq \theta \leq \beta$; take an arbitrary θ from $[\alpha, \beta]$, make a half plane to intersect V, on this half plane if we treat the z-axis as the polar axis and if the polar coordinate (ρ, φ) of the cross section σ_θ satisfies

$$\varphi_1(\theta) \leq \varphi \leq \varphi_2(\theta), \rho_1(\theta, \varphi) \leq \rho \leq \rho_1(\theta, \varphi)$$

then

$$\iiint_V f(\rho\sin\varphi\cos\theta, \rho\sin\varphi\sin\theta, \rho\cos\varphi)\rho^2 \sin\varphi \, d\rho d\varphi d\theta$$
$$= \int_\alpha^\beta d\theta \int_{\varphi_1(\theta)}^{\varphi_2(\theta)} \sin\varphi d\varphi \int_{\rho_1(\theta,\varphi)}^{\rho_2(\theta,\varphi)} f(\rho\sin\varphi\cos\theta, \rho\sin\varphi\sin\theta, \rho\cos\varphi)\rho^2 \sin\varphi \, d\rho \tag{7}$$

When the integral region V is a ball centered at the origin or centered at a point on coordinate axis and passing through the origin; or it's part of a ball; or it's a cone with vertex the origin and with axis one of the coordinate axis, and the integrant is a function about $x^2 + y^2 + z^2$, then it is better to use spherical coordinate.

Example 7 Evaluate $\iiint_{x^2+y^2+z^2 \leq 1} (ax + by)^2 \, dv$.

Solution By symmetries of the integral region we get

$$\iiint_\Omega xy \, dv = 0, \iiint_\Omega x^2 \, dv = \iiint_\Omega y^2 \, dv = \iiint_\Omega z^2 \, dv = \frac{1}{3}\iiint_\Omega (x^2 + y^2 + z^2) \, dv$$

Then

the Original Integral $= \dfrac{1}{3}(a^2 + b^2) \iiint_\Omega (x^2 + y^2 + z^2) \, dv$

Chapter 9 Multiple Integrals

$$= \frac{a^2+b^2}{3} \int_0^{2\pi} d\theta \int_0^{\pi} \sin\varphi d\varphi \int_0^1 \rho^4 d\rho = \frac{4}{15}\pi(a^2+b^2)$$

Example 8 Find the volume of the ball with radius R.

Solution Take the center of the ball as the origin, then

$$V: 0 \leqslant \theta \leqslant 2\pi, 0 \leqslant \varphi \leqslant \pi, 0 \leqslant \rho \leqslant R$$

$$V = \iiint_V dV = \int_0^{2\pi} d\theta \int_0^{\pi} \sin\varphi d\varphi \int_0^R \rho^2 d\rho = \frac{4}{3}\pi R^3$$

Example 9 Evaluate $I = \iiint_V \sqrt{x^2+y^2+z^2}\, dV$, where

$$V: x^2 + y^2 + z^2 \geqslant 2Rz, \text{ and } x^2+y^2+z^2 \leqslant 2R^2, z \geqslant 0$$

Solution We sketch the integral region as shown in Fig. 9.34, and notice that the integrant is a function of $x^2+y^2+z^2$, so we choose the spherical coordinate system

$$V: 0 \leqslant \theta \leqslant 2\pi, \frac{\pi}{4} \leqslant \varphi \leqslant \frac{\pi}{2}, 2R\cos\varphi \leqslant \rho \leqslant \sqrt{2}R$$

$$I = \iiint_V \rho^3 \sin\varphi d\rho d\varphi d\theta = \int_0^{2\pi} d\theta \int_{\pi/4}^{\pi/2} \sin\varphi d\varphi \int_{2R\cos\varphi}^{\sqrt{2}R} \rho^3 d\rho = \frac{4}{5}\sqrt{2}\pi R^4$$

Example 10 Suppose there is a right circular cone with height h and a generatrix of length l. Suppose the mass density is a constant μ. There is a particle of mass m on the vertex of the cone. Find the universal gravitation between the cone and the particle.

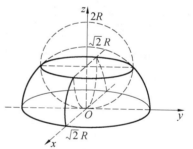

Fig. 9.34

Solution Set the coordinate system as shown in Fig. 9.35. By symmetries, the component forces of the gravitation F on the x and y-axes are both zero, we just need to find the component force on the z-axis, F_z, obviously

$$V: 0 \leqslant \theta \leqslant 2\pi, 0 \leqslant \varphi \leqslant \arccos\frac{h}{l}, 0 \leqslant \rho \leqslant \frac{h}{\cos\varphi}$$

Similar as for the application of definite integral, we usually use the method of differential element for applications of multiple integral. For any point (ρ,φ,θ) of V, we take the volume element

$$dV = \rho^2 \sin\varphi d\rho d\varphi d\theta$$

The component force of it to the particle m on the z-axis is

$$dF_z = \frac{km\mu dV}{\rho^2}\cos\varphi = km\mu\sin\varphi\cos\varphi d\rho d\varphi d\theta$$

where k is the constant of gravitation. Then

$$F_z = \iiint_V km\mu\sin\varphi\cos\varphi d\rho d\varphi d\theta$$

$$= \int_0^{2\pi} d\theta \int_0^{\arccos(h/l)} km\mu\sin\varphi\cos\varphi d\varphi \int_0^{h/\cos\varphi} d\rho$$

$$= 2\pi k\mu mh\left(1 - \frac{h}{l}\right)$$

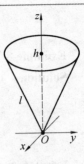

Fig. 9.35

Then the universal gravitation we want is

$$F = \left\{0, 0, 2\pi k\mu mh\left(1 - \frac{h}{l}\right)\right\}$$

Example 11 Suppose V is the rotation body obtained by rotating the logarithmic spiral

$$r = ae^{\theta/4}, 0 \leqslant \theta \leqslant \pi (a > 0)$$

about the polar axis a full circle under the polar coordinate system (Fig. 9.36). The mass density at a point equals to the distance from this point to the polar point. Find the mass m of V.

Solution Take the positive direction of the z-axis as the polar axis, then r, θ is equivalent to ρ, φ in the spherical coordinate system. Then the equation of the rotation body under the spherical coordinate system is

$$\rho = ae^{\varphi/4}, 0 \leqslant \varphi \leqslant \pi$$

Then

$$V: 0 \leqslant \rho \leqslant 2\pi, 0 \leqslant \varphi \leqslant \pi, 0 \leqslant \rho \leqslant ae^{\varphi/4}$$

$$m = \iiint_V \rho dV = \iiint_V \rho\rho^2\sin\varphi d\rho d\varphi d\theta$$

$$= \int_0^{2\pi} d\theta \int_0^\pi \sin\varphi d\varphi \int_0^{ae^{\varphi/4}} \rho^3 d\rho$$

$$= \frac{\pi}{2} \int_0^\pi a^4 e^\varphi \sin\varphi d\varphi$$

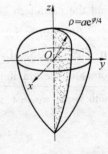

Fig. 9.36

$$= \frac{\pi a^4}{2} \cdot \frac{1}{2} e^\varphi(\sin\varphi - \cos\varphi)\Big|_0^\pi = \frac{\pi a^4}{4}(e^\pi + 1)$$

From this example we see that the triple integrals on a solid enclosed by a rotation body obtained by a space curve rotating around the polar axis a full circle under the po-

Chapter 9 Multiple Integrals

lar coordinate system can also be treated as in spherical coordinate system.

When calculating multiple integrals, we usually need to sketch the integral region first, then we choose the suitable coordinate system according to both the integral region and the integrant. We have to pay attention to the area element or volume element under the chosen coordinate system. Then we need to express the integral region by inequalities in order to determine the upper and lower limit of the integral. At last, we transform the multiple integral into an iterated integral to calculate.

9.4 Concepts and Calculations of The First Type Curve Integral

9.4.1 Concepts and Properties of the First Type Curve Integral

Mass of curvilinear component When designing curvilinear components, in order to use the material more reasonably, we need to design the different parts of the component to have different thickness according to force bearing status. Then the linear density of the component is a variable. Suppose the component lies on a curve arc L in the Oxy plane, its end points are A, B. For any point (x, y) on L, the linear density is $\mu(x, y)$. Find the mass M of the component.

If the linear density of a component is a constant, then the mass of this component equals to the product of the linear density and the length of the component. When the linear density of the component is a variable, to overcome this difficulty, we could divide the curve L into n pieces using the points $M_1, M_2, \cdots, M_{n-1}$ (Fig. 9.37), we take a small piece $\widehat{M_{i-1} M_i}$ to analysis.

When the linear density changing continuously, as long as this piece is short enough, we can use the linear density of an arbitrary point (ξ_i, η_i) to replace the linear density of other points on the piece, then the mass of the piece is

$$\Delta m_i \approx \mu(\xi_i, \eta_i) \Delta s_i$$

where Δs_i is the length of $\widehat{M_{i-1} M_i}$. Then the total mass of the whole component is approximately

$$M \approx \sum_{i=1}^{n} \mu(\xi_i, \eta_i) \Delta s_i$$

Let $\lambda = \max(\Delta s_1, \Delta s_2, \cdots, \Delta s_n)$, to calculate the accurate value of the mass we take the limit of the right hand side of the above equation as $\lambda \to 0$ and we get

$$M = \lim_{\lambda \to 0} \sum_{i=1}^{n} \mu(\xi_i, \eta_i) \Delta s_i$$

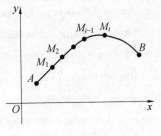

Fig. 9.37

We give the following definition

Definition 9.3 Suppose L is a piece of smooth curve arc on the Oxy plane and the function $f(x,y)$ is bounded on L. Arbitrarily plot $n-1$ points $M_1, M_2, \cdots, M_{n-1}$ to divide L into n pieces. Suppose the length of the ith piece is Δs_i and (ξ_i, η_i) is an arbitrary point in the ith piece. We take the product $f(\xi_i, \eta_i) \Delta s_i$, $i = 1, 2, \cdots, n$ and make the sum $\sum_{i=1}^{n} f(\xi_i, \eta_i) \Delta s_i$. Let $\lambda = \max(\Delta s_1, \Delta s_2, \cdots, \Delta s_n)$. If the limit of the sum always exists as $\lambda \to 0$, then we call the limit value the curve integral of the function $f(x,y)$ on the curve L with respect to arc length (or the first type curve integral), denoted by $\int_L f(x,y) \, ds$, where ds is the arc length element of L, i.e.

$$\int_L f(x,y) \, ds = \lim_{\lambda \to 0} \sum_{i=1}^{n} f(\xi_i, \eta_i) \Delta s_i$$

where $f(x,y)$ is the integrant, L is the integral arc.

If L is a closed curve, then we denote the curve integral of $f(x,y)$ on the closed curve L with respect to the arc length as $\oint_L f(x,y) \, ds$.

The above definition can be generalized to the case that the integrant is $f(x,y,z)$ and the integral arc is the space curve Γ, i.e. the curve integral of the function $f(x,y,z)$ on the curve Γ with respect to arc length

$$\int_\Gamma f(x,y,z) \, ds = \lim_{\lambda \to 0} \sum_{i=1}^{n} f(\xi_i, \eta_i, \zeta_i) \Delta s_i$$

By the definition of curve integral with respect to arc length, the curve integral with respect to arc length possesses all the properties for double integrals showed in Section 9.1.

9.4.2 Calculation The First Type Curve Integral

Theorem 9.2 Suppose l is a plane curve segment with end points A, B given by parametric equations $x = x(t)$, $y = y(t)$, $\alpha \leqslant t \leqslant \beta$, where $x(t)$, $y(t)$ are continuously differentiable on $[\alpha, \beta]$ (i.e. the curve l is smooth). If the function $f(x, y)$ is continuous on l, then the curve integral with respect to arc length (i.e. the first type curve integral) exists, and

$$\int_l f(x,y)\,ds = \int_\alpha^\beta f(x(t), y(t))\sqrt{x'^2(t) + y'^2(t)}\,dt$$

Proof The integrability is obvious. We will study the calculating formula. Suppose the point A corresponds to $t = \alpha$ and the point B corresponds to $t = \beta$. Since the point (x, y) is on the curve l, then the integrant $f(x, y)$ is a function of t, i.e. $f(x(t), y(t))$. And by the formula of arc length and the mean value theorem for integrals

$$\Delta s_i = \int_{t_i}^{t_{i+1}} \sqrt{x'^2(t) + y'^2(t)}\,dt = \sqrt{x'^2(\tau_i) + y'^2(\tau_i)}\,\Delta t_i$$

Notice that in the definition of curve integral with respect to arc length, the point (ξ_i, η_i) in $\int_l f(x,y)\,ds = \lim_{\lambda \to \infty} \sum_{i=1}^n f(\xi_i, \eta_i)\Delta s_i$ is arbitrary, then base on the assumption that the function is integrable, we could choose $(\xi_i, \eta_i) = (x(\tau_i), y(\tau_i))$ in order to calculate the integral, i.e. (Fig. 9.38)

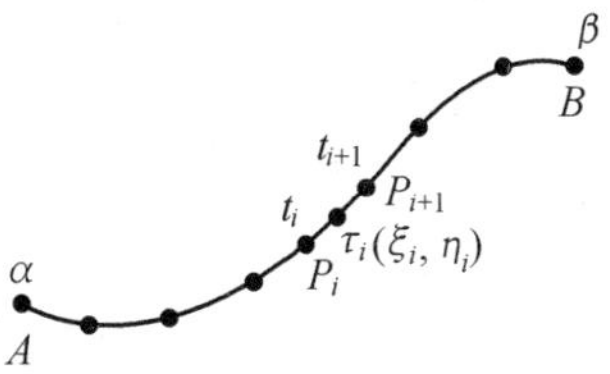

Fig. 9.38

$$\int_l f(x,y)\,ds = \lim_{\lambda \to \infty} \sum_{i=1}^n f(\xi_i, \eta_i)\Delta s_i$$

$$= \lim_{\lambda \to 0} \sum_{i=1}^n f(x(\tau_i), y(\tau_i))\sqrt{x'^2(\tau_i) + y'^2(\tau_i)}\,\Delta t_i$$

$$= \int_\alpha^\beta f(x(t), y(t))\sqrt{x'^2(t) + y'^2(t)}\,dt$$

Then the curve integral with respect to arc length can be transformed into definite integral

$$\int_l f(x,y)\,ds = \int_\alpha^\beta f(x(t), y(t))\sqrt{x'^2(t) + y'^2(t)}\,dt \tag{1}$$

Notice here, the arc length element ds is exactly differential of arc length

Calculus(II)

$$ds = \sqrt{x'^2(t) + y'^2(t)}\, dt$$

Since when $ds > 0$, $dt > 0$, then the upper limit of the definite integral in equation (1) must bigger than the lower limit. So

$$\int_{\widehat{AB}} f(P)\, ds = \int_{\widehat{BA}} f(P)\, ds$$

This is a feature of the first type curve integral, it's different than the definite integral.

If l is a space curve segment $\widehat{AB}$

$$x = x(t), y = y(t), z = z(t), \alpha \leq t \leq \beta$$

then we have the following formula

$$\int_l f(x,y,z)\, ds = \int_\alpha^\beta f(x(t), y(t), z(t)) \sqrt{x'^2(t) + y'^2(t) + z'^2(t)}\, dt \qquad (2)$$

Example 1 Evaluate $\int_l y\, ds$, where: 1° l is the arc segment of the curve $y^2 = 4x$ between $(0,0)$ and $(1,2)$; 2° l is the lower half of the cardioid $r = a(1 + \cos \theta)$ (Fig. 9.39).

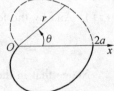

Fig. 9.39

Solution 1° Since $l: x = \dfrac{1}{4} y^2, 0 \leq y \leq 2$

$$ds = \sqrt{x'^2_y + 1}\, dy = \sqrt{1 + \dfrac{y^2}{4}}\, dy$$

then by equation (1), we get

$$\int_l y\, ds = \int_0^2 y \sqrt{1 + \dfrac{y^2}{4}}\, dy = \dfrac{4}{3}(2\sqrt{2} - 1)$$

2° Since $l: r = a(1 + \cos \theta), \pi \leq \theta \leq 2\pi$, then

$$x = r\cos \theta = a(1 + \cos \theta) \cos \theta, y = r\sin \theta = a(1 + \cos \theta) \sin \theta$$

(θ is the parameter), and the differential of arc length under polar coordinate system is

$$ds = \sqrt{r^2(\theta) + r'^2(\theta)}\, d\theta = \sqrt{a^2 (1 + \cos \theta)^2 + a^2 \sin^2 \theta}\, d\theta = a\sqrt{2(1 + \cos \theta)}\, d\theta$$

Then by formula (1), we get

$$\int_l y\, ds = \int_\pi^{2\pi} \sqrt{2} a^2 (1 + \cos \theta)^{3/2} \sin \theta\, d\theta = -\dfrac{16}{5} a^2$$

Example 2 Evaluate $\int_{\widehat{BB'}} x |y|\, ds$, where $\widehat{BB'}$ is the right half of the ellipse $x = a\cos t, y = b\sin t (a > b > 0)$ (Fig. 9.40).

Chapter 9 Multiple Integrals

Solution Since $\widehat{BB'}: x = a\cos t, y = b\sin t, -\dfrac{\pi}{2} \leqslant t \leqslant \dfrac{\pi}{2}$, and $ds = \sqrt{x'^2_t + y'^2_t}\, dt = \sqrt{a^2\sin^2 t + b^2\cos^2 t}\, dt$, then by equation (1), we get

$$\int_{\widehat{BB'}} x \mid y \mid ds = \int_{-\frac{\pi}{2}}^{\frac{\pi}{2}} a\cos t \mid b\sin t \mid \sqrt{a^2\sin^2 t + b^2\cos^2 t}\, dt$$

$$= 2ab \int_0^{\frac{\pi}{2}} \cos t\sin t \sqrt{a^2 - (a^2 - b^2)\cos^2 t}\, dt$$

$$= \dfrac{2ab}{3(a+b)}(a^2 + ab + b^2)$$

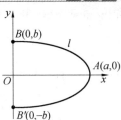

Fig. 9.40

Example 3 Suppose l is a segment of the cylindrical helix (Fig. 9.41)

$$l: x = a\cos t, y = a\sin t, z = bt, 0 \leqslant t \leqslant 2\pi$$

1° Evaluate the arc length of l. 2° Evaluate $\displaystyle\int_l \dfrac{ds}{x^2 + y^2 + z^2}$.

Solution The differential of arc length is

$$ds = \sqrt{(-a\sin t)^2 + (a\cos t)^2 + b^2}\, dt = \sqrt{a^2 + b^2}\, dt$$

1° By equation (2) we get the arc length

$$s = \int_l ds = \int_0^{2\pi} \sqrt{a^2 + b^2}\, dt = 2\pi\sqrt{a^2 + b^2}$$

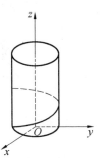

Fig. 9.41

2° From equation (2)

$$\int_l \dfrac{ds}{x^2 + y^2 + z^2} = \int_0^{2\pi} \dfrac{\sqrt{a^2 + b^2}}{a^2 + b^2 t^2}\, dt = \dfrac{\sqrt{a^2 + b^2}}{ab}\arctan\dfrac{b\pi}{2a}$$

Finally we point out, when $f(x,y) \geqslant 0$, the geometric interpretation of the first type curve integral on the plane curve l, $\displaystyle\int_l f(x,y)\,ds$, is the area of the portion of the cylinder with directrix l and generatrix parallel to the z-axis between the plane $z = 0$ and the surface $z = f(x,y)$ (Fig. 9.42).

Example 4 Find the area of the part of the cylinder $\left(x - \dfrac{R}{2}\right)^2 + y^2 = \left(\dfrac{R}{2}\right)^2$ inside the sphere $x^2 + y^2 + z^2 = R^2$ (Fig. 9.43).

Solution By the symmetry of the figure, we only need to find the area of the part in the first octant then times four. The intersection of the cylinder and the Oxy plane is

$$l: r = R\cos\theta, 0 \leqslant \theta \leqslant \frac{\pi}{2}$$

The arc length element is

$$ds = \sqrt{r^2 + r'^2}\,d\theta = R\,d\theta$$

so the area we want is

$$S = 4\int_l \sqrt{R^2 - x^2 - y^2}\,ds = 4\int_0^{\pi/2} \sqrt{R^2 - R^2\cos^2\theta}\,R\,d\theta = 4R^2$$

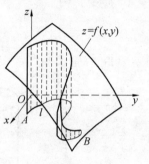

Fig. 9.42

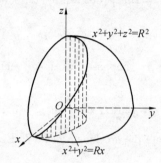

Fig. 9.43

9.5 The First Type Surface Integral

9.5.1 Definition of Surface Integral with Respect to Area

In solving engineering problems, sometimes we need to calculate the mass of a thin plate, i.e. given a space surface segment S and its area density $\mu(x,y,z)$, find its mass M.

Apply the similar procedure as we did in 9.4.1, we get

$$M = \lim_{\lambda \to 0} \sum_{i=1}^{n} \mu(x_i, y_i, z_i)\Delta S_i$$

We will encounter limits like this in other problems. We get the definition for the first type surface integral by abstracting specific significance of it.

Definition 9.4 Suppose Σ is a smooth surface and $f(x,y,z)$ is a bounded function on Σ. Divide Σ into n small closed sub-regions

$$\Delta S_1, \Delta S_2, \cdots, \Delta S_n$$

Chapter 9 Multiple Integrals

and use those to denote the areas. Let $d_i = \sup\limits_{P_1,P_2 \in \Delta S_i} (P_1, P_2)$ be the diameter of ΔS_i, and denote $\lambda = \max\limits_{1 \leq i \leq n}(d_i)$. Take an arbitrary point $P_i(\xi_i, \eta_i, \zeta_i) \in \Delta S_i (i = 1, 2, \cdots, n)$ and make the sum of the products

$$\sum_{i=1}^{n} f(\xi_i, \eta_i, \zeta_i) \Delta S_i$$

If no matter how to divide the region Σ and how to pick the points (ξ_i, η_i, ζ_i), the limit

$$\lim_{\lambda \to 0} \sum_{i=1}^{n} f(\xi_i, \eta_i, \zeta_i) \Delta S_i$$

always exists and gives the same value, we call this limit the surface integral of the function $f(x,y,z)$ on the surface Σ with respect to area (or the first type surface integral), and denoted by $\iint_\Sigma f(x,y,z) \, dS$, i. e.

$$\iint_\Sigma f(x,y,z) \, dS = \lim_{\lambda \to 0} \sum_{i=1}^{n} f(\xi_i, \eta_i, \zeta_i) \Delta S_i$$

where dS is the area element of the surface, $f(x,y,z)$ is the integrant, Σ is the integral surface.

By the definition of the surface integral with respect to area, we know that it has similar properties as double integral.

9.5.2 Calculating Surface Integrals with Respect to Area

Suppose the equation of the space surface Σ is

$$z = z(x,y), (x,y) \in \sigma_{xy}$$

where σ_{xy} is the projection of the surface Σ on the Oxy plane. The function $f(x,y,z)$ is continuous on the surface Σ, then the surface integral with respect to area (or the first type surface integral)

$$\iint_\Sigma f(x,y,z) \, dS = \lim_{\lambda \to 0} \sum_{i=1}^{n} f(x_i, y_i, z_i) \Delta S_i$$

exists.

If $z(x,y)$ has continuous first order partial derivatives on σ_{xy}, then by equation (6) from Section 9.2 and the mean value theorem of integral from Section 9.1, we get

$$\Delta S_i = \iint_{\Delta \sigma_i} \sqrt{1 + \left(\frac{\partial z}{\partial x}\right)^2 + \left(\frac{\partial z}{\partial y}\right)^2} \, d\sigma = \sqrt{1 + \left(\frac{\partial z}{\partial x}\right)_i^2 + \left(\frac{\partial z}{\partial y}\right)_i^2} \, \Delta \sigma_i$$

where $\Delta\sigma_i$ is the projection of ΔS_i on the Oxy plane (Fig. 9.44), $\left(\dfrac{\partial z}{\partial x}\right)_i$, $\left(\dfrac{\partial z}{\partial y}\right)_i$ are the partial derivatives at some point (x_i, y_i) in $\Delta\sigma_i$. Since the point (x, y, z) is on the surface S, then the integrant $f(x, y, z)$ is a function of two variables x, y, $f(x, y, z(x, y))$. Then by the definition of double integral

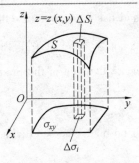

Fig. 9.44

$$\lim_{\lambda \to 0} \sum_{i=1}^{n} f(x_i, y_i, z_i) \Delta S_i$$
$$= \lim_{\lambda \to 0} \sum_{i=1}^{n} f(x_i, y_i, z(x_i, y_i)) \sqrt{1 + \left(\dfrac{\partial z}{\partial x}\right)_i^2 + \left(\dfrac{\partial z}{\partial y}\right)_i^2} \Delta\sigma_i$$
$$= \iint_{\sigma_{xy}} f(x, y, z(x, y)) \sqrt{1 + \left(\dfrac{\partial z}{\partial x}\right)^2 + \left(\dfrac{\partial z}{\partial y}\right)^2} \, d\sigma$$

Then the surface integral with respect to area can be transformed into double integral

$$\iint_S f(x, y, z) \, dS = \iint_{\sigma_{xy}} f(x, y, z(x, y)) \sqrt{1 + \left(\dfrac{\partial z}{\partial x}\right)^2 + \left(\dfrac{\partial z}{\partial y}\right)^2} \, d\sigma \qquad (1)$$

According to the different status of the surface Σ, we can transform the surface integral with respect to area into a double integral on the projection of Σ on other coordinate planes. So, when calculating surface integrals with respect to area, we first need to choose a proper projection plane according to the surface Σ, find the projection of Σ and write out the equation of Σ, then we find the area element, at last we substitute the equation of the surface into the integrant, and calculate the double integral on the right hand side of equation (1).

Example 1 Evaluate $\iint_S (x^2 + y^2 + z^2) \, dS$, where S is the part of the cone $z = \sqrt{x^2 + y^2}$ between the planes $z = 0$ and $z = 1$ (Fig. 9.45).

Solution The equation of the surface S is
$$z = \sqrt{x^2 + y^2}$$
$$\dfrac{\partial z}{\partial x} = \dfrac{x}{\sqrt{x^2 + y^2}}, \quad \dfrac{\partial z}{\partial y} = \dfrac{y}{\sqrt{x^2 + y^2}}$$
$$\sqrt{1 + \left(\dfrac{\partial z}{\partial x}\right)^2 + \left(\dfrac{\partial z}{\partial y}\right)^2} = \sqrt{2}$$

The projection of S on the Oxy plane is a disk $\sigma_{xy} : x^2 + y^2 \leqslant 1$, so

$$\iint_S (x^2 + y^2 + z^2)\,\mathrm{d}S = \iint_{\sigma_{xy}} 2\sqrt{2}\,(x^2 + y^2)\,\mathrm{d}\sigma$$

$$= 2\sqrt{2} \int_0^{2\pi} \mathrm{d}\theta \int_0^1 r^3\,\mathrm{d}r = \sqrt{2}\,\pi$$

Example 2 Evaluate $\iint_S (x^3 + x^2 y + z)\,\mathrm{d}S$, where S is the part of the sphere $z = \sqrt{a^2 - x^2 - y^2}$ above the plane $z = h\,(0 < h < a)$ (Fig. 9.46).

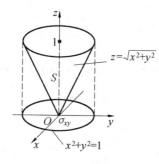

Fig. 9.45   Fig. 9.46

Solution By the symmetries

$$\iint_S (x^3 + x^2 y + z)\,\mathrm{d}S = \iint_S z\,\mathrm{d}S$$

The equation of S is $z = \sqrt{a^2 - x^2 - y^2}$, so

$$\frac{\partial z}{\partial x} = \frac{-x}{\sqrt{a^2 - x^2 - y^2}},\quad \frac{\partial z}{\partial y} = \frac{-y}{\sqrt{a^2 - x^2 - y^2}}$$

$$\sqrt{1 + \left(\frac{\partial z}{\partial x}\right)^2 + \left(\frac{\partial z}{\partial y}\right)^2} = \frac{a}{\sqrt{a^2 - x^2 - y^2}}$$

The projection of S on the Oxy plane is a disk σ_{xy}

$$x^2 + y^2 \leqslant a^2 - h^2$$

Thus

$$\iint_S (x^3 + x^2 y + z)\,\mathrm{d}S = \iint_S z\,\mathrm{d}S$$

$$= \iint_{\sigma_{xy}} \sqrt{a^2 - x^2 - y^2}\,\frac{a}{\sqrt{a^2 - x^2 - y^2}}\,\mathrm{d}\sigma$$

$$= \pi a(a^2 - h^2)$$

Calculus(Ⅱ)

What noticeable is that, it's sometimes convenient to calculate the first type surface integral on sphere using spherical coordinate system, such as the above example. The surface S is

$$\rho = a, 0 \leqslant \theta \leqslant 2\pi, 0 \leqslant \varphi \leqslant \arccos \frac{h}{a}$$

The area element of the surface is

$$dS = a^2 \sin\varphi \, d\varphi \, d\theta$$

So

$$\iint_S (x^3 + x^2 y + z) \, dS = \iint_S z \, dS = \int_0^{2\pi} d\theta \int_0^{\arccos\frac{h}{a}} a^3 \cos\varphi \sin\varphi \, d\varphi = \pi a(a^2 - h^2)$$

Example 3 Evaluate $I = \oiint_S xyz \, dS$, where S is the surface of the tetrahedron enclosed by the plane $x + y + z = 1$ and the three coordinate planes.

Solution If we use S_1, S_2, S_3 to denote the three surfaces of the tetrahedron on the coordinate planes Oxy, Oyz, Ozx respectively (Fig. 9.47), then since the equation of S_1 is $z = 0$, so

$$\iint_{S_1} xyz \, dS = 0$$

Similarly, we have

$$\iint_{S_2} xyz \, dS = 0, \iint_{S_3} xyz \, dS = 0$$

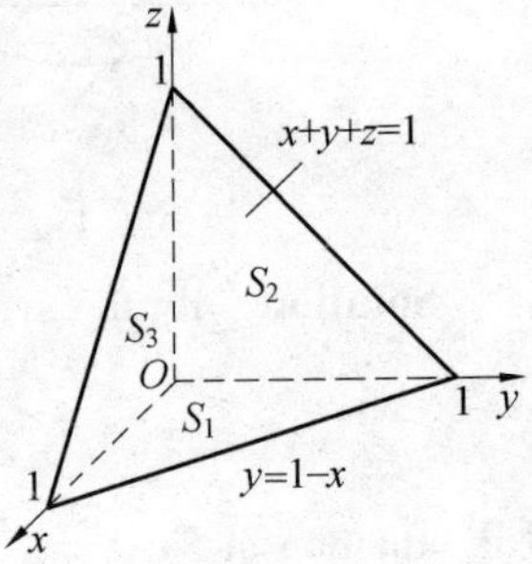

Fig. 9.47

The equation of the fourth surface S_4 is

$$z = 1 - x - y$$
$$0 \leqslant x \leqslant 1, 0 \leqslant y \leqslant 1 - x$$

so $z'_x = -1, z'_y = -1, \sqrt{1 + z'^2_x + z'^2_y} = \sqrt{3}$. then

$$I = \oiint_S xyz \, dS = \iint_{S_4} xyz \, dS = \iint_{\sigma_{xy}} xy(1-x-y) \sqrt{3} \, dx \, dy$$

$$= \int_0^1 dx \int_0^{1-x} \sqrt{3} xy(1 - x - y) \, dy = \frac{\sqrt{3}}{120}$$

Chapter 9 Multiple Integrals

9.6 Application of Integrals

We already learned double integrals, triple integrals, curve integral with respect to arc length, and surface integral with respect to area. These integrals have extensive applications, such as, finding the areas of plane regions, areas of surfaces, volumes of solids, lengths of curves, and masses of objects. In this section we will introduce the methods to calculate the centers of mass and moments of inertia of objects. For simplicity of writing, we use $\int_\Omega f(P) \mathrm{d}\Omega$ to denote the integrals we just mentioned.

9.6.1 Center of Mass

By statics, when the mass of the point in the system of particles $P_i(x_i,y_i,z_i)$ ($i = 1,2,\cdots,n$) is m_i, the coordinate of the center of mass of the system is

$$\bar{x} = \frac{\sum_{i=1}^{n} m_i x_i}{\sum_{i=1}^{n} m_i}, \bar{y} = \frac{\sum_{i=1}^{n} m_i y_i}{\sum_{i=1}^{n} m_i}, \bar{z} = \frac{\sum_{i=1}^{n} m_i z_i}{\sum_{i=1}^{n} m_i}$$

Suppose the mass distribution density function on a geometric body Ω, $\mu(P)$, $P \in \Omega$, is a continuous function. We divide Ω into n parts with very small diameters, $\Delta\Omega_i$ ($i = 1,2,\cdots,n$), arbitrarily take a point $P_i \in \Delta\Omega_i$, think of $\Delta\Omega_i$ as a particle of mass $\mu(P_i)\Delta\Omega_i$ at P_i. We got a system of n particles, take the center of mass of this system, then refine the partition infinitely and take the limit, we obtain the coordinate of the center of mass

$$\bar{x} = \frac{\int_\Omega \mu(P)x\mathrm{d}\Omega}{\int_\Omega \mu(P)\mathrm{d}\Omega}, \bar{y} = \frac{\int_\Omega \mu(P)y\mathrm{d}\Omega}{\int_\Omega \mu(P)\mathrm{d}\Omega}, \bar{z} = \frac{\int_\Omega \mu(P)z\mathrm{d}\Omega}{\int_\Omega \mu(P)\mathrm{d}\Omega} \qquad (1)$$

Here Ω may be a solid, a surface, a curve, a plane segment or a planar curve. By equation (1), the x coordinate of the center of mass equals to the ratio of the static distance to the plane $x = 0$, $\int_\Omega \mu(P)x\mathrm{d}\Omega$, and the total mass $\int_\Omega \mu(P)\mathrm{d}\Omega$, and we have similar results for the y coordinate and z coordinate. When the density $\mu(P)$ is a constant, the center of mass is also called centroid. From equation (1), we can easily obtained the coordinate of the centroid by deleting μ.

Example 1 Find the center of mass of the homogeneity plate σ enclosed by two circles $r = 2\sin\theta, r = 4\sin\theta$ (Fig. 9.48).

Solution Since the plate σ is symmetric with respect to the y-axis, and is homogeneity (area density μ is a constant), then $\bar{x} = 0$, so we only need to find $\bar{y}$. Since

$$m = \iint_\sigma \mu d\sigma = \mu(\pi \cdot 2^2 - \pi \cdot 1^2) = 2\pi\mu$$

$$M_y = \iint_\sigma \mu y d\sigma = \int_0^\pi d\theta \int_{2\sin\theta}^{4\sin\theta} \mu r^2 \sin\theta dr$$

$$= \frac{112}{3}\mu \int_0^{\pi/2} \sin^4\theta d\theta = 7\pi\mu$$

Then $\bar{y} = \dfrac{7}{3}$, i.e. the center of mass is $\left(0, \dfrac{7}{3}\right)$.

Example 2 Given the volume density $\mu = k(x^2 + y^2 + z^2)$ of the cone shown in Fig. 9.49, where k is a constant, find the center of mass.

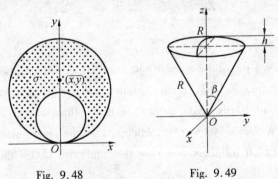

Fig. 9.48   Fig. 9.49

Solution From Fig. 9.49 we know that the cone V is symmetric with respect to the coordinate planes Oyz and Ozx, and the density function is an even function with respect to x, y. So the center of mass must lie on z-axis, i.e. we just need to find $\bar{z}$. Since

$$m = \iiint_V k(x^2 + y^2 + z^2) dV = \int_0^{2\pi} d\theta \int_0^\beta k\sin\varphi d\varphi \int_0^R \rho^4 d\rho$$

$$= \frac{2\pi}{5} kR^5 (1 - \cos\beta)$$

and

Chapter 9 Multiple Integrals

$$M_z = \iiint_V k(x^2 + y^2 + z^2)z\,dV = \int_0^{2\pi} d\theta \int_0^{\beta} k\sin\varphi\cos\varphi\,d\varphi \int_0^R \rho^5 d\rho$$

$$= \frac{\pi}{6}kR^6(1 - \cos^2\beta)$$

then

$$\bar{z} = \frac{5}{12}R(1 + \cos\beta) = \frac{5}{12}(2R - h)$$

where $h = R(1 - \cos\beta)$. Then the center of mass of the cone is $(0, 0, \frac{5}{12}(2R-h))$.

9.6.2 Moment of Inertia

Moment of inertia is an important concept in mechanism, we need to use it when study rigid rotation. By mechanism, the moment of inertia of n particles to a fixed axis

$$\sum_{i=1}^n r_i^2 m_i$$

where m_i and r_i are the mass and the distance from the i th particle to the fixed axis respectively.

For a rigid body Ω, suppose the mass distribution density $\mu(P)$ is a continuous function on Ω, then how to find the moment of inertia of Ω to some fixed axis? Here we use the method of differential element. Take an arbitrary small rigid body $\Delta\Omega$ in Ω, and take an arbitrary point $P \in \Delta\Omega$. Suppose the distance from P to the fixed axis is r, then we have the mass element $\mu(P)\Delta\Omega$, and the corresponding moment of inertial element

$$r^2\mu(P)\Delta\Omega$$

then the moment of inertia of the rigid body Ω to the fixed axis is

$$I = \int_\Omega r^2\mu(P)\,d\Omega$$

If we set up the rectangular coordinate system $Oxyz$, then the moments of inertia of the rigid body Ω to the x-axis, y-axis and z-axis are

$$I_x = \int_\Omega (y^2 + z^2)\mu\,d\Omega, \quad I_y = \int_\Omega (x^2 + z^2)\mu\,d\Omega \qquad (2)$$

$$I_z = \int_\Omega (x^2 + y^2)\mu\,d\Omega$$

where $\mu = \mu(P)$ is the density function. Here Ω could be a solid, a surface, a curve, a plane segment or a planar curve.

Example 3 Find the moment of inertia I of the rotating parabolic body $x^2 + y^2 \leq z \leq 1$ (denoted by Ω) with density 1 to the z-axis.

Solution
$$I = \iiint_\Omega (x^2 + y^2) dv = \int_0^1 dz \iint_{x^2+y^2 \leq z} (x^2 + y^2) dxdy$$
$$= \int_0^1 dz \int_0^{2\pi} d\theta \int_0^{\sqrt{z}} r^3 dr = \frac{\pi}{2} \int_0^1 z^2 dz = \frac{\pi}{6}$$

Example 4 Given a homogeneity cylindrical helix l
$$x = a\cos t, y = a\sin t, z = bt, 0 \leq t \leq 2\pi$$
1° Find the center of mass of l; 2° Find the moment of inertia of l to the z-axis, I_z.

Solution The arc length element of the space curve l is
$$ds = \sqrt{x_t'^2 + y_t'^2 + z_t'^2} dt = \sqrt{(-a\sin t)^2 + (a\cos t)^2 + b^2} dt = \sqrt{a^2 + b^2} dt$$

1° Suppose the linear density is a constant μ, since
$$m = \int_l \mu ds = \int_0^{2\pi} \mu \sqrt{a^2 + b^2} dt = 2\pi\mu\sqrt{a^2 + b^2}$$
$$M_x = \int_l \mu x ds = \int_0^{2\pi} \mu a\cos t \sqrt{a^2 + b^2} dt = 0$$
$$M_y = \int_l \mu y ds = \int_0^{2\pi} \mu a\sin t \sqrt{a^2 + b^2} dt = 0$$
$$M_z = \int_l \mu z ds = \int_0^{2\pi} \mu bt \sqrt{a^2 + b^2} dt = 2\pi^2 b\mu \sqrt{a^2 + b^2}$$

Then the center of mass of l is $(0, 0, \pi b)$.

2° The moment of inertial of l to the z-axis is
$$I_z = \int_l (x^2 + y^2) \mu ds = \int_0^{2\pi} a^2 \mu \sqrt{a^2 + b^2} dt = 2\pi\mu a^2 \sqrt{a^2 + b^2} = a^2 m$$

where $m = 2\pi\mu\sqrt{a^2 + b^2}$ is the mass of l.

9.7 Examples

Example 1 Calculate $\iint_D |\cos(x + y)| d\sigma$ where D is the region bounded by the lines $x = \frac{\pi}{2}, y = 0, y = x$.

Solution Sketch the graph of the region D, as shown in Fig. 9.50. Use the line

Chapter 9 Multiple Integrals

$y + x = \dfrac{\pi}{2}$ to divide D into two parts: D_1 and D_2. Hence

$$|\cos(x+y)| = \begin{cases} \cos(x+y), (x,y) \in D_1 \\ -\cos(x+y), (x,y) \in D_2 \end{cases}$$

So

$$\iint_D |\cos(x+y)|\, d\sigma$$
$$= \iint_{D_1} \cos(x+y)\, d\sigma - \iint_{D_2} \cos(x+y)\, d\sigma$$

Since

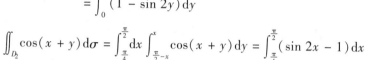

Fig. 9.50

$$\iint_{D_1} \cos(x+y)\, d\sigma = \int_0^{\frac{\pi}{4}} dy \int_y^{\frac{\pi}{2}-y} \cos(x+y)\, dx$$
$$= \int_0^{\frac{\pi}{4}} (1 - \sin 2y)\, dy$$

$$\iint_{D_2} \cos(x+y)\, d\sigma = \int_{\frac{\pi}{4}}^{\frac{\pi}{2}} dx \int_{\frac{\pi}{2}-x}^{x} \cos(x+y)\, dy = \int_{\frac{\pi}{4}}^{\frac{\pi}{2}} (\sin 2x - 1)\, dx$$

we get

$$\iint_D |\cos(x+y)|\, d\sigma = \int_0^{\frac{\pi}{4}} (1 - \sin 2t)\, dt + \int_{\frac{\pi}{4}}^{\frac{\pi}{2}} (1 - \sin 2t)\, dt$$
$$= \int_0^{\frac{\pi}{2}} (1 - \sin 2t)\, dt = \dfrac{\pi}{2} - 1$$

When the integrand contains absolute value sign, we usually divide the integral region into several parts according to the additivity property, in order to remove the absolute value.

Example 2 Find the area of the graph enclosed by the lemniscates $r^2 = 2a^2 \sin 2\theta$.

Solution The graph of the lemniscates is shown in Fig. 9.51. As the pole angle θ increases from 0 to $\dfrac{\pi}{2}$, sketch one petal of the lemniscates; as the pole angle θ increases from π to $\dfrac{3\pi}{2}$, sketch another petal of the lemniscates.

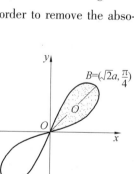

Fig. 9.51

Since the graph is symmetric about the pole, the area is

$$S = 2\iint_\sigma 1 d\sigma = 2\int_0^{\frac{\pi}{2}} d\theta \int_0^{a\sqrt{2\sin 2\theta}} r dr$$

$$= 2\int_0^{\frac{\pi}{2}} a^2 \sin 2\theta d\theta = 2a^2$$

Example 3 Calculate the integral $\int_0^1 \dfrac{x^b - x^a}{\ln x} dx \, (a, b > 0)$.

Solution Note that

$$\frac{x^b - x^a}{\ln x} = \int_a^b x^y dy$$

so we can convert the original integral to an iterated integral. Then by interchanging the order of integration of the iterated integral yields

$$\int_0^1 \frac{x^b - x^a}{\ln x} dx = \int_0^1 dx \int_a^b x^y dy = \int_a^b dy \int_0^1 x^y dx = \ln \frac{1+b}{1+a}$$

Example 4 Verify that

$$\left[\int_a^b f(x)g(x) dx\right]^2 \leq \int_a^b f^2(x) dx \int_a^b g^2(x) dx$$

Solution In the square region $D: a \leq x \leq b, a \leq y \leq b$, we have

$$\iint_D [f(x)g(y) - f(y)g(x)]^2 d\sigma \geq 0$$

Hence

$$\iint_D f^2(x)g^2(y) d\sigma + \iint_D f^2(y)g^2(x) d\sigma \geq 2\iint_D f(x)g(x)f(y)g(y) d\sigma$$

Since the integral domain D is a square, x and y appear symmetrically in the domain, the two integrals of the left side of the above equality equal. Converting the double integrals to iterated integrals yields

$$2\int_a^b f^2(x) dx \int_a^b g^2(y) dy \geq 2\int_a^b f(x)g(x) dx \int_a^b f(y)g(y) dy$$

so

$$\left[\int_a^b f(x)g(x) dx\right]^2 \leq \int_a^b f^2(x) dx \int_a^b g^2(x) dx$$

Example 5 A cylinder of radius a is stitched on the great circle of a homogeneous hemispheroid of radius a, and they are made of the same material. For what value of the height of the cylinder the position of the center of mass of the jointing solid is located

Chapter 9 Multiple Integrals

at the centre of spheroid?

Solution Suppose that the cylinder has height H, and setup the coordinate system as illustrated in Fig. 9.52. By symmetry the center of mass of the jointing solid must lie on the z-axis. In order to let the center of mass lie on the centre of spheroid (i.e. the origin of coordinate), all that we need is

$$\bar{z} = \frac{\iiint_V \mu z dV}{\iiint_V \mu dV} = \frac{\iiint_V z dV}{\iiint_V dV} = 0$$

where the constant μ is the mass volume density. Because

$$\iiint_V z dV = \iiint_{\text{hemispheroid}} z dV + \iiint_{\text{cylinder}} z dV$$

$$= \int_0^{2\pi} d\theta \int_{\frac{\pi}{2}}^{\pi} \cos\varphi \sin\varphi d\varphi \int_0^a \rho^3 d\rho + \int_0^{2\pi} d\theta \int_0^a r dr \int_0^H z dz$$

$$= -\frac{1}{4}\pi a^4 + \frac{1}{2}\pi a^2 H^2$$

we have

$$-\frac{1}{4}\pi a^4 + \frac{1}{2}\pi a^2 H^2 = 0$$

Solving the equation yields $H = \frac{\sqrt{2}}{2}a$, i.e. the height of the cylinder should equal to $\frac{\sqrt{2}}{2}a$.

Example 6 Find the volume of the solid obtained by rotating around the polar axis the region enclosed by the cardioid $r = a(1 - \cos\theta) (a > 0, 0 \leq \theta \leq \pi)$.

Solution Take the polar axis as the positive direction of oz-axis, then the equation of the boundary surface of the rotation body in spherical coordinates is

$$\rho = a(1 - \cos\varphi) \quad (\text{as shown in Fig. 9.53})$$

So the volume is

$$V = \iiint_V 1 dV = \int_0^{2\pi} d\theta \int_0^\pi d\varphi \int_0^{a(1-\cos\varphi)} \rho^2 \sin\varphi d\rho$$

$$= 2\pi \int_0^\pi \sin\varphi \frac{a^3(1-\cos\varphi)^3}{3} d\varphi = \frac{8}{3}\pi a^3$$

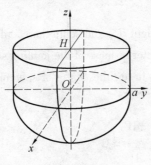

Fig. 9.52　　　　　　　　　　Fig. 9.53

Example 7　Calculate the space curve integral $\oint_c (z + y^2) ds$, where c is the intersection between the sphere $x^2 + y^2 + z^2 = R^2$ and the plane $x + y + z = 0$.

Solution　Because x, y and z appear symmetrically in the expressions of the sphere $x^2 + y^2 + z^2 = R^2$ and the plane $x + y + z = 0$, so x, y, z are interchangeable in the equation of the curve c. So we can use the rotation properties of variable x, y, z to compute the curve integral. Since

$$\oint_c x ds = \oint_c y ds = \oint_c z ds$$

and

$$\oint_c x^2 ds = \oint_c y^2 ds = \oint_c z^2 ds$$

we get

$$\oint_c z ds = \frac{1}{3} \oint_c (x + y + z) ds = \frac{1}{3} \oint_c 0 ds = 0$$

and

$$\oint_c y^2 ds = \frac{1}{3} \oint_c (x^2 + y^2 + z^2) ds = \frac{1}{3} R^2 \oint_c ds = \frac{1}{3} R^2 2\pi R$$

Thus $\oint_c (z + y^2) ds = \oint_c z ds + \oint_c y^2 ds = \frac{2}{3} \pi R^3$.

It is an ingenious method to simplify the integrand by using the curve and surface equations when calculating the curve and surface integrals.

Example 8　A cylinder with radius R and height h has uniform electric charge, and the area density of charge is a constant σ. Find the electric field intensity $\boldsymbol{E}(0) = \{E_x(0), E_y(0), E_z(0)\}$ of the center of the base circle.

Chapter 9 Multiple Integrals

Solution Set up the rectangular coordinate system as shown in Fig. 9.54. By symmetry we obtain $E_x(0) = 0$, $E_y(0) = 0$. So we only need to find $E_z(0)$. Take an area element dS on the cylinder, which has charge element σdS. The component about the z-axis of the electric field intensity produced by σdS at the origin O is

$$dE_z(0) = \frac{-k\sigma dS}{\rho^2}\cos\varphi = \frac{-k\sigma z}{\rho^3}dS = \frac{-k\sigma z}{(R^2 + z^2)^{3/2}}dS$$

where k is a constant, and ρ denotes the distance from the origin O to the area element. Therefore

$$E_z(0) = \iint_S \frac{-k\sigma z}{(R^2 + z^2)^{3/2}}dS$$

where the integral surface is the cylinder.

Sometimes it is very convenient to calculate surface integrals under cylindrical coordinates.

In cylindrical coordinates the equation of the surface S is

$$r = R, 0 \leq \theta \leq 2\pi, 0 \leq z \leq H$$

Obviously, the area element in the cylinder is

$$dS = Rd\theta dz$$

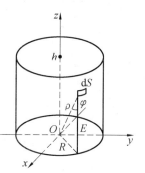

Fig. 9.54

Thus

$$E_z(0) = \iint_S \frac{-k\sigma z}{(R^2 + z^2)^{3/2}}Rd\theta dz = -k\sigma R\int_0^{2\pi}d\theta\int_0^H \frac{zdz}{(R^2 + z^2)^{3/2}}$$

$$= -2\pi k\sigma\left(1 - \frac{R}{\sqrt{R^2 + H^2}}\right)$$

Namely

$$E(0) = \left\{0, 0, -2\pi k\sigma\left(1 - \frac{R}{\sqrt{R^2 + H^2}}\right)\right\}$$

Exercises 9

9.1

1. Evaluate the following integrals

(1) $I = \iint_\sigma (x + y + 10)\,d\sigma$, where σ is the disk $x^2 + y^2 \leq 4$.

(2) $I = \iint_D |y|\,d\sigma$, where the integral region $D: 0 \leq x \leq 1, -1 \leq y \leq 1$.

2. Let D be a triangular region in the Oxy plane with vertices $(1,1), (-1,1)$ and $(-1,-1)$, and D_1 be the first quadrant part of D. Prove that
$$\iint_D (xy + \cos x \sin y)\,d\sigma = 2\iint_{D_1} \cos x \sin y\,d\sigma$$

3. Compare the values of the integrals in each of the following group.

(1) $\iint_D (x+y)^2\,d\sigma$ and $\iint_D (x+y)^3\,d\sigma$, where $D: (x-2)^2 + (y-2)^2 \leq 2^2$;

(2) $\iint_D \ln(x+y)\,d\sigma$ and $\iint_D xy\,d\sigma$, where D is bounded by the lines $x = 0, y = 0$, $x + y = \dfrac{1}{2}, x + y = 1$.

4. The double integral of the function $\dfrac{\sin(\pi\sqrt{x^2+y^2})}{x^2+y^2}$ on the annulus $D: 1 \leq x^2 + y^2 \leq 4$ ( ).

(A) doesn't exist (B) exists and is positive

(C) exists and is negative (D) exists and equals to zero

9.2

1. Sketch the graph of the integral domain σ, and convert the double integral $\iint_\sigma f(x,y)\,d\sigma$ to an iterated integral with different orders of integration.

(1) σ is bounded by the lines $x + y = 1, x - y = 1, x = 0$;

(2) σ is enclosed by the lines $y = 0, y = a, y = x, y = x - 2a\ (a > 0)$;

(3) $\sigma: xy \geq 1, y \leq x, 0 \leq x \leq 2$;

(4) $\sigma: x^2 + y^2 \leq 1, x \geq y^2$;

(5) $\sigma: 4x^2 + 9y^2 \geq 36, y^2 \leq x + 4$.

2. Evaluate the following double integrals.

(1) $\iint_D \dfrac{x^2}{1+y^2}\,d\sigma$, where the region $D: 0 \leq x \leq 1, 0 \leq y \leq 1$;

(2) $\iint_D (x+y)\,d\sigma$, where D is the triangular region with vertices $O(0,0), A(1,$

$0)$, $B(1,1)$;

(3) $\iint_D \dfrac{x^2}{y^2} d\sigma$, where D is the region enclosed by $y = 2$, $y = x$, $xy = 1$;

(4) $\iint_D \cos(x + y) dxdy$, where D is the region bounded by the lines $x = 0$, $y = x$, $y = \pi$;

(5) $\iint_D \dfrac{x\sin y}{y} dxdy$, where D is the region enclosed by $y = x$, $y = x^2$;

(6) $\iint_D y^2 dxdy$, where D is the region bounded by the x-axis and one arch of the cycloid $x = a(t - \sin t)$, $y = a(1 - \cos t)$ $(0 \leqslant t \leqslant 2\pi, a > 0)$.

3. Calculate the following double integrals:

(1) $\iint_\sigma [x^2 y + \sin(xy^2)] d\sigma$, where σ is the region enclosed by $x^2 - y^2 = 1$, $y = 0$, $y = 1$;

(2) $\iint_\sigma x |y| dxdy$, $D: y \leqslant x, x \leqslant 1, y \geqslant -\sqrt{2 - x^2}$;

(3) $\iint_\sigma (1 - 2x + \sin y^3) dxdy$, $D: x^2 + y^2 \leqslant R^2$.

4. Sketch the integral region of the following iterated integral and interchange the order of integration.

(1) $\displaystyle\int_1^e dx \int_0^{\ln x} f(x,y) dy$;

(2) $\displaystyle\int_0^1 dx \int_x^{2x} f(x,y) dy$;

(3) $\displaystyle\int_0^1 dy \int_{\sqrt{y}}^{\sqrt[3]{y}} f(x,y) dx$;

(4) $\displaystyle\int_0^1 dy \int_{\sqrt{1-y^2}}^{-\sqrt{1-y^2}} f(x,y) dx$;

(5) $\displaystyle\int_{1/2}^{1/\sqrt{2}} dx \int_{1/2}^{x} f(x,y) dy + \int_{1/\sqrt{2}}^{1} dx \int_{x^2}^{x} f(x,y) dy$;

(6) $\displaystyle\int_0^{\frac{a}{2}} dy \int_{\sqrt{a^2-2ay}}^{\sqrt{a^2-y^2}} f(x,y) dx + \int_{\frac{a}{2}}^{a} dy \int_0^{\sqrt{a^2-y^2}} f(x,y) dx$.

5. Evaluate $\displaystyle\int_0^1 dx \int_{x^2}^1 \dfrac{xy}{\sqrt{1 + y^3}} dy$.

6. Find the volume of the solid enclosed by the surfaces $z = x^2 + y^2$, $y = x^2$, $y = 1$ and $z = 0$.

7. A planar thin lamina bounded by the curve $xy = 1$ and the line $x + y = \dfrac{5}{2}$ has mass

area density $\dfrac{1}{x}$. Find the mass of the thin lamina.

8. Evaluate the following double integrals:

(1) $\iint_D \ln(1 + x^2 + y^2)\,d\sigma$, where D is a disk: $x^2 + y^2 \leq 1$;

(2) $\iint_D \sqrt{a^2 - x^2 - y^2}\,d\sigma$, $D: x^2 + y^2 \leq ay, |y| \geq |x|\, (a > 0)$;

(3) $\iint_D \sin\sqrt{x^2 + y^2}\,d\sigma$, $D: \pi^2 \leq x^2 + y^2 \leq 4\pi^2$;

(4) $\iint_D (x^2 + y^2)\,d\sigma$, $D: x^2 + y^2 \geq 2x, x^2 + y^2 \leq 4x$;

(5) $\iint_D (x^2 + y^2)^{3/2}\,d\sigma$, $D: x^2 + y^2 \leq 1, x^2 + y^2 \leq 2x$;

(6) $\iint_D \arctan\dfrac{y}{x}\,dxdy$, $D: 1 \leq x^2 + y^2 \leq 4, x \geq 0, y \geq 0$;

(7) $\iint_D |x^2 + y^2 - 4|\,dxdy$, $D: x^2 + y^2 \leq 16$;

(8) $\iint_D \sqrt{x^2 + y^2}\,dxdy$, $D: 0 \leq x \leq a, 0 \leq y \leq a$.

9. Calculate the area of the given plane region by means of double integrals.

(1) The common region inside the cardioid $r = a(1 - \cos\theta)$ and outside the circle $r = a$;

(2) The region enclosed by the curve $(x^2 + y^2)^2 = 8a^2xy\, (a > 0)$.

10. Find the area of the given surface.

(1) The part of the conic surface $y^2 + z^2 = x^2$ which lies within the cylinder $x^2 + y^2 = a^2$;

(2) The part of the conic surface $z = \sqrt{x^2 + y^2}$ cut off by the parabolic cylindrical surface $z^2 = 2x$;

(3) The part of the paraboloid of revolution $2z = x^2 + y^2$ cut off by the cylinder $x^2 + y^2 = 1$;

(4) The part of the hyperbolic paraboloid $z = xy$ cut off by the cylinder $x^2 + y^2 = a^2$;

(5) The part of the sphere $x^2 + y^2 + z^2 = 3a^2$ which lies within the paraboloid of revolution $x^2 + y^2 - 2az = 0\,(a > 0)$.

11. Suppose that the center of the sphere Σ with radius R lies on a fixed sphere $x^2 +$

Chapter 9 Multiple Integrals

$y^2 + z^2 = a^2 (a > 0)$. For what value of R does the part of the sphere Σ inside the fixed sphere attains a maximal area?

12. Let $f(x) \in C[0,1]$, and $\int_0^1 f(x)dx = a$, find $\int_0^1 dx \int_x^1 f(x)f(y) dy$.

9.3

1. Convert the triple integral $\iiint_V f(x,y,z)dV$ into an iterated integral in rectangular coordinate system, where the integral region V is

(1) enclosed by the surfaces $z = x^2 + 2y^2$ and $z = 2 - x^2$;

(2) enclosed by the surface $z = 1 - \sqrt{x^2 + y^2}$ and the planes $z = x (x \geq 0), x = 0$;

(3) enclosed by the system of inequalities $0 \leq x \leq \sin z, x^2 + y^2 \leq 1, 0 \leq z \leq \pi$.

2. Compute the following triple integrals in rectangular coordinates.

(1) $\iiint_V xy^2 z^3 dV$, where V is the region enclosed by the surfaces $z = xy, y = x, x = 1, z = 0$;

(2) $\iiint_V y\cos(x + z)dV$, where V is the region enclosed by the cylindrical surface $y = \sqrt{x}$ and the planes $y = 0, z = 0, x + z = \dfrac{\pi}{2}$;

(3) $\iiint_V z^2 dxdydz$, where V is the region enclosed by $\dfrac{x}{a} + \dfrac{y}{b} + \dfrac{z}{c} = 1, x = 0, y = 0$ and $z = 0$;

(4) $\iiint_V (x + y + z)dV$, where V is the region defined by the system of inequalities $0 \leq x \leq a, 0 \leq y \leq b, 0 \leq z \leq c$;

(5) $\iiint_V y[1 + xf(z)]dV$, where V is the region determined by the system of inequalities $-1 \leq x \leq 1, x^3 \leq y \leq 1, 0 \leq z \leq x^2 + y^2$, and $f(z)$ is any continuous function.

3. Convert the following iterated integrals to the iterated integrals in cylindrical or spherical coordinates and compute the integral values.

(1) $\int_0^1 dx \int_0^{\sqrt{1-x^2}} dy \int_0^{\sqrt{1-x^2-y^2}} (x^2 + y^2) dz$;

Calculus(II)

(2) $\int_0^2 dx \int_0^{\sqrt{2x-x^2}} dy \int_0^a z\sqrt{x^2+y^2}\, dz$.

4. Evaluate the given triple integrals.

(1) $\iiint_V (z + x^2 + y^2)\, dV$, where V is the solid enclosed by the plane $z = 4$ and the lateral surface of solid of revolution generated by rotating the curve $\begin{cases} y^2 = 2z \\ x = 0 \end{cases}$, around the z-axis.

(2) $\iiint_V \dfrac{1}{1 + x^2 + y^2}\, dV$, where V is the space region bounded by the conic surface $x^2 + y^2 = z^2$ and the plane $z = 1$;

(3) $\iiint_V (x^2 + y^2)\, dV$, where V is the space region bounded by the paraboloid of revolution $2z = x^2 + y^2$ and the planes $z = 2, z = 8$;

(4) $\iiint_V (x^2 + y^2)\, dV$, where V is the space region bounded by two hemispheres $z = \sqrt{A^2 - x^2 - y^2}$, $z = \sqrt{a^2 - x^2 - y^2}$ ($A > a$) and the plane $z = 0$;

(5) $\iiint_V (x + z)\, dV$, where V is the space region bounded by the conic surface $z = \sqrt{x^2 + y^2}$ and the hemisphere $z = \sqrt{1 - x^2 - y^2}$;

(6) $\iiint_V \dfrac{x^2 + y^2}{z^2}\, dV$, where V is the space region defined by the system of inequalities $x^2 + y^2 + z^2 \geq 1, x^2 + y^2 + (z-1)^2 \leq 1$;

(7) $\iiint_V (x^3 y - 3xy^2 + 3xy)\, dV$, where V is the spheroid
$$(x-1)^2 + (y-1)^2 + (z-2)^2 \leq 1$$

5. A solid enclosed by the surfaces $x = \sqrt{y - z^2}$, $\dfrac{1}{2}\sqrt{y} = x$ and the plane $y = 1$ has volume density $|z|$. Find the mass m of the solid.

6. Using triple integrals to find the volume V of each of the given solids.

(1) A solid enclosed by the surfaces $az = x^2 + y^2$, $2az = a^2 - x^2 - y^2$ ($a > 0$);

(2) A solid determined by the system of inequalities
$$x^2 + y^2 - z^2 \leq 0, x^2 + y^2 + z^2 \leq a^2$$

Chapter 9 Multiple Integrals

7. Let $f(x)$ be continuous, $F(t) = \iiint_V [z^2 + f(x^2 + y^2)] dV$, where V is defined by the system of inequalities $0 \leq z \leq h, x^2 + y^2 \leq t^2$. Find $\dfrac{dF}{dt}$.

8. A snow pile of volume V and exposed surface area S melts at a rate given by $dV/dt = -0.9S$. Assume that during melting the pile maintains the shape
$$z = h - \frac{2(x^2 + y^2)}{h}$$
where $h = h(t)$ (t denotes the time). How long does it take for a snow pile of height $h(0) = 130$ cm to disappear? (Hint: Find $V = V(h)$ and $S = S(h)$ and a differential equation for $h(t)$)

9.4

1. Evaluate the following first type curve integrals.

(1) $\int_l \sqrt{2y} \, ds$, where l is one arch of the cycloid
$$x = a(t - \sin t), y = a(1 - \cos t)$$

(2) $\int_l (x^{\frac{4}{3}} + y^{\frac{4}{3}}) ds$, where l is the arc in the first quadrant of the asteroid
$$x = a\cos^3 t, y = a\sin^3 t \, (0 \leq t \leq \frac{\pi}{2})$$

(3) $\oint_c \sqrt{x^2 + y^2} \, ds$, where c is the circle $x^2 + y^2 = ax$;

(4) $\int_l x \, ds$, where l is the arc segment of the hyperbolic curve $xy = 1$ from the point $\left(\dfrac{1}{2}, 2\right)$ to the point $(1,1)$;

(5) $\int_l |y| \, ds$, where l is the curve $x = \sqrt{1 - y^2}$;

(6) $\oint_c e^{\sqrt{x^2+y^2}} \, ds$, where c is the boundary curve of the first quadrant plane region bounded by the curve $x^2 + y^2 = a^2$, the line $y = x$ and the positive half of the x-axis;

(7) $\int_L z \, ds$, where L is the arc segment of the space curve $x = t\cos t, y = t\sin t, z = t$ as t varies from 0 to t_0;

(8) $\int_L \dfrac{z^2}{x^2 + y^2} ds$, where L is the arc segment of the spiral $x = a\cos t, y = a\sin t, z =$

Calculus(Ⅱ)

at as t varies from 0 to 2π;

(9) $\oint_c (2xy + 2x^2 + 4y^2) ds$, where c is the ellipse $\dfrac{x^2}{4} + \dfrac{y^2}{3} = 1$ with perimeter a;

(10) $\oint_L (2yz + 2zx + 2xy) ds$, where L the space circle $\begin{cases} x^2 + y^2 + z^2 = a^2 \\ x + y + z = \dfrac{3}{2}a \end{cases}$;

(11) $\oint_L (x^2 + y^2) ds$, where L the space circle $\begin{cases} x^2 + y^2 + z^2 = 1 \\ x + y + z = 0 \end{cases}$;

(12) $\oint_C (2x^2 + 3y^2) ds$, where C is the curve $x^2 + y^2 = 2(x + y)$.

2. Determine the area of the part of the cylindrical surface.

(1) The area of the part of the cylindrical surface $x^2 + y^2 = R^2$ bounded by the Oxy plane and the cylindrical surface $z = R + \dfrac{x^2}{R}$;

(2) The area of the part of the cylindrical surface $x^2 + y^2 = 1$ cut off by the parabolic cylindrical surface $x = z^2$ (Express the area as a definite integral, do not calculate the value).

3. Use curve integration to evaluate the area of the surface of revolution obtained by rotating the curve $l: y = \dfrac{x^2}{4} - \dfrac{1}{2}\ln x (1 \leq x \leq 2)$ about the line $y = \dfrac{3}{4}x - \dfrac{9}{8}$.

4. Suppose that the density of each point on the catenary $y = \dfrac{a}{2}(e^{\frac{x}{a}} - e^{-\frac{x}{a}})$ ($a > 0$) is inversely proportional to the coordinate of this point, and that the density at the point $(0, a)$ equals to δ. Find the mass of one segment of the catenary when x varies from $x_1 = 0$ to $x_2 = a$.

9.5

1. Evaluate the given first type surface integrals.

(1) $\iint_S (2x + \dfrac{4}{3}y + z) dS$, where S is the part in the first octant of the plane $\dfrac{x}{2} + \dfrac{y}{3} + \dfrac{z}{4} = 1$;

(2) $\iint_S x^2 y^2 dS$, where S is the upper hemisphere $z = \sqrt{R^2 - x^2 - y^2}$;

Chapter 9 Multiple Integrals

(3) $\iint_S \dfrac{1}{x^2 + y^2 + z^2} dS$, where S is the lower hemisphere $z = -\sqrt{R^2 - x^2 - y^2}$;

(4) $\iint_S |y| \sqrt{z}\, dS$, where S is the surface $z = x^2 + y^2 (z \leqslant 1)$;

(5) $\iint_S (xy + yz + zx)\, dS$, where S is the part of the conical surface $z = \sqrt{x^2 + y^2}$ cut off by the surface $x^2 + y^2 = 2ax (a > 0)$;

(6) $\oiint_\Sigma (3x^2 + y^2 + 2z^2)\, dS$, where Σ is the sphere $(x - 1)^2 + (y - 1)^2 + (z - 1)^2 = 3$.

2. Find the mass of a paraboloid thin shell $z = \dfrac{1}{2}(x^2 + y^2)\, (0 \leqslant z \leqslant 1)$ with mass area density $\mu(x, y, z) = z$.

3. Prove the inequality
$$\oiint_\Sigma (x + y + z + \sqrt{3}\, a)^3 dS \geqslant 108\pi\, a^5 (a > 0)$$
where Σ is the sphere $x^2 + y^2 + z^2 - 2ax - 2ay - 2az + 2a^2 = 0$.

4. Let S be the upper half of the ellipsoidal surface $\dfrac{x^2}{2} + \dfrac{y^2}{2} + z^2 = 1$, the point $P(x, y, z) \in S$, π be the tangent plane to S at the point P, and $\rho(x, y, z)$ be the distance from the origin to the plane π. Find
$$\iint_S \dfrac{z}{\rho(x, y, z)} dS$$

9.6

1. A plane thin lamina enclosed by the parabola $y = x^2$ and the line $y = x$ has area density $\mu = x^2 y$. Find the center of mass of the thin lamina.

2. A homogeneous solid is bounded by the paraboloid of revolution $z = x^2 + y^2$ and the plane $z = 1$. Find the center of mass of the solid.

3. A homogeneous solid is bounded by the parabolic cylindrical surface $y = \sqrt{x}$, $y = 2\sqrt{x}$, and the planes $z = 0$, $x + z = 6$. Find the center of mass of the solid.

4. Find the center of mass of a cone shaped thin shell $z = \dfrac{h}{R}\sqrt{x^2 + y^2}\, (0 \leqslant z \leqslant h$, R, h are constants) with area density $\mu = 1$.

5. Fine the center of mass of the boundary curve for one eighth sphere $x^2 + y^2 + z^2 = R^2, x \geq 0, y \geq 0, z \geq 0$ given the linear density of the curve $\rho = 1$.

6. Find the moment of inertia of the right circular cylinder with radius r, height h, uniform mass distribution and volume density $\mu = 1$ about its axial line.

7. Find the moment of inertia I_z about the z-axis of the homogeneous solid with volume density μ_0 enclosed by the upper sphere $x^2 + y^2 + z^2 = 2$ and the conical surface $z = \sqrt{x^2 + y^2}$.

8. Find the moment of inertia I_z about z-axis of the homogeneous hemispherical shell $z = \sqrt{a^2 - x^2 - y^2}$ with surface density μ_0 (Use spherical coordinates to calculate I_z).

9. Let the linear density of the material curve
$$\begin{cases} x^2 + y^2 + z^2 = R^2 \\ x^2 + y^2 = Rx \end{cases} (z \geq 0)$$
be $\sqrt{x}$. Find the sum $I_x + I_y + I_z$, where I_x, I_y, I_z are the moments of inertia of the curve about the x-axis, y-axis and z-axis, respectively.

9.7

1. Find the volume of the solid enclosed by the surface $\sqrt{x} + \sqrt{y} + \sqrt{z} = \sqrt{a}\ (a > 0)$ and the three coordinate planes.

2. Calculate $\iint_D dxdy$, where D is the region determined by a set of inequalities: $x \geq 0, y \geq 0, (x^2 + y^2)^3 \leq 4a^2x^2y^2\ (a > 0)$.

3. Suppose that f has continuous third order partial derivatives with $f(0) = f'(0) = f''(0) = -1, f(2) = -\dfrac{1}{2}$. Evaluate the iterated integral
$$I = \int_0^2 dx \int_0^x \sqrt{(2-x)(2-y)}\, f'''(y)\, dy$$

4. Compute the double integral $\iint_\sigma \sqrt{|y - x^2|}\, dxdy$, where σ is the region bounded by the lines $x = -1, x = 1, y = 0$ and $y = 2$.

5. Let the function $f(x)$ be continuous in the interval $[0,1]$, and $\int_0^1 f(x)dx = A$. Find $\int_0^1 dx \int_x^1 f(x)f(y)dy$.

Chapter 9 Multiple Integrals

6. Evaluate $\int_{-\infty}^{+\infty}\int_{-\infty}^{+\infty} \min\{x,y\} e^{-(x^2+y^2)} dxdy$.

7. Show that the volume of the solid enclosed by the tangent plane to the paraboloid $z = x^2 + y^2 + 1$ at any point and the paraboloid $z = x^2 + y^2$ must be a fixed value. Find the fixed value.

8. Find the tangent plane to the paraboloid $z = x^2 + y^2 + 1$ such that the volume of the solid bounded by the tangent plane, the paraboloid and the cylindrical surface $(x - 1)^2 + y^2 = 1$ is minimized, and find the minimum value.

9. A homogeneous plane lamina with density $\mu = 1$ occupies the region bounded by the curve $y = \ln x$ and the lines $y = 0$, $x = e$. Find the moment of inertia $I(t)$ of the lamina around the line $x = t$ and find the minimum value of $I(t)$.

10. A cylindrical container with radius R and height H which holds water of height $\frac{2}{3}H$, is placed on a centrifuge with high-speed rotation. Under the action of the centrifugal force, the water surface of the container has a shape of paraboloid of revolution. Find the lowest point of the liquid level when the water is just about to overflow the container.

11. Let $f(t)$ be continuous. Prove that
$$\iint_D f(x-y) dxdy = \int_{-A}^{A} f(t)(A - |t|) dt$$
where A is a positive constant, $D: |x| \leq A/2, |y| \leq A/2$.

12. Let the function $f(x)$ be continuous, positive and monotonically decreasing in the interval $[0,1]$. Show that
$$\frac{\int_0^1 xf^2(x) dx}{\int_0^1 xf(x) dx} \leq \frac{\int_0^1 f^2(x) dx}{\int_0^1 f(x) dx}$$

13. Verify that
$$\iiint_{x^2+y^2+z^2 \leq 1} f(z) dxdydz = \pi \int_{-1}^{1} f(u)(1 - u^2) du$$
Applying the equality to compute
$$\iiint_{x^2+y^2+z^2 \leq 1} (z^4 + z^2 \sin^3 z) dxdydz$$

14. Let $F(t) = \iiint_\Omega f(x^2 + y^2 + z^2) dxdydz$, where f is a differentiable function, and

Calculus(II)

the domain of integration Ω is a spheroid $x^2 + y^2 + z^2 \leq t^2$. Determine $F'(t)$.

15. Compute $\iiint_V | \sqrt{x^2 + y^2 + z^2} - 1 | \, dV$, where V is the solid enclosed by the conical surface $z = \sqrt{x^2 + y^2}$ and the plane $z = 1$.

16. Determine $\iiint_V (x + 2y + 3z) \, dV$, where V is a cone whose base is a circular region with radius 1 and center at the point $(1,1,1)$ and is located on the plane $x + y + z = 3$

17. Prove that the moment of inertia around the x-axis of the solid of revolution with volume density $\mu = 1$ generated by rotating the curved Trapezoid about the x-axis bounded by the continuous curve $y = f(x) > 0$, the lines $x = a, x = b$ and x-axis equals to

$$I_x = \frac{\pi}{2} \int_a^b f^4(x) \, dx$$

18. Given

$$f(x,y,z) = \begin{cases} x^2 + y^2, & \text{as } z \geq \sqrt{x^2 + y^2} \\ 0, & \text{as } z < \sqrt{x^2 + y^2} \end{cases}$$

Calculate the surface integral $\iint_{x^2+y^2+z^2 = R^2} f(x,y,z) \, dS$.

19. Using the spherical coordinates to evaluate the moment of inertia I_z about the z-axis of the homogeneous spherical shell $x^2 + y^2 + z^2 = R^2$ with area density μ_0.

20. Use curve integration to evaluate the area S of the surface of revolution obtained by rotating the plane curve $l: y = \frac{1}{3}x^3 + 2x, 0 \leq x \leq 1$ about the line $L: y = \frac{4}{3}x$.

21. Compute the curve integral with respect to arc

$$\int_l (|x| + |y|)^2 (1 + \sin xy) \, ds$$

where the curve l is the unit circle with center at the origin.

Chapter 10 The Second Type Curve Integral, Surface Integral, and Vector Field

In chapter 9, we generalized the method of integral to a function of several variables $f(P)$ on a geometric body Ω. In this chapter, we will generalize the method of integral to directed curves and directed surfaces in vector fields as needed, i. e. the second type curve integrals and the second type surface integrals. Meanwhile, we will introduce the concepts of divergence and curl of vector fields.

10.1 The Second Type Curve Integral

There are all kinds of fields in physics, such as height field, temperature field, density field, electropotential field, velocity field, force field, magnetic field, and so on. In general, we call a space region filled with some physical quantity a field. In mathematics, this means a numerical function or a vector function defined on a space region. When it is a numerical function, we usually call it a scalar field; when it is a vector function, we call it a vector field. According to this classification, in the above fields we just mentioned, the height field, the temperature field, the density field and the electropotential field are scalar fields, and the velocity field, the force field, the magnetic field are vector fields.

If the physical quantity of a field at a point P is only relevant to the position of P, but is irrelevant to time, then we call this field a stationary field, and denoted by $u(P)$, $P \in D, D \subseteq \mathbf{R}^3$ (if it is a scalar field) or $A(P), P \in D, D \subseteq \mathbf{R}^3$ (if it is a vector field); if the physical quantity of a field at a point is relevant to both the position of the point and the time t, then we call it a time-varying field, and denoted by $u(P,t), P \in D, D \subseteq \mathbf{R}^3, t \in \mathbf{R}^+$ (if it is a scalar field) or $A(P,t), P \in D, D \subseteq \mathbf{R}^3, t \in \mathbf{R}^+$ (if it is a vector field). In this book we only discuss stationary fields.

10.1.1 Work Done by Variable Force and The Concepts of The Second Type Curve Integral

We first discuss the problem of work done by field force, and then introduce the concept of the second type curve integral.

Example 1 Suppose there is a force field of continuous plane force
$$F(x,y) = P(x,y)i + Q(x,y)j, (x,y) \in D$$
A particle in the field is moving from a point A to a point B along a smooth curve l. Find the work done to the particle by the field force F.

Solution When F is a constant force, l is the directed line segment $\overrightarrow{AB}$, the work done by the force is
$$W = F \cdot \overrightarrow{AB}$$

In general, we apply the method of definite integral. Firstly, we divide $\widehat{AB}$ into n pieces using $n+1$ arbitrary points on the curve arc l
$$A = M_0, M_1, M_2, \cdots, M_{n-1}, M_n = B$$
Set $M_k(x_k, y_k)$, $\Delta x_k = x_k - x_{k-1}$, $\Delta y_k = y_k - y_{k-1}(k=1,2,\cdots,n)$. Then, take an arbitrary typical directed sub-arc $\widehat{M_{k-1}M_k}$ (Fig. 10.1). Since it is smooth and very short, we can use the displacement vector $\overrightarrow{M_{k-1}M_k} = \Delta x_k i + \Delta y_k j$ to approximate $\widehat{M_{k-1}M_k}$, and since $P(x,y), Q(x,y)$ are continuous, we can use the force $F(\xi_k, \eta_k) = P(\xi_k, \eta_k)i + Q(\xi_k, \eta_k)j$ at an arbitrary point (ξ_k, η_k) of $\widehat{M_{k-1}M_k}$ to approximate the variable force.

Then the work done by the variable force $F(x,y)$ along the directed sub-arc $\widehat{M_{k-1}M_k}$ is
$$\Delta W_k \approx F(\xi_k, \eta_k) \cdot \overrightarrow{M_{k-1}M_k}$$
i.e.
$$\Delta W_k \approx P(\xi_k, \eta_k)\Delta x_k + Q(\xi_k, \eta_k)\Delta y_k$$
Then
$$W = \sum_{k=1}^{n} \Delta W_k \approx \sum_{k=1}^{n} F(\xi_k, \eta_k) \cdot \overrightarrow{M_{k-1}M_k}$$

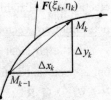

Fig. 10.1

Chapter 10 The Second Type Curve Integral, Surface Integral, and Vector Field

$$= \sum_{k=1}^{n} [P(\xi_k, \eta_k) \Delta x_k + Q(\xi_k, \eta_k) \Delta y_k]$$

Finally, we let the number of dividing points increase infinitely to make the length of the longest sub-arc $\lambda \to 0$, then take the limit, we get the work done

$$W = \lim_{\lambda \to 0} \sum_{k=1}^{n} F(\xi_k, \eta_k) \cdot \overrightarrow{M_{k-1} M_k}$$

$$= \lim_{\lambda \to 0} \sum_{k=1}^{n} [P(\xi_k, \eta_k) \Delta x_k + Q(\xi_k, \eta_k) \Delta y_k]$$

From the above example, we abstract the following important concept.

Definition 10.1 Suppose l is a smooth directed curve segment from the point A to the point B on the Oxy plane, and the vector function

$$F(x,y) = P(x,y)i + Q(x,y)j$$

is defined on l. Divide $\widehat{AB}$ into n pieces using points on l

$$A = M_0, M_1, M_2, \cdots, M_{n-1}, M_n = B$$

Set $M_k(x_k, y_k)$, $\Delta x_k = x_k - x_{k-1}, \Delta y_k = y_k - y_{k-1}(k = 1, 2, \cdots, n)$. Take an arbitrary point (ξ_k, η_k) from every directed sub-arc $\widehat{M_{k-1} M_k}$, and make the sum of the products

$$\sum_{k=1}^{n} F(\xi_k, \eta_k) \overrightarrow{M_{k-1} M_k} = \sum_{k=1}^{n} [P(\xi_k, \eta_k) \Delta x_k + Q(\xi_k, \eta_k) \Delta y_k]$$

Let $\lambda = \max_k (\text{arc length of } \widehat{M_{k-1} M_k})$, if the limit

$$\lim_{\lambda \to 0} \sum_{k=1}^{n} F(\xi_k, \eta_k) \cdot \overrightarrow{M_{k-1} M_k} = \lim_{\lambda \to 0} \sum_{k=1}^{n} [P(\xi_k, \eta_k) \Delta x_k + Q(\xi_k, \eta_k) \Delta y_k]$$

exists, and is irrelevant to the choices of $M_k, (\xi_k, \eta_k)(k = 1, 2, \cdots, n)$, then we call this limit value the curve integral of the vector function $F(x,y)$ on the directed arc l, or the second type curve integral of the functions $P(x,y), Q(x,y)$ along the directed curve $l(\widehat{AB})$, and denoted by

$$\int_l F \cdot ds \text{ or } \int_l P(x,y) dx + Q(x,y) dy$$

where ds is called the arc length element vector, and its direction is coincident with the direction of the tangent line to the positive direction of the directed curve l (Fig. 10.2)

$$ds = dx i + dy j$$

We call

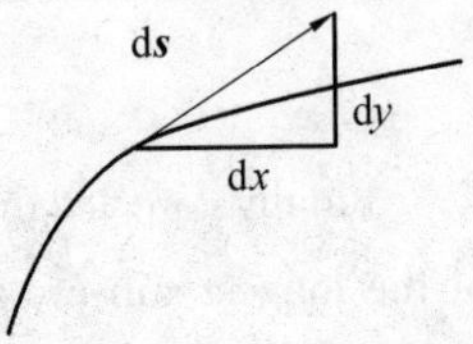

$$\int_l P(x,y)\,\mathrm{d}x = \lim_{\lambda\to 0}\sum_{k=1}^{n} P(\xi_k,\eta_k)\Delta x_k \qquad (1)$$

Fig. 10.2

the curve integral of the function $P(x,y)$ along the directed arc l with respect to x, and call

$$\int_l Q(x,y)\,\mathrm{d}y = \lim_{\lambda\to 0}\sum_{k=1}^{n} Q(\xi_k,\eta_k)\Delta y_k \qquad (2)$$

the curve integral of the function $Q(x,y)$ along the directed arc l with respect to y.

Similarly, we can define the curve integral of a vector function
$$\boldsymbol{F}(x,y,z) = P(x,y,z)\boldsymbol{i} + Q(x,y,z)\boldsymbol{j} + R(x,y,z)\boldsymbol{k}$$
along a directed space curve Γ

$$\int_\Gamma \boldsymbol{F}\cdot\mathrm{d}\boldsymbol{s} = \int_\Gamma P\mathrm{d}x + Q\mathrm{d}y + R\mathrm{d}z$$

where $\mathrm{d}\boldsymbol{s} = \mathrm{d}x\boldsymbol{i} + \mathrm{d}y\boldsymbol{j} + \mathrm{d}z\boldsymbol{k}$, and

$$\int_\Gamma P\mathrm{d}x = \lim_{\lambda\to 0}\sum_{k=1}^{n} P(\xi_k,\eta_k,\zeta_k)\Delta x_k$$

$$\int_\Gamma Q\mathrm{d}y = \lim_{\lambda\to 0}\sum_{k=1}^{n} Q(\xi_k,\eta_k,\zeta_k)\Delta y_k$$

$$\int_\Gamma R\mathrm{d}z = \lim_{\lambda\to 0}\sum_{k=1}^{n} R(\xi_k,\eta_k,\zeta_k)\Delta z_k$$

are called the curve integrals of the functions P,Q,R along the directed curve Γ with respect to x,y,z, respectively.

So, the work done by the force $\boldsymbol{F}$ in Example 1 is $W = \int_{AB}\boldsymbol{F}\cdot\mathrm{d}\boldsymbol{s}$.

When the integrant is continuous on the path of integration, the second type curve integral exists.

By the definition of the second type curve integral we get the following properties (assuming all the integrals involved exist):

$1°\ \int_{AB}(k_1 f_1 + k_2 f_2)\mathrm{d}x = k_1\int_{AB} f_1\mathrm{d}x + k_2\int_{AB} f_2\mathrm{d}x$ (k_1,k_2 are constants); (Linearity)

$2°\ \int_{AB} f\mathrm{d}x = \int_{AC} f\mathrm{d}x + \int_{CB} f\mathrm{d}x$ (C lies on $\widehat{AB}$); (Additivity of integral paths)

$3°\ \int_{AB} f\mathrm{d}x = -\int_{BA} f\mathrm{d}x$.

Property $3°$ tells us: The second type curve integral is relevant to the orientation of

Chapter 10 The Second Type Curve Integral, Surface Integral, and Vector Field

the integral path, and if we change the orientation (turn $\widehat{AB}$ into $\widehat{BA}$), then the value of the integral changes sign. This is coincident with the definite integral, but different than the first type curve integral. Why is that?

10.1.2 Calculating The Second Type Curve Integral

Suppose the parametric equation of a plane curve $\widehat{AB}$ started at A ended at B is
$$x = x(t), y = y(t), t \text{ is between } \alpha \text{ and } \beta$$
the initial point A corresponds to $t = \alpha$, the terminal point B corresponds to $t = \beta$, functions $x(t), y(t) \in C^1$ (i.e. the curve segment $\widehat{AB}$ is smooth), and their derivatives are not zero at the same time. Functions $P(x,y), Q(x,y)$ are continuous on $\widehat{AB}$. Based on these assumptions, the second type curve integral of the functions $P(x,y), Q(x,y)$ along the directed curve segment $\widehat{AB}$ exist. Next, we will see how to calculate this integral.

Assume the dividing point M_k in Definition 10.1 corresponds to $t = t_k$, by the Lagrange mean value theorem we have
$$\Delta x_k = x_k - x_{k-1} = x(t_k) - x(t_{k-1}) = x'(\tau_k) \Delta t_k$$
where $\Delta t_k = t_k - t_{k-1}, \tau_k$ is between t_k, t_{k-1}. Then by Definition 10.1 we have
$$\int_{\widehat{AB}} P(x,y) \, dx = \lim_{\lambda \to 0} \sum_{k=1}^{n} P(x(\tau_k), y(\tau_k)) x'(\tau_k) \Delta t_k$$
Since when $\lambda \to 0$, we have $\max_k |\Delta t_k| \to 0$, so the limit of the sum of the right hand side of the above equation is exactly the definite integral of the function $P(x(t), y(t)) x'(t)$ from α to β, $\int_{\alpha}^{\beta} P(x(t), y(t)) x'(t) \, dt$, then we have

$$\int_{\widehat{AB}} P(x,y) \, dx = \int_{\alpha}^{\beta} P(x(t), y(t)) x'(t) \, dt \tag{1}$$

similarly, we have

$$\int_{\widehat{AB}} Q(x,y) \, dx = \int_{\alpha}^{\beta} Q(x(t), y(t)) y'(t) \, dt \tag{2}$$

We have similar results for the second type curve integrals of smooth space curves. In a word, the second type curve integrals can be transformed into definite integrals. We just need to substitute the parametric equation of the curve into the integral expression, the parameter corresponds to the initial point of the curve is the lower limit of the integral, and the one corresponds to the terminal point is the upper limit, then we transform the integral into a definite integral.

For example, the equation of $\widehat{AB}$ is
$$y = y(x)$$
when the coordinate of A is $(a, y(a))$ and the coordinate of B is $(b, y(b))$, treat x as the parameter, then we have
$$\int_{\widehat{AB}} P(x,y) dx = \int_a^b P(x, y(x)) dx$$
$$\int_{\widehat{AB}} Q(x,y) dx = \int_a^b Q(x, y(x)) y'(x) dx$$

Example 2 Evaluate $\int_{\widehat{AB}} xy dx$, where $\widehat{AB}$ is the directed arc segment of the parabola $y^2 = x$ from the point $A(1, -1)$ to the point $B(1, 1)$.

Solution By Fig. 10.3, if we want to express $\widehat{AB}$ as a function of x, we need to divide $\widehat{AB}$ into two pieces: $\widehat{AO}: y = -\sqrt{x}$, x varies from 1 to 0; $\widehat{OB}: y = \sqrt{x}$, x varies from 0 to 1, then
$$\int_{\widehat{AB}} xy dx = \int_{\widehat{AO}} xy dx + \int_{\widehat{OB}} xy dx = \int_1^0 -x\sqrt{x} dx + \int_0^1 x\sqrt{x} dx$$
$$= 2\int_0^1 x^{3/2} dx = \frac{4}{5}$$

If we express the equation of $\widehat{AB}$ as a function of y
$$x = y^2, y \text{ varies from } -1 \text{ to } 1$$
then
$$\int_{\widehat{AB}} xy dx = \int_{-1}^1 y^2 y \, dy^2 = 2\int_{-1}^1 y^4 dy = \frac{4}{5}$$

What noticeable is: The arc segment $\widehat{AB}$ is Example 2 is symmetric about the x-axis, the integrant xy is an odd function of y, but this second type curve integral is not zero. This is because that the second type curve integrals are along directed curves, and the orientation matters, so the second type curve integrals are different than the first type curve integrals.

Chapter 10 The Second Type Curve Integral, Surface Integral, and Vector Field

Example 3 Evaluate $\int_c (x^2 + y^2) dx + (x^2 - y^2) dy$, where c is the part of the curve $y = 1 - |1 - x|$ from the point with $x = 0$ to the point with $x = 2$ (Fig. 10.4).

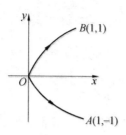

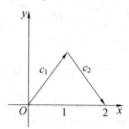

Fig. 10.3   Fig. 10.4

Solution Divide c into two pieces c_1, c_2

$$\int_c = \int_{c_1} + \int_{c_2}$$

on c_1, $y = x$, $dx = dy$

$$\int_{c_1} (x^2 + y^2) dx + (x^2 - y^2) dy = 2\int_0^1 x^2 dx = \frac{2}{3}$$

and on c_2, $x = 2 - y$, $dx = -dy$, then

$$\int_{c_2} (x^2 + y^2) dx + (x^2 - y^2) dy = -2\int_1^0 y^2 dy = \frac{2}{3}$$

thus

$$\int_c (x^2 + y^2) dx + (x^2 - y^2) dy = \frac{4}{3}$$

Example 4 Evaluate $\int_l x^2 dx + (y - x) dy$, where

1° l is the upper half circle $y = \sqrt{a^2 - x^2}$, oriented counterclockwisely;
2° l is the line segment on the x-axis from the point $A(a, 0)$ to the point $B(-a, 0)$.

Solution 1° By Fig. 10.5, the parametric equation of l is

$$x = a\cos t, y = a\sin t$$

A corresponds to $t = 0$, B corresponds to $t = \pi$, then

$$\int_l x^2 dx + (y - x) dy = \int_0^\pi a^2 \cos^2 t \, d(a\cos t) + (a\sin t - a\cos t) d(a\sin t)$$

$$= \int_0^\pi a^3 \cos^2 t \, d\cos t + \int_0^\pi a^2 \sin t \sin t - \int_0^\pi a^2 \cos^2 t \, dt$$

Calculus(Ⅱ)

$$=-\frac{2}{3}a^3-\frac{\pi}{2}a^2$$

2° The equation of l is

$$y=0$$

where x is from a to $-a$, then

$$\int_l x^2\,dx+(y-x)\,dy=\int_a^{-a}x^2\,dx=-\frac{2}{3}a^3$$

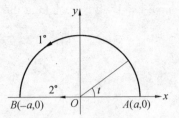

Fig. 10.5

Example 5 In an electrostatic field generated by an electric charge q at the origin $(0,0,0)$, a unit positive charge moves along a smooth curve Γ

$$x=x(t),y=y(t),z=z(t)$$

from point A to point B, suppose the point A corresponds to $t=\alpha$, B corresponds to $t=\beta$, find the work done by the electrostatic field, W.

Solution Suppose the radius vector of the point $M(x,y,z)$ is $\overrightarrow{OM}=\boldsymbol{r}$, i.e.

$$\boldsymbol{r}=x\boldsymbol{i}+y\boldsymbol{j}+z\boldsymbol{k},r=|\boldsymbol{r}|=\sqrt{x^2+y^2+z^2}$$

By the Coulomb's law, the unit positive charge at the point M bears the electric field force

$$\boldsymbol{F}=\frac{q}{r^3}\boldsymbol{r}=\frac{q}{r^3}\{x,y,z\}$$

thus, the work done is

$$W=\int_\Gamma \boldsymbol{F}\cdot d\boldsymbol{s}=\int_\Gamma \frac{q}{r^3}\boldsymbol{r}\cdot d\boldsymbol{s}=q\int_\Gamma \frac{x\,dx+y\,dy+z\,dz}{(x^2+y^2+z^2)^{3/2}}$$

$$=q\int_\alpha^\beta \frac{xx'+yy'+zz'}{(x^2+y^2+z^2)^{3/2}}dt=q\int_{r(\alpha)}^{r(\beta)}\frac{dr}{r^2}=q\left[\frac{1}{r(\alpha)}-\frac{1}{r(\beta)}\right]$$

where $r(\alpha),r(\beta)$ are the distances from the points A and B to the origin, respectively.

This example shows that the work done by the electric field force of an electrostatic field is only relevant to the initial point and terminal point of the moving unit positive charge, and is irrelevant to the path of the motion. A force field possesses this property is called a conservative force field, such as the gravitational field.

10.1.3 The Relation Between The Second Type Curve Integral and The First Type Curve Integral

The second type curve integral and the first type curve integral have different physi-

Chapter 10 The Second Type Curve Integral, Surface Integral, and Vector Field

cal backgrounds and features, but they are actually related.

Suppose L is a directed curve segment on the Oxy plane from the point A to the point B with equation

$$x = x(t), y = y(t)$$

when t varies monotonically from α to β, the corresponding point $M(x,y)$ moves from the initial point A to the terminal point B. Assume $x(t), y(t)$ have continuous derivatives, and not equal zero at the same time, and $P(x,y), Q(x,y)$ are continuous on L. Then by formulas (1), (2) of 10.1.2, we have

$$\int_L P(x,y)\,dx + Q(x,y)\,dy = \int_\alpha^\beta \{P(x(t),y(t))x'(t) + Q(x(t),y(t))y'(t)\}\,dt$$

We could assume $\alpha < \beta$ (if $\alpha > \beta$, then let $s = -t$, the equation of L can be written as $x = x(-s), y = y(-s)$, and when s varies monotonically from $-\alpha$ to $-\beta$, the corresponding point $M(x,y)$ moves from the initial point A to the terminal point B). By Section 9.6 we know, the vector $t = \{x'(t), y'(t)\}$ is a tangent vector to L at the point $M(x(t), y(t))$, and its direction is coincident with the moving direction of $M(x,y)$ as t increases. So when $\alpha < \beta$, this direction is actually the orientation of the directed arc L. We call such a tangent vector whose orientation coincident with the directed arc L the tangent vector of the directed arc. So the tangent vector of the directed arc L is $t = \{x'(t), y'(t)\}$, and its directional cosines are

$$\cos \alpha = \frac{x'(t)}{\sqrt{x'^2(t) + y'^2(t)}}$$

$$\cos \beta = \frac{y'(t)}{\sqrt{x'^2(t) + y'^2(t)}}$$

then by the calculating formula for the curve integral with respect to arc length, we get

$$\int_L \{P(x,y)\cos\alpha + Q(x,y)\cos\beta\}\,ds$$

$$= \int_\alpha^\beta \{P(x(t),y(t))\cos\alpha + Q(x(t),y(t))\cos\beta\}\sqrt{x'^2(t) + y'^2(t)}\,dt$$

$$= \int_\alpha^\beta \{P(x(t),y(t))x'(t) + Q(x(t),y(t))y'(t)\}\,dt$$

so

$$\int_L P(x,y)\,dx + Q(x,y)\,dy = \int_L \{P(x,y)\cos\alpha + Q(x,y)\cos\beta\}\,ds \tag{3}$$

Similarly, we get the relation between the two types of curve integrals on a space curve Γ

$$\int_\Gamma P(x,y,z)\,dx = \int_\Gamma P(x,y,z)\cos\alpha\,ds$$

$$\int_\Gamma Q(x,y,z)\,dy = \int_\Gamma Q(x,y,z)\cos\beta\,ds$$

$$\int_\Gamma R(x,y,z)\,dz = \int_\Gamma R(x,y,z)\cos\gamma\,ds$$

where $\alpha = \alpha(x,y,z)$, $\beta = \beta(x,y,z)$, $\gamma = \gamma(x,y,z)$ are the directional angles of the tangent vectors to the directed arc Γ at the point $M(x,y,z)$.

10.2　The Green's Theorem

The Newton-Leibniz formula $\int_a^b F'(x)\,dx = F(b) - F(a)$ connects the definite integral with the values of the original function of the integrant at the upper and lower limits of the integral. Similarly, the Green's theorem which will be introduced in this section relates the double integral on a plane region to the curve integral on the boundary of this region. In fluid mechanics, the Green's Theorem and the Stokes' Theorem revel the relation between the circulation and the curl, which are important concepts in both mathematics and physical field theory.

Firstly, we need to introduce some concepts about plane region. Suppose D is a plane region, if the region enclosed by any closed curve in D still belongs to D, then, we call D a planar simply connected region, otherwise, we call D a planar complex connected region. The simply connected region can also be described as the following way: Any closed curve in D can continuously contract to a point in D without passing any point outside D, or more intuitively, simply connected regions are the regions with no "hole". For example, the planar disk $\{(x,y)\,|\,x^2 + y^2 < 4\}$ and the upper half plane $\{(x,y)\,|\,y > 0\}$ are both simply connected regions, but the planar annulus and the disk with center removed are both complex connected regions.

Suppose D is a plane region, L is the boundary curve, then we define the positive direction of L with respect to D as: When the observer walks along L following this direction, the part of D that close to him is always on his left hand side. For example, for the

Chapter 10 The Second Type Curve Integral, Surface Integral, and Vector Field

region $\{(x,y) \mid x^2 + y^2 < 4\}$, the circle $x^2 + y^2 = 4$ oriented conterclockwisely is its positive boundary; for the region $\{(x,y) \mid x^2 + y^2 > 1\}$, the circle $x^2 + y^2 = 1$ oriented clockwisely is its positive boundary; and for the region $\{(x,y) \mid 1 < x^2 + y^2 < 4\}$, the circle $x^2 + y^2 = 4$ oriented counterclockwisely and the circle $x^2 + y^2 = 1$ oriented clockwisely together form its positive boundary.

Theorem 10.1 Suppose the closed plane region D on the Oxy plane is enclosed by a piecewise smooth simple closed curve C, the functions $P(x,y), Q(x,y)$ have continuous partial derivatives on D, then we have the following Green's formula

$$\oint_C P(x,y) \, dx + Q(x,y) \, dy = \iint_D \left(\frac{\partial Q}{\partial x} - \frac{\partial P}{\partial y} \right) dx dy \qquad (1)$$

where the integral on the closed curve C follows the positive orientation.

Proof Suppose the region D is x-type, i.e. can be expressed as

$$a \leqslant x \leqslant b, y_1(x) \leqslant y \leqslant y_2(x)$$

(Fig. 10.6), then by the method of calculating double integrals

$$-\iint_D \frac{\partial P}{\partial y} dx dy = -\int_a^b dx \int_{y_1(x)}^{y_2(x)} \frac{\partial P}{\partial y} dy = \int_a^b [P(x,y_1(x)) - P(x,y_2(x))] dx$$

and by the method of calculating the curve integrals

$$\oint_C P(x,y) dx = \int_{\overline{AB}+BE+\overline{EF}+FA} P(x,y) dx$$

$$= \int_a^b P(x,y_1(x)) dx + \int_b^a P(x,y_2(x)) dx$$

$$= \int_a^b [P(x,y_1(x)) - P(x,y_2(x))] dx$$

thus

$$-\iint_D \frac{\partial P}{\partial y} dx dy = \oint_C P(x,y) dx$$

When D is not an x-type region, we just need to divide D into several x-type regions using some piecewise smooth curves, then we can get the above equation. For a region D as shown in Fig. 10.7, we use an arc $\stackrel{\frown}{AB}$ to divide D into two x-type regions D_1, D_2. By applying the above results, the properties of multiple integrals and the properties of the second type curve integrals, we get

$$-\iint_D \frac{\partial P}{\partial y} dx dy = -\iint_{D_1} \frac{\partial P}{\partial y} dx dy - \iint_{D_2} \frac{\partial P}{\partial y} dx dy = \oint_{AEBA} P dx + \oint_{ABFA} P dx$$

$$= \int_{AEB} Pdx + \int_{BA} Pdx + \int_{AB} Pdx + \int_{BFA} Pdx$$

$$= \oint_{AEBFA} Pdx = \oint_C Pdx$$

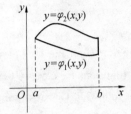

Fig. 10.6   Fig. 10.7

Similarly, we can prove

$$\iint_D \frac{\partial Q}{\partial x} dxdy = \oint_C Qdy$$

Specially, when D is a complex connected region (Fig. 10.8), Theorem 10.1 still holds.

The Green's formula has another form. Suppose t is a tangent vector having the same direction as the curve C. Let n be the external normal vector of C. Rotate n counterclockwisely 90 degree to get t. By Fig. 10.9, the included angles of them and two coordinate axes satisfy

$$\widehat{(t,x)} = \pi - \widehat{(n,y)}$$

$$\widehat{(t,y)} = \widehat{(n,x)}$$

then by the relation between two types of curve integrals

$$\oint_C Pdx + Qdy = \oint_C [P\cos\widehat{(t,x)} + Q\cos\widehat{(t,y)}]ds$$

we get

$$\oint_C [-P\cos\widehat{(n,y)} + Q\cos\widehat{(n,x)}]ds = \iint_D (\frac{\partial Q}{\partial x} - \frac{\partial P}{\partial y})dxdy$$

Substitute the Q in the above equation by P, and P by $-Q$, we get another form of the Green's formula

$$\oint_C [P\cos\widehat{(n,x)} + Q\cos\widehat{(n,y)}]ds = \iint_D (\frac{\partial P}{\partial x} + \frac{\partial Q}{\partial y})dxdy \qquad (2)$$

Chapter 10 The Second Type Curve Integral, Surface Integral, and Vector Field

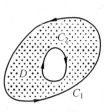
Fig. 10.8

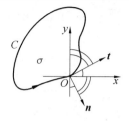
Fig. 10.9

The physical significance of equation (2) reveals the relation between the net flux and the divergence passing through the boundary of a planar velocity field. Next, we will introduce a simple application of the Green's formula (1), if we let $P(x,y) = -y$, $Q(x,y) = x$, then we have

$$\oint_C x\,dy - y\,dx = 2\iint_D dx\,dy = 2S$$

where S is the area of D, so the area S of the region D enclosed by the closed curve C can be calculated by curve integral

$$S = \frac{1}{2}\oint_C x\,dy - y\,dx$$

Example 1 Find the area S of the region enclosed by the ellipse $x = a\cos t, y = b\sin t, 0 \leqslant t \leqslant 2\pi$.

Solution By equation (3)

$$S = \frac{1}{2}\oint_C x\,dy - y\,dx = \frac{1}{2}\int_0^{2\pi} ab(\cos^2 t + \sin^2 t)\,dt = \pi ab$$

Example 2 Estimate the integral $I_R = \oint_{x^2+y^2=R^2} \frac{y\,dx - x\,dy}{(x^2 + xy + y^2)^2}$, and prove $\lim_{R \to \infty} I_R = 0$.

Solution Firstly, we take a look of the denominator of the integrant, since

$$x^2 + xy + y^2 \geqslant \frac{x^2 + y^2}{2} = \frac{R^2}{2}$$

and
$$I_R = \oint_{x^2+y^2=R^2} \frac{y\,dx - x\,dy}{(x^2 + xy + y^2)^2} = \oint_{x^2+y^2=R^2} \frac{y\cos\alpha - x\cos\beta}{(x^2 + xy + y^2)^2}\,ds$$

on $x^2 + y^2 = R^2$, $|y\cos\alpha - x\cos\beta| \leqslant \sqrt{x^2+y^2}\sqrt{\cos^2\alpha + \cos^2\beta} = \sqrt{x^2+y^2}$, then

$$|I_R| \leqslant \frac{4}{R^4}\oint_{x^2+y^2=R^2} \sqrt{x^2+y^2}\,ds = \frac{4}{R^4} \cdot 2\pi R^2 = \frac{8\pi}{R^2}$$

Since $0 \leq |I_R| \leq \dfrac{8\pi}{R^2}$, then $\lim\limits_{R \to \infty} |I_R| = 0$, so $\lim\limits_{R \to \infty} I_R = 0$.

For the second type curve integrals on plane closed curves, when $\dfrac{\partial Q}{\partial x} - \dfrac{\partial P}{\partial y}$ is simple, we usually need to transform it into double integrals by using the Green's formula.

Example 3 Evaluate $J = \int_{\overset{\frown}{AO}} (e^x \sin y - my) dx + (e^x \cos y - m) dy$, where $\overset{\frown}{AO}$ is the upper half circle $x^2 + y^2 = ax$ form the point $A(a, 0)$ to the point $O(0, 0)$.

Solution Here the integral path $\overset{\frown}{AO}$ is not a closed curve, but since
$$P = e^x \sin y - my, \quad Q = e^x \cos y - m$$
$$\frac{\partial Q}{\partial x} = e^x \cos y, \quad \frac{\partial P}{\partial y} = e^x \cos y - m$$

we get $\dfrac{\partial Q}{\partial x} - \dfrac{\partial P}{\partial y} = m$. To use the Green's formula, based on the $\overset{\frown}{AO}$, we add another curve segment to complete a closed curve. Since we need to calculate curve integral on the additional curve, the curve should be simple, usually we take the line segments or polygonal lines parallel to the coordinate axes. Here we add a line segment $\overline{OA}$, then by the Green's formula

$$\oint_{\overset{\frown}{AO} + \overline{OA}} (e^x \sin y - my) dx + (e^x \cos y - m) dy = \iint_D m dx dy = \frac{1}{8} m\pi a^2$$

Since the equation of $\overline{OA}$ is $y = 0, 0 \leq x \leq a$, then

$$\int_{\overline{OA}} (e^x \sin y - my) dx + (e^x \cos y - m) dy = \int_0^a 0 dx = 0$$

thus
$$J = \frac{1}{8} m\pi a^2 - 0 = \frac{1}{8} m\pi a^2$$

Example 4 Evaluate $\oint_C \dfrac{(x + 4y) dy + (x - y) dx}{x^2 + 4y^2}$, where C is an arbitrary closed curve oriented positively and not passes through the origin.

Solution Since
$$\frac{\partial Q}{\partial x} = \frac{4y^2 - x^2 - 8xy}{(x^2 + 4y^2)^2} = \frac{\partial P}{\partial y}, (x, y) \neq (0, 0)$$

When the region enclosed by C does not contain the origin, by the Green's formula

Chapter 10 The Second Type Curve Integral, Surface Integral, and Vector Field

$$\oint_C \frac{(x+4y)\,dy + (x-y)\,dx}{x^2 + 4y^2} = 0$$

When the region enclosed by C contains the origin, we can apply the parametric equation of the curve C to transform the curve integral into definite integral. Since the curve C is not given specifically in this problem, we apply the method of eliminating the origin. Since the denominator of the integrant is $x^2 + 4y^2$, for simplicity of calculation, we add an additional ellipse $C_1 : x = 2\varepsilon\cos t$, $y = \varepsilon\sin t$, t varies from 0 to 2π, $\varepsilon > 0$ is small enough (Fig. 10.10). Apply the Green's formula to the complex connected region enclosed by the curves C and C_1, we get

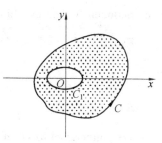

Fig. 10.10

$$\oint_{C+C_1^-} \frac{(x+4y)\,dy + (x-y)\,dx}{x^2+4y^2} = 0$$

thus

$$\oint_C \frac{(x+4y)\,dy + (x-y)\,dx}{x^2+4y^2} = \oint_{C_1} \frac{(x+4y)\,dy + (x-y)\,dx}{x^2+4y^2}$$

$$= \frac{1}{4\varepsilon^2}\oint_{C_1}(x+4y)\,dy + (x-y)\,dx$$

$$= \frac{1}{4}\int_0^{2\pi}[(2\cos t + 4\sin t)\cos t -$$

$$(2\cos t - \sin t)2\sin t]\,dt$$

$$= \frac{1}{2}\int_0^{2\pi} dt = \pi$$

We could also first simplify the integrant by using the equation of the curve, then apply the Green's formula. For example

$$\oint_C \frac{(x+4y)\,dy + (x-y)\,dx}{x^2+4y^2} = \frac{1}{4\varepsilon^2}\oint_{C_1}(x+4y)\,dy + (x-y)\,dx$$

$$= \frac{1}{4\varepsilon^2}\iint_{D_1}(1+1)\,dxdy = \pi$$

where D_1 is the region enclosed by the ellipse C_1.

Here, the curve integral on C is transformed into a curve integral on C_1, can you explain this from the viewpoint of curve integral independent of the integral path?

Can we say "In the region satisfies $\dfrac{\partial Q}{\partial x} = \dfrac{\partial P}{\partial y}$, the path of a closed curve integral can continuously deform arbitrarily"?

Example 5 Evaluate $I = \oint_c (z-y)dx + (x-z)dy + (x-y)dz$, where c is the curve

$$\begin{cases} x^2 + y^2 = 1 \\ x - y + z = 2 \end{cases}$$

oriented counterclockwisely if saw from the positive direction of the z-axis.

Solution Suppose the projection of c on the Oxy plane is c_1 (Fig. 10.11), since $x - y + z = 2$, we get $z = 2 - x + y$, substitute into the original integral

$$I = \oint_{c_1} (2-x)dx + (2x - y - 2)dy + (x - y)d(2 - x + y)$$

$$= \oint_{c_1} (y - 2)dx + (3x - 2y - 2)dy$$

$$= \iint_D (3 - 2)dxdy = 2\pi$$

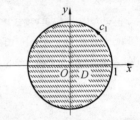

Fig. 10.11

10.3 Conditions for Plane Curve Integrals Being Independent of Path, Conservative Fields

10.3.1 Conditions for Plane Curve Integrals Being Independent of Path

Finding the antiderivative is an important problem in the theory of integration of single variable functions. If $f(x)$ is continuous on the interval I, then $F(x) = \int_{x_0}^{x} f(t)dt$ is an antiderivative of $f(x)$, i.e.

$$dF = f(x)dx, x \in I$$

If $Pdx + Qdy$ is the total differential of a function u, i.e.

$$du = Pdx + Qdy$$

then we call u the original function of $Pdx + Qdy$.

For two continuous binary functions $P(x,y), Q(x,y)$ on the plane region G, or for

Chapter 10 The Second Type Curve Integral, Surface Integral, and Vector Field

the vector field $P(x,y)\boldsymbol{i} + Q(x,y)\boldsymbol{j}$ on G, it's obvious that the curve integral from a fixed point (x_0, y_0) on G to another point (x,y) along the curve l

$$u(x,y) = \int_l P(x,y)\,dx + Q(x,y)\,dy$$

might be the original function of $P(x,y)\,dx + Q(x,y)\,dy$, i. e.

$$du = P(x,y)\,dx + Q(x,y)\,dy, (x,y) \in G$$

Unfortunately, the function $u(x,y)$ not just depends on x,y but also depends on the integral path l. So, if we follow the above curve integral, we usually cannot find a definite function in G. Unless the integral is independent of path, then if we fix the initial point, we can find a definite solution on G. Next, we will show that this is the only condition we need.

If for any two points A, B in the region G, and any two curves l_1, l_2 from A to B, we always have

$$\int_{l_1} P\,dx + Q\,dy = \int_{l_2} P\,dx + Q\,dy$$

then we say the curve integral $\int_l P\,dx + Q\,dy$ is independent of path on G.

Theorem 10.2 Suppose the functions $P(x,y), Q(x,y)$ have continuous partial derivatives on a simply connected region G, then the following four statements are equivalent.

(i) For any closed curve C on G, the intergral

$$\oint_C P\,dx + Q\,dy = 0$$

(ii) In G, the curve integral

$$\int_{AB} P\,dx + Q\,dy$$

is independent of path.

(iii) In G, the expression $P\,dx + Q\,dy$ is the total differential of some function $u(x, y)$, i. e.

$$du = P\,dx + Q\,dy$$

(iv) In G, P, Q satisfy

$$\frac{\partial P}{\partial y} = \frac{\partial Q}{\partial x}$$

Proof (i)⇒(ii). Suppose A,B are two arbitrary points in G, $\overset{\frown}{AMB}$ and $\overset{\frown}{ANB}$ are two arbitrary curve segments from A to B in G, then we have

$$\int_{AMB} Pdx + Qdy - \int_{ANB} Pdx + Qdy = \oint_{AMBNA} Pdx + Qdy = 0$$

thus

$$\int_{AMB} Pdx + Qdy = \int_{ANB} Pdx + Qdy$$

(ii)⇒(iii) Since the curve integral is independent of path, when the initial point $A(x_0,y_0)$ is fixed, it is a binary function of the terminal point $B(x,y)$, denoted by

$$u(x,y) = \int_{(x_0,y_0)}^{(x,y)} Pdx + Qdy \tag{1}$$

To prove

$$du = Pdx + Qdy$$

first we need to prove

$$\frac{\partial u}{\partial x} = P(x,y), \quad \frac{\partial u}{\partial y} = Q(x,y)$$

Since

$$u(x + \Delta x, y) = \int_{(x_0,y_0)}^{(x+\Delta x,y)} Pdx + Qdy$$

is independent of path, for simplicity, we integrate first from $A(x_0,y_0)$ to $B(x,y)$ following the curve l, then integrate along the horizontal line segment from $B(x,y)$ to $B'(x + \Delta x, y)$, as shown in Fig. 10.12. Then

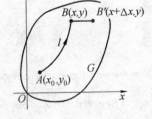

Fig. 10.12

$$u(x + \Delta x, y) = u(x,y) + \int_{(x,y)}^{(x+\Delta x,y)} Pdx + Qdy$$

Since on the horizontal line segment $\overline{BB'}$, the vertical coordinate y does not change, so $dy = 0$, then

$$\Delta_x u = u(x + \Delta u, y) - u(x,y) = \int_x^{x+\Delta x} P(x,y)dx$$

By the mean value theorem of integral, we get

$$\Delta_x u = P(x + \theta \Delta x, y) \Delta x, 0 \leqslant \theta \leqslant 1$$

Since $P(x,y)$ is continuous, so

$$\frac{\partial u}{\partial x} = \lim_{\Delta x \to 0} \frac{\Delta_x u}{\Delta x} = \lim_{\Delta x \to 0} P(x + \theta \Delta x, y) = P(x,y)$$

Chapter 10 The Second Type Curve Integral, Surface Integral, and Vector Field

Similarly, we can prove
$$\frac{\partial u}{\partial y} = Q(x,y)$$

Since the partial derivatives are continuous, so u is differentiable, and
$$du = Pdx + Qdy$$

(iii) $\Rightarrow$ (iv). Since $du = Pdx + Qdy$, then
$$\frac{\partial u}{\partial x} = P, \quad \frac{\partial u}{\partial y} = Q$$

And since P, Q have continuous partial derivatives, then
$$\frac{\partial P}{\partial y} = \frac{\partial^2 u}{\partial x \partial y} = \frac{\partial^2 u}{\partial y \partial x} = \frac{\partial Q}{\partial x}$$

(iv) $\Rightarrow$ (i). For any closed curve C in G, since G is simply connected, so the region D enclosed by C belongs to G. By the Green's formula and (iv), we have
$$\oint_C Pdx + Qdy = \iint_D \left(\frac{\partial Q}{\partial x} - \frac{\partial P}{\partial y}\right) dx dy = 0$$

This theorem is very important, it not only shows the necessary and sufficient condition for curve integrals being independent of path, but also shows that for the expression $Pdx + Qdy$ being the total differential of a function. Among these conditions, (iv) is the one that easiest to be checked.

The simply connected hypothesis of the region G in Theorem 10.2 is necessary. For example, the functions
$$P(x,y) = \frac{-y}{x^2 + y^2}, \quad Q(x,y) = \frac{x}{x^2 + y^2}$$

on the complex connected region
$$\frac{1}{2} \leq x^2 + y^2 \leq 2$$

always satisfies
$$\frac{\partial P}{\partial y} = \frac{y^2 - x^2}{(x^2 + y^2)^2} = \frac{\partial Q}{\partial x}$$

but the curve integral along the unit circle $C: x^2 + y^2 = 1$ in the region
$$\oint_C Pdx + Qdy = \oint_C \frac{xdy - ydx}{x^2 + y^2} = \oint_C xdy - ydx = 2\pi \neq 0$$

For a continuously differentiable vector field on a complex connected region, the

Calculus(II)

condition (iv) cannot guarantee the conditions (i),(ii),(iii) hold, but the conditions (i),(ii),(iii) are still equivalent.

When the curve integral is independent of path, we can substitute the integral path by a simpler one.

Example 1 Evaluate $\int_l (x^4 + 4xy^3)dx + (6x^2y^2 - 5y^4)dy$, where l is part of the sine curve $y = \sin x$ from the point $O(0,0)$ to the point $A\left(\dfrac{\pi}{2}, 1\right)$.

Solution 1 Since

$$\dfrac{\partial Q}{\partial x} = 12xy^2, \quad \dfrac{\partial P}{\partial y} = 12xy^2, \quad \dfrac{\partial Q}{\partial x} = \dfrac{\partial P}{\partial y}$$

so the integral is independent of path, for convenience of calculation, substitute the path l by the polygonal line formed by the line segment from the point $O(0,0)$ to the point $B\left(\dfrac{\pi}{2}, 0\right)$ followed by the line segment from B to A (Fig. 10.13). Then

$$\int_l (x^4 + 4xy^3)dx + (6x^2y^2 - 5y^4)dy$$

$$= \left(\int_{\overline{OB}} + \int_{\overline{BA}}\right)(x^4 + 4xy^3)dx + (6x^2y^2 - 5y^4)dy$$

$$= \int_0^{\pi/2} x^4 dx + \int_0^1 \left[6\left(\dfrac{\pi}{2}\right)^2 y^2 - 5y^4\right]dy = \dfrac{\pi^5}{160} + \dfrac{\pi^2}{2} - 1$$

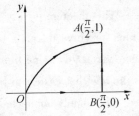

Fig. 10.13

By the way, when the curve integral is independent of path, i.e. the integral expression is the total differential of some function $u(x,y)$, then by equation (1) we get

$$\int_A^B Pdx + Qdy = u\big|_A^B \tag{2}$$

Solution 2 Since

$$(x^4 + 4xy^3)dx + (6x^2y^2 - 5y^4)dy$$

$$= d\left(\dfrac{1}{5}x^5 - y^5\right) + 2(2xy^3 dx + 3x^2y^2 dy)$$

$$= d\left(\dfrac{1}{5}x^5 - y^5 + 2x^2y^3\right)$$

$$I = \left[\dfrac{1}{5}x^5 - y^5 + 2x^2y^3\right]_{(0,0)}^{(\pi/2,0)} = \dfrac{\pi^5}{160} + \dfrac{\pi^2}{2} - 1$$

Example 2 Suppose when $x > 0$, $f(x)$ is differentiable, and $f(1) = 2$. On an arbi-

Chapter 10 The Second Type Curve Integral, Surface Integral, and Vector Field

trary closed curve C in the right half plane $(x > 0)$, we always have

$$\oint_C 4x^3 y\,dx + xf(x)\,dy = 0$$

Find $\int_{\widehat{AB}} 4x^3 y\,dx + xf(x)\,dy$, where $\widehat{AB}$ is the arc from the point $A(1,0)$ to the point $B(2,3)$.

Solution By the given conditions, the curve integral is independent of path on the right half plane. Thus, $\dfrac{\partial Q}{\partial x} = \dfrac{\partial P}{\partial y}$, i. e.

$$xf'(x) + f(x) = 4x^3$$

Solve this first order linear differential equation, and using the condition $f(1) = 2$, we get

$$f(x) = \frac{1}{x} + x^3$$

Since the curve integral is independent of path, we use the polygonal line formed by the line segment from the point $A(1,0)$ to the point $D(2,0)$, followed by the line segment from $D(2,0)$ to $B(2,3)$, then

$$\int_{\widehat{AB}} 4x^3 y\,dx + xf(x)\,dy = \int_{\widehat{AB}} 4x^3 y\,dx + (1 + x^4)\,dy$$

$$= \left(\int_{\widehat{AD}} + \int_{\widehat{DB}}\right) 4x^3 y\,dx + (1 + x^4)\,dy$$

$$= \int_1^2 0\,dx + \int_0^3 (1 + 2^4)\,dy = 51$$

10.3.2 The Conservative Field, Original Function and Total Differential Equations

By the knowledge of physics we know that the work done by the electric field force of an electrostatic field is only relevant to the initial point and terminal point of the moving unit positive charge, and is irrelevant to the path of the motion. A force field possesses this property is called a conservative force field, such as the gravitational field.

In the continuous vector field $\mathbf{F}(x,y) = P(x,y)\mathbf{i} + Q(x,y)\mathbf{j}$, $(x,y) \in D$, if the second type curve integral

$$\int_l \mathbf{F} \cdot d\mathbf{s} = \int_l P(x,y)\,dx + Q(x,y)\,dy$$

is independent of path, then we call the vector field $\boldsymbol{F}$ a conservative field.

In the continuous vector field $\boldsymbol{F}(x,y) = P(x,y)\boldsymbol{i} + Q(x,y)\boldsymbol{j}, (x,y) \in D$, if there exists a differentiable single valued function $u(x,y)$, such that
$$\boldsymbol{F} = \text{grad } u$$
then $\boldsymbol{F}$ is the gradient field of the scalar field u, then we call the vector field $\boldsymbol{F}$ a potential field, and call $v(x,y) = -u(x,y)$ the potential function of the field $\boldsymbol{F}$.

Theorem 10.2 tells us, if a continuously differentiable simply connected vector field is a conservative field, then it must be a potential field, and vice versa, i.e.

$$\boxed{\text{Conservative field}} \Leftrightarrow \boxed{\text{Potential field}}$$

The above discussion for plane vector fields can easily be generalized to space vector fields.

Now the question is for a potential field $\boldsymbol{F} = P\boldsymbol{i} + Q\boldsymbol{j}$, how to find the potential function v? Since $v = -u$, the question is actually the one to find the original function u of the expression $P\mathrm{d}x + Q\mathrm{d}y$. The equation (1) in the proof of Theorem 10.2 already gave us the answer

$$u(x,y) = \int_{(x_0,y_0)}^{(x,y)} P\mathrm{d}x + Q\mathrm{d}y + C$$

Since at this moment the curve integral is independent of path, we usually take the polygonal line that parallel to the coordinate axes as the integral path, and transform the curve integral into a definite integral (Fig. 10.14). When we take the polygonal line ARB as the integral path

$$u(x,y) = \int_{x_0}^{x} P(x,y_0)\mathrm{d}x + \int_{y_0}^{y} Q(x,y)\mathrm{d}y + C \qquad (3)$$

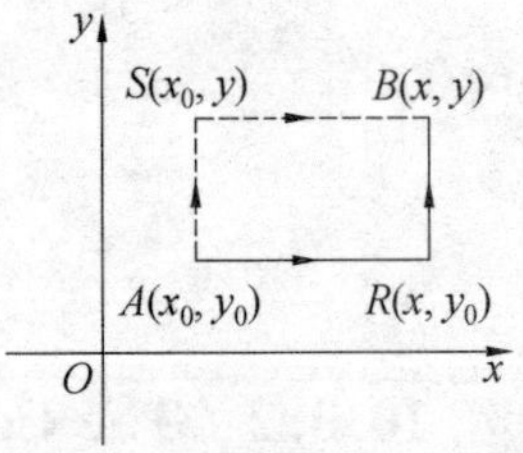

When we take the polygonal line ASB

$$u(x,y) = \int_{x_0}^{x} P(x,y)\mathrm{d}x + \int_{y_0}^{y} Q(x_0,y)\mathrm{d}y + C \qquad (4)$$

since the point $A(x_0,y_0)$ is taken arbitrarily in G, so when choose it we need to let the $P(x,y_0)$ or $Q(x_0,y)$ in equations (3) and (4) be easy to integrate.

Fig. 10.14

Example 3 Prove $(4x^3 + 10xy^3 - 3y^4)\mathrm{d}x + (15x^2y^2 - 12xy^3 + 5y^4)\mathrm{d}y$ is a total differential, and find its original function.

Solution Since

Chapter 10 The Second Type Curve Integral, Surface Integral, and Vector Field

$$\frac{\partial Q}{\partial x} = 30xy^2 - 12y^3 = \frac{\partial P}{\partial y}$$

then, $(4x^3 + 10xy^3 - 3y^4)dx + (15x^2y^2 - 12xy^3 + 5y^4)dy$ is the total differential of some function $u(x,y)$. Take the point $A(x_0, y_0) = (0,0)$, by equation (3) we get

$$u(x,y) = \int_0^x 4x^3 dx + \int_0^y (15x^2y^2 - 12xy^3 + 5y^4)dy + C$$
$$= x^4 + 5x^2y^2 - 3xy^4 + y^5 + C$$

If the left hand side of the first order differential equation

$$P(x,y)dx + Q(x,y)dy = 0 \qquad (5)$$

is the total differential of the function $u(x,y)$, then we call equation (5) a total differential equation. Then the equation (5) can be written as

$$du(x,y) = 0$$

then

$$u(x,y) = C$$

is a general solution of equation (5). Here $u(x,y)$ can be determined by (3) or (4).

Example 4 Solve the equation

$$(4x^3y^3 - 3y^2 + 5)dx + (3x^4y^2 - 6xy - 4)dy = 0$$

Solution Since

$$\frac{\partial Q}{\partial x} = 12x^3y^2 - 6y = \frac{\partial P}{\partial y}$$

so the original equation is a total differential equation. Take $(x_0, y_0) = (0,0)$, by equation (3), we get

$$u(x,y) = \int_0^x 5dx + \int_0^y (3x^4y^2 - 6xy - 4)dy$$
$$= 5x + x^4y^3 - 3xy^2 - 4y$$

so the general solution of the equation is

$$5x + x^4y^3 - 3xy^2 - 4y = C$$

Example 5 Solve the equation

$$[\cos(x + y^2) + 3y]dx + [2y\cos(x + y^2) + 3x]dy = 0$$

Solution Rewrite the left hand side of the equation as

$$\cos(x + y^2)(dx + 2ydy) + (3ydx + 3xdy)$$
$$= \cos(x + y^2)d(x + y^2) + d(3xy) = d[\sin(x + y^2) + 3xy]$$

so the general solution of the equation is

$$\sin(x + y^2) + 3xy = C$$

Example 6 Solve the equation
$$(2xy^2 + y)dx - xdy = 0$$

Solution Since
$$\frac{\partial Q}{\partial x} = -1, \frac{\partial P}{\partial y} = 4xy + 1$$

so this is not a total differential equation. But if we rewrite the equation as
$$2xy^2 dx + ydx - xdy = 0$$

if multiplying $\dfrac{1}{y^2}$ to both sides of the equation, we get

$$2xdx + \frac{ydx - xdy}{y^2} = 0$$

i. e.
$$d\left(x^2 + \frac{x}{y}\right) = 0$$

which is a total differential equation, and its general solution is
$$x^2 + \frac{x}{y} = C$$

For an equation which is not total differential
$$P(x,y)dx + Q(x,y)dy = 0$$

if there exists a non-zero function $\mu = \mu(x,y)$ such that
$$\mu P(x,y)dx + \mu Q(x,y)dy = 0$$

is a total differential equation, then we call $\mu = \mu(x,y)$ the integrating factor of the equation. For example, $\dfrac{1}{y^2}$ is the integrating factor of the differential equation in Example 6. The method of solving differential equations by multiplying integrating factor is called the method of integrating factor, which is an elementary method of solving first order differential equations.

10.4 The Second Type Surface Integral

10.4.1 Oriented Surface

Smooth surfaces usually have two sides, for example, the surface determined by $z =$

Chapter 10 The Second Type Curve Integral, Surface Integral, and Vector Field

$f(x,y)$ has upward side and downward side, the surface determined by $x = x(y,z)$ has front side and back side, and a closed surface has inner side and outer side. A characteristic of two-sided surfaces is: If a bug on a point of the surface wants to get to the other side of the surface, it must cross the boundary of the surface. Since the normal vector of the surface at any point has two different directions, so we can distinguish the two sides of the surface by regulate the direction of the normal vector. For example, for the surface $z = z(x,y)$, we let the included angle between the normal vector of the upward side and the positive side of the z-axis be smaller than $\dfrac{\pi}{2}$, and that of the normal vector of the downward side and the positive side of the z-axis be bigger than $\dfrac{\pi}{2}$. For a closed surface, we let the normal vector of the outer side be pointing outward and that of the inner side be pointing inward. A surface with regulated normal vectors is called an oriented surface.

By the way, a surface has two sides is called a two-sided surface. But there are surfaces have only one side, which are called one-sided surfaces, such as the Möbius band, which is obtained from a rectangular paper strip $ABCD$, by a half twist and pasting the two sides together (Fig. 10. 15).

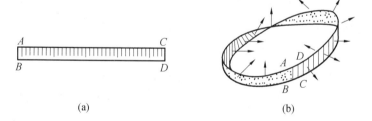

Fig. 10. 15

Next, we will study the projections of an oriented surface onto the coordinate planes. Suppose Σ is a space oriented surface with area S, and the directional cosines of the normal vector are $\cos\alpha, \cos\beta, \cos\gamma$. At this moment, α, β, γ are exactly the included angles between Σ and the coordinate planes Oyz, Ozx, Oxy, respectively (Fig. 10. 16). We call the values $\Sigma_{yz} = S\cos\alpha$, $\Sigma_{zx} = S\cos\beta$, $\Sigma_{xy} = S\cos\gamma$ the projective values of the oriented surface Σ on to the coordinate planes Oyz, Ozx, Oxy.

Let σ_{xy} be the area of the projection of Σ on to the Oxy plane. Then when $\gamma < \dfrac{\pi}{2}$, $\boldsymbol{n}$ points up, $S\cos\gamma$ is positive, $\Sigma_{xy} = \sigma_{xy}$; when $\gamma > \dfrac{\pi}{2}$, $\boldsymbol{n}$ points down, $S\cos\gamma$ is negative, $\Sigma_{xy} = -\sigma_{xy}$.

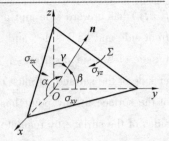

Fig. 10.16

10.4.2 Concepts for The second Type Surface Integral

Example 1 Suppose in the region G, there is a continuous incompressible fluid velocity field
$$\boldsymbol{v} = P(x,y,z)\boldsymbol{i} + Q(x,y,z)\boldsymbol{j} + R(x,y,z)\boldsymbol{k}$$
Find the net flow Φ in the unit time that through the oriented surface Σ in G in the given direction.

Solution 1° For the case that $\boldsymbol{v}$ is a constant vector, Σ is an oriented surface. Suppose the area of Σ is S, the unit normal vector is $\boldsymbol{n}^0$, then
$$\Phi = S|\boldsymbol{v}|\cos(\widehat{\boldsymbol{v},\boldsymbol{n}^0}) = \boldsymbol{v} \cdot \boldsymbol{n}^0 S = \boldsymbol{v} \cdot \boldsymbol{S}$$
where $\boldsymbol{S} = S\boldsymbol{n}^0$ (Fig. 10.17).

2° The general case.

Divide the surface Σ into $\Delta S_1, \Delta S_2, \cdots, \Delta S_n$ (Fig. 10.18), and also use these to denote their areas, Let $\lambda = \max\limits_{1 \leqslant i \leqslant n}\{\text{diameter of } \Delta S_i\}$. Take an arbitrary point $M_i(\xi_i, \eta_i, \zeta_i) \in \Delta S_i$, let $\boldsymbol{n}_i^0 = \{\cos\alpha_i, \cos\beta_i, \cos\gamma_i\}$ be the given unit normal vector of the oriented surface Σ at the point M_i. Use $\boldsymbol{v}(M_i)$ to take place of the velocity of all the points on ΔS_i, and treat ΔS_i as a surface piece that passes through M_i with normal vector $\boldsymbol{n}_i^0$, then the flow through ΔS_i
$$\Delta \Phi_i \approx \boldsymbol{v}(M_i) \cdot \boldsymbol{n}_i^0 \Delta S_i = \boldsymbol{v}(M_i) \cdot \Delta \boldsymbol{S}_i$$
where $\Delta \boldsymbol{S}_i = \Delta S_i \boldsymbol{n}_i^0$, then the net flow Φ in the unit time that through the oriented surface Σ in G in the given direction is
$$\Phi = \lim_{\lambda \to 0}\sum_{i=1}^n \boldsymbol{v}(M_i) \cdot \boldsymbol{n}_i^0 \Delta S_i = \lim_{\lambda \to 0}\sum_{i=1}^n \boldsymbol{v}(M_i) \cdot \Delta \boldsymbol{S}_i$$
$$= \lim_{\lambda \to 0}\sum_{i=1}^n [P(\xi_i,\eta_i,\zeta_i)\cos\alpha_i + Q(\xi_i,\eta_i,\zeta_i)\cos\beta_i + R(\xi_i,\eta_i,\zeta_i)\cos\gamma_i]\Delta S_i$$

Chapter 10 The Second Type Curve Integral, Surface Integral, and Vector Field

$$= \lim_{\lambda \to 0} \sum_{i=1}^{n} [P(M_i)\Delta\Sigma_{iyz} + Q(M_i)\Delta\Sigma_{izx} + R(M_i)\Delta\Sigma_{ixy}]$$

 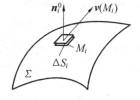

Fig. 10.17　　　　　　　　Fig. 10.18

Problems similar to this are very common in practice. By abstracting, we obtain the following important concept.

Definition 10.2 Suppose Σ is a smooth oriented piece of surface, the vector function

$$F(x,y,z) = \{P(x,y,z), Q(x,y,z), R(x,y,z)\}$$

is defined on Σ. Divide Σ into $\Delta S_1, \Delta S_2, \cdots, \Delta S_n$, and also use these to denote their areas, Let $\lambda = \max_{1 \leq i \leq n} \{\text{diameter of } \Delta S_i\}$. Take an arbitrary point $M_i(\xi_i, \eta_i, \zeta_i) \in \Delta S_i$, let $n_i^0 = \{\cos\alpha_i, \cos\beta_i, \cos\gamma_i\}$ be the given unit normal vector of the oriented surface Σ at the point M_i. Let $\Delta\Sigma_{iyz}, \Delta\Sigma_{izx}, \Delta\Sigma_{ixy}$ be the projections of ΔS_i onto the Oyz, Ozx, Oxy planes. If no matter how to divide the surface Σ and how to choose the points M_i, the limit

$$\lim_{\lambda \to 0} \sum_{i=1}^{n} F(M_i) \cdot \Delta S_i = \lim_{\lambda \to 0} \sum_{i=1}^{n} F(M_i) \cdot n_i^0 \Delta S_i$$

$$= \lim_{\lambda \to 0} \sum_{i=1}^{n} [P(M_i)\cos\alpha_i + Q(M_i)\cos\beta_i + R(M_i)\cos\gamma_i]\Delta S_i$$

$$= \lim_{\lambda \to 0} \sum_{i=1}^{n} [P(M_i)\Delta\Sigma_{iyz} + Q(M_i)\Delta\Sigma_{izx} + R(M_i)\Delta\Sigma_{ixy}] \quad (1)$$

always exists, and gives the same value, then we call this limit value the surface integral of the vector function F on the oriented surface Σ, or the second type surface integral of the functions P, Q, R on the oriented surface Σ. Denoted by

$$\iint_\Sigma F(M) \cdot dS \quad \text{or} \quad \iint_\Sigma P dy \wedge dz + Q dz \wedge dx + R dx \wedge dy$$

and call

$$\iint_\Sigma P(x,y,z) dy \wedge dz, \iint_\Sigma Q(x,y,z) dz \wedge dx, \iint_\Sigma R(x,y,z) dx \wedge dy$$

the surface integral of the functions P, Q, R on the oriented surface Σ with respect to yz, zx, xy, respectively. Where $d\mathbf{S}$ is the area element vector of the surface

$$d\mathbf{S} = dS\{\cos \alpha, \cos \beta, \cos \gamma\}, dy \wedge dz = \cos \alpha dS$$
$$dz \wedge dx = \cos \beta dS, dx \wedge dy = \cos \gamma dS$$

are the projections of $d\mathbf{S}$ on the Oyz, Ozx, Oxy planes, respectively. Traditionally, we omit the outer product symbol " $\wedge$ ", and denote $dy \wedge dz$ by $dydz$ for short, and denote $\iint_\Sigma P dy \wedge dz$ by $\iint_\Sigma P dydz$, and so on.

According to this, we can defined the net flow of Example 1 as $\Phi = \iint_\Sigma v(M) \cdot d\mathbf{S}$.

When P, Q, R are continuous on Σ, the second type surface integral exist. And by the definition formula(1) of the second type surface integral, we get the following properties:

(i) The relationship between the second type surface integral and the first type surface integral. Suppose $\cos \alpha, \cos \beta, \cos \gamma$ are the directional cosines of the given normal vector of the oriented surface Σ, then

$$\iint_\Sigma Pdydz + Qdzdx + Rdxdy = \iint_\Sigma (P\cos \alpha + Q\cos \beta + R\cos \gamma) dS$$

(ii) If we use $-\Sigma$ to denote the surface with the opposite orientation as the oriented surface Σ, then

$$\iint_\Sigma Rdxdy = - \iint_{-\Sigma} Rdxdy \quad \text{(Orientability)}$$

(iii) If k_1, k_2 are constants, then

$$\iint_\Sigma (k_1 R_1 + k_2 R_2) dxdy = k_1 \iint_\Sigma R_1 dxdy + k_2 \iint_\Sigma R_2 dxdy \quad \text{(Linearity)}$$

(iv) If we divide the oriented surface Σ into two pieces Σ_1, Σ_2, then

$$\iint_\Sigma Rdxdy = \iint_{\Sigma_1} Rdxdy + \iint_{\Sigma_2} Rdxdy \quad \text{(Additivity of integral region)}$$

(v) When Σ is the cylinder with generatrix parallel to the z-axis, then

$$\iint_\Sigma Rdxdy = 0$$

We have the similar results for surface integrals with respect to yz, zx.

Chapter 10 The Second Type Curve Integral, Surface Integral, and Vector Field

10.4.3 Calculating The Second Type Surface Integral

As it for the first type surface integral, the second type integrals are also transformed into double integrals to compute. Suppose the equation of the surface Σ is

$$z = z(x,y), (x,y) \in \sigma_{xy}$$

where σ_{xy} is the projection of the surface Σ on the Oxy plane, $z(x,y)$ has continuous partial derivatives on σ_{xy} (i.e. the surface Σ is smooth); the functions $P(x,y,z), Q(x,y,z)$ and $R(x,y,z)$ are continuous on Σ. Since

$$\boldsymbol{n} = \pm \{-z'_x, -z'_y, 1\}$$

are normal vectors of the surface Σ on the different direction, then the unit normal vector

$$\boldsymbol{n}^0 = \{\cos\alpha, \cos\beta, \cos\gamma\}$$

$$= \frac{\pm 1}{\sqrt{1 + z'^2_x + z'^2_y}} \{-z'_x, -z'_y, 1\}$$

And

$$dS = \sqrt{1 + z'^2_x + z'^2_y}\, dxdy$$

so by applying the relation between these two types of surface integrals (i) and the calculation of the first type surface integral, we get the calculating formula for the second type surface integral

$$\iint_{\Sigma(\frac{up}{down})} P(x,y,z)\,dydz + Q(x,y,z)\,dzdx + R(x,y,z)\,dxdy$$

$$= \pm \iint_{\sigma_{xy}} [-P(x,y,z(x,y))z'_x - Q(x,y,z(x,y))z'_y + R(x,y,z(x,y))]\,dxdy$$

(2)

It transform the second type surface integral into a double integral on the projection σ_{xy} of the surface Σ on the Oxy plane. When we take the upward side of the surface Σ, the sign before the double integral is positive; when we take the downward side of Σ, the sign before the double integral is negative. If the equation of the surface Σ is not single valued on σ_{xy}, we can divide Σ into several single valued pieces. Other formulas for the projections of the surface Σ on other coordinate planes are similar.

The vector version of the equation (2) is

$$\iint_{\Sigma(\frac{up}{down})} \boldsymbol{F} \cdot d\boldsymbol{S} = \pm \iint \boldsymbol{F} \cdot \boldsymbol{n}\,dxdy$$

Calculus(Ⅱ)

where $n = \{-z'_x, -z'_y, 1\}$.

Specially, by equation(2) we have

$$\iint_{\Sigma(\substack{up\\down})} R(x,y,z)\,dxdy = \pm \iint R(x,y,z(x,y))\,dxdy \qquad (3)$$

Similarly, if we express the equation of Σ as $x = x(y,z), (y,z) \in \sigma_{yz}$, then we have

$$\iint_{\Sigma(\substack{front\\back})} P(x,y,z)\,dydz = \pm \iint_{\sigma_{yz}} P(x(y,z),y,z)\,dydz \qquad (4)$$

If we express the equation of Σ as $y = y(x,z), (x,z) \in \sigma_{zx}$, then we have

$$\iint_{\Sigma(\substack{right\\left})} Q(x,y,z)\,dzdx = \pm \iint_{\sigma_{zx}} Q(x,y(x,z),z)\,dzdx \qquad (5)$$

So, for surface integrals for different coordinates, we can project the surface Σ onto different coordinate planes, then transform the integral into double integrals on different projections. We have to pay attention when using this method: 1° Determine the independent variables of the integral, express the surface Σ as a function of these two variables, and determine the projection of Σ; 2° Plug the equation of Σ into the integrant, and transform the integral into a double integral on the projection; 3° Determine the sign in front of the double integral according to the orientation of the surface Σ.

Example 2 Evaluate $\iint_{\Sigma} xyz\,dxdy$, where Σ is part of the sphere $x^2 + y^2 + z^2 = 1$ satisfies $x \geq 0, y \geq 0$, and oriented outward.

Solution As showed in Fig. 10.19, divide Σ into Σ_1, Σ_2

$$\Sigma_1: z = \sqrt{1 - x^2 - y^2} \quad \text{(upwards)}$$

$$\Sigma_2: z = -\sqrt{1 - x^2 - y^2} \quad \text{(downwards)}$$

There projections on the Oxy plane are both

$$\sigma_{xy}: x \geq 0, y \geq 0, x^2 + y^2 \leq 1$$

so

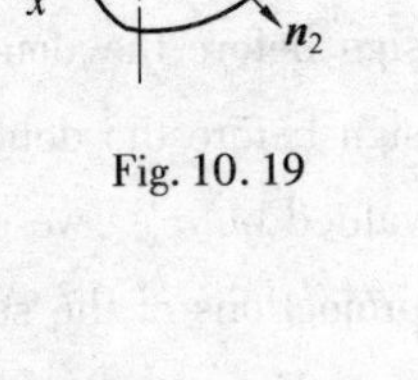

$$\iint_{\Sigma} xyz\,dxdy = \left(\iint_{\Sigma_1} + \iint_{\Sigma_2}\right) xyz\,dxdy$$

$$= \iint_{\sigma_{xy}} xy\sqrt{1 - x^2 - y^2}\,dxdy - \iint_{\sigma_{xy}} xy(-\sqrt{1 - x^2 - y^2})\,dxdy$$

$$= 2\int_0^{\pi/2} d\theta \int_0^1 r^3 \sqrt{1 - r^2} \sin\theta\cos\theta\,dr = \frac{2}{15}$$

Fig. 10.19

Notice that the oriented surface Σ in this problem is symmetric with respect to $z = 0$,

Chapter 10 The Second Type Curve Integral, Surface Integral, and Vector Field

and the integrant is an odd function of z, but the integral with respect to the coordinates x, y is not zero. If the oriented surface Σ is symmetric with respect to $z = 0$, and the integrant is an even function of z, then the integral with respect to x, y is zero.

Example 3 Suppose Σ is the part of the cylinder $x^2 + y^2 = R^2$ between the planes $z = 0$ and $z = h$, oriented outwards. Find the flow Q in the unit time of the velocity field $v = \{x^2, y^3, z\}$ through Σ.

Solution The flow $Q = \iint_\Sigma x^2 dydz + y^3 dzdx + zdxdy$, since the projection of Σ on the yOz plane is a rectangular region (Fig. 10.20, 10.21), but the included angle between the normal vector of the surface $x = \sqrt{R^2 - y^2}$ and the x-axis is acute, that of $x = -\sqrt{R^2 - y^2}$ and the x-axis is obtuse. Thus $\iint_\Sigma x^2 dydz = 0$; since $\Sigma \perp xOy$, so $\iint_\Sigma zdxdy = 0$; so only need to calculate

$$\iint_\Sigma y^3 dzdx = 2\iint_D (R^2 - x^2)^{3/2} dxdz$$

where D is the projection of Σ on the xOz plane: $0 \leq z \leq h$, $-R \leq x \leq R$. Thus

$$Q = 4h \int_0^R (R^2 - x^2)^{3/2} dx = 4\pi R^4 \cdot \int_0^{\pi/2} \cos^4 t dt = \frac{3}{4}\pi R^4$$

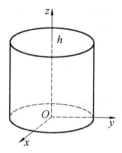
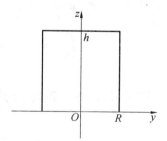

Fig. 10.20   Fig. 10.21

Example 4 Evaluate $J = \iint_\Sigma (x^2 + y^2) dzdx + zdxdy$, where Σ is the part of the cone $z = \sqrt{x^2 + y^2}$ ($0 \leq z \leq 1$) in the first octant, oriented downwards.

Solution 1 See Fig. 10.22, the projection of Σ on the Ozx plane is positive, and the projective region is

$$\sigma_{zx}: 0 \leq z \leq 1, 0 \leq x \leq z$$

Calculus(II)

and the projection of Σ on the Oxy plane is negative, and the projective region is

$$\sigma_{xy}: 0 \leq \theta \leq \frac{\pi}{2}, 0 \leq r \leq 1$$

so

$$\iint_\Sigma (x^2 + y^2)\,dzdx + z\,dxdy = \iint_{\sigma_{zx}} z^2\,dzdx - \iint_{\sigma_{xy}} \sqrt{x^2 + y^2}\,dxdy$$

$$= \int_0^1 z^2\,dz \int_0^z dx - \int_0^{\pi/2} d\theta \int_0^1 r^2\,dr$$

$$= \frac{1}{4} - \frac{\pi}{6}$$

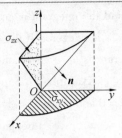
Fig. 10.22

To avoid project the surface onto different coordinate planes, we can also calculate the second type surface integral following equation (2).

Solution 2 Since the equation of the surface is $z = \sqrt{x^2 + y^2}$, so

$$z'_y = \frac{y}{\sqrt{x^2 + y^2}}$$

by equation(2), we know

$$J = \iint_{\sigma_{xy}} [(x^2 + y^2)z'_y - \sqrt{x^2 + y^2}]\,dxdy$$

$$= \iint_{\sigma_{xy}} [\sqrt{x^2 + y^2}\,y - \sqrt{x^2 + y^2}]\,dxdy$$

$$= \int_0^{\pi/2} d\theta \int_0^1 (r^2 \sin\theta - r)r\,dr = \frac{1}{4} - \frac{\pi}{6}$$

10.5 The Gauss Formula, The Flux and Divergence

The Green's formula connects the curve integral on a planar closed curve with the double integral on the region enclosed by the curve. In this section, we will introduce the Gauss formula, which will show us the relation between the surface integral on a space closed surface and the triple integral on the solid enclosed by the surface. The Gauss formula has its physical significance—the flux and divergence.

10.5.1 The Gauss Formula

Theorem 10.3 Suppose the space region V is enclosed by a piecewise smooth

Chapter 10 The Second Type Curve Integral, Surface Integral, and Vector Field

closed surface Σ, the functions $P(x,y,z), Q(x,y,z), R(x,y,z)$ have continuous partial derivatives on V, then we have the Gauss formula

$$\oiint_{\Sigma\text{outwards}} P\,dydz + Q\,dzdx + R\,dxdy = \iiint_V \left(\frac{\partial P}{\partial x} + \frac{\partial Q}{\partial y} + \frac{\partial R}{\partial z}\right)dxdydz \qquad (1)$$

i. e.

$$\oiint_\Sigma (P\cos\alpha + Q\cos\beta + R\cos\gamma)\,dS = \iiint_V \left(\frac{\partial P}{\partial x} + \frac{\partial Q}{\partial y} + \frac{\partial R}{\partial z}\right)dxdydz \qquad (2)$$

where $\cos\alpha, \cos\beta, \cos\gamma$ are the directional cosines of the external normal vector of Σ.

Proof Suppose the projection of the space region V on the Oxy plane is σ_{xy}. Suppose the straight lines paralle to the z-axis and pass through V intersect the boundary of V, Σ, in exactly two points, then in general, the boundary surface Σ consists of $\Sigma_1, \Sigma_2, \Sigma_3$ (Fig. 10.23):

$\Sigma_1 : z = z_1(x,y), (x,y) \in \sigma_{xy}$;

$\Sigma_2 : z = z_2(x,y), (x,y) \in \sigma_{xy}$;

Σ_3: Part of the cylinder with the boundary of σ_{xy} as the directrix and generatrix parallel to the z-axis, between z_1, z_2.

By the calculation of surface integral, we have

$$\oiint_{\Sigma\text{outwards}} R(x,y,z)\,dxdy = \left(\iint_{\Sigma\text{downwards}} + \iint_{\Sigma\text{upwards}} + \iint_{\Sigma\text{outwards}}\right) R(x,y,z)\,dxdy$$

$$= -\iint_{\sigma_{xy}} R(x,y,z_1(x,y))\,dxdy + \iint_{\sigma_{xy}} R(x,y,z_2(x,y))\,dxdy$$

On the other hand, by the calculation of triple integrals, we have

$$\iiint_V \frac{\partial R}{\partial z}dxdydz = \iint_{\sigma_{xy}} dxdy \int_{z_1(x,y)}^{z_2(x,y)} \frac{\partial R}{\partial z}dz$$

$$= \iint_{\sigma_{xy}} [R(x,y,z_2(x,y)) - R(x,y,z_1(x,y))]\,dxdy$$

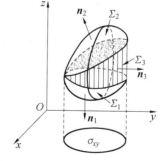

Fig. 10.23

then, we have

$$\oiint_{\Sigma\text{outwards}} R(x,y,z)\,dxdy = \iiint_V \frac{\partial R}{\partial z}dxdydz$$

If V is very complicate, and does not satisfies the assumption, then we just need to divide V into several pieces by using smooth surfaces, such that each piece satisfies the assumption. Notice that on the two opposite sides of an interface the two integrals cancel out, so the above equation still holds for general complex connected region V.

Similarly, we can prove

$$\oiint_{\Sigma_{outwards}} P(x,y,z) \, dydz = \iiint_V \frac{\partial P}{\partial x} dxdydz$$

$$\oiint_{\Sigma_{outwards}} Q(x,y,z) \, dzdx = \iiint_V \frac{\partial Q}{\partial y} dxdydz$$

so equation (1) holds.

The Gauss formula presents us a new way to calculate surface integrals on closed surfaces.

Example 1 Evaluate $I = \oiint_\Sigma (x - y) \, dxdy + (y - z) \, ydydy$, where Σ is the boundary surface of the solid Ω enclosed by $z = 0, z = 3$ and $x^2 + y^2 = 1$, oriented outwards.

Solution By the Gauss formula

$$I = \oiint_\Sigma (x - y) \, dxdy + (y - z) \, ydydy$$

$$= \iiint_\Omega (0 + y - z) \, dV = \int_0^{2\pi} d\theta \int_0^1 \rho d\rho \int_0^3 (\rho\sin\theta - z) \, dz$$

$$= \int_0^{2\pi} d\theta \left(\int_0^1 3\rho^2 \sin\theta - \frac{9}{2}\rho \right) d\rho = -\frac{9\pi}{2}$$

Example 2 Evaluate $\oiint_S \frac{xdydz + ydzdx + zdxdy}{\sqrt{x^2 + y^2 + z^2}}$, where S is the sphere $x^2 + y^2 + z^2 = a^2$ oriented outwards.

Solution Since the integrant has no definition at the origin, so we cannot use the Gauss formula directly. But since the point (x,y,z) in the integrant is on the surface, we can simplify the integrant by using the equation of the surface, so that we can apply the Gauss formula

$$\oiint_S \frac{xdydz + ydzdx + zdxdy}{\sqrt{x^2 + y^2 + z^2}} = \frac{1}{a} \oiint_S xdydz + ydzdx + zdxdy$$

$$= \frac{3}{a} \iiint_V dxdydz = 4\pi a^2$$

Example 3 Evaluate $I = \iint_\Sigma xdydz + ydzdx + (z^2 - 2z) \, dxdy$, where Σ is part of the cone $z = \sqrt{x^2 + y^2}$ between $0 \leqslant z \leqslant 1$, oriented upwards.

Chapter 10 The Second Type Curve Integral, Surface Integral, and Vector Field

Solution Here
$$\frac{\partial P}{\partial x} + \frac{\partial Q}{\partial y} + \frac{\partial R}{\partial z} = 2z$$
is easy, but the surface Σ is not closed, in order to use the Gauss formula, we add an additional oriented surface
$$\Sigma_1 : z = 1 (x^2 + y^2 \leqslant 1)$$
oriented downwards (Fig. 10.24), then

$$I = (\oiint_{\Sigma+\Sigma_1} - \oiint_{\Sigma_1}) x\,dy\,dz + y\,dz\,dx + (z^2 - 2z)\,dx\,dy$$

$$= -\iiint_V 2z\,dV + \iint_{\sigma_{xy}} (1 - 2)\,dx\,dy$$

$$= -\int_0^{2\pi} d\theta \int_{-0}^{1} r\,dr \int_r^1 2z\,dz - \pi = -\frac{3\pi}{2}$$

Example 4 Suppose the function $f(u)$ has continuous derivatives, evaluate

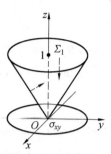

Fig. 10.24

$$J = \oiint_\Sigma x^3\,dz\,dx + [y^3 + yf(yz)]\,dz\,dx + [z^3 - zf(yz)]\,dx\,dy$$

where Σ is the boundary surface of the solid enclosed by the cone $x = \sqrt{y^2 + z^2}$ and the spheres $x = \sqrt{1 - y^2 - z^2}$ and $x = \sqrt{4 - x^2 - y^2}$, oriented outwards (Fig. 10.25).

Solution There is an abstract function in the integrant, so we cannot calculate the integral directly.

Since
$$P = x^3, Q = y^3 + yf(yz), R = z^3 - zf(yz)$$
$$\frac{\partial P}{\partial x} = 3x^2, \frac{\partial Q}{\partial y} = 3y^2 + f(yz) + yzf'(yz)$$
$$\frac{\partial R}{\partial z} = 3z^2 - f(yz) - yzf'(yz)$$

Then by the Gauss formula

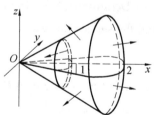

Fig. 10.25

$$J = \iiint_V 3(x^2 + y^2 + z^2)\,dV$$

$$= 3\iiint_V \rho^4 \sin\varphi\,d\rho\,d\varphi\,d\theta$$

$$= 3\int_0^{2\pi} d\theta \int_0^{\pi/4} \sin\varphi\,d\varphi \int_1^2 \rho^4\,d\rho = \frac{93\pi}{5}(2 - \sqrt{2})$$

10.5.2 The Flux and Divergence of Vector Fields

In Section 10.4, we introduced the second type surface integral, and we knew that for the velocity field of an irreducible fluid $v(M)$, the net flow passing through an oriented surface Σ in a given direction is

$$\Phi = \iint_\Sigma v \cdot n^0 \mathrm{d}S = \iint_\Sigma v \cdot \mathrm{d}S$$

where $\mathrm{d}S = \mathrm{d}Sn^0$, n^0 is the unit normal vector of the oriented surface Σ.

In the electric intensity field E, the electric flow through the oriented surface Σ is

$$\Phi_E = \iint_\Sigma E \cdot n^0 \mathrm{d}S = \iint_\Sigma E \cdot \mathrm{d}S$$

which can be thought of as the number of electric power lines through Σ.

In the magnetic induction intensity filed B, the magnetic flow through the oriented surface Σ is

$$\Phi_B = \iint_\Sigma B \cdot n^0 \mathrm{d}S = \iint_\Sigma B \cdot \mathrm{d}S$$

which can be thought of as the number of magnetic power lines through Σ.

When we study vector fields, we usually need to think of this type of surface integrals. They are very important.

Definition 10.3 In the vector field $F(M)$, suppose Σ is an oriented surface, we call the surface integral

$$\Phi = \iint_\Sigma F \cdot n^0 \mathrm{d}S = \iint_\Sigma F \cdot \mathrm{d}S \tag{3}$$

the flux of the vector field $F(M)$ through the oriented surface Σ in a given direction.

In the rectangular coordinate system, if

$$F(M) = \{P(x,y,z), Q(x,y,z), R(x,y,z)\}$$

then the flux

$$\Phi = \iint_\Sigma P(x,y,z) \mathrm{d}y\mathrm{d}z + Q(x,y,z) \mathrm{d}z\mathrm{d}x + R(x,y,z) \mathrm{d}x\mathrm{d}y$$

Now take the velocity field as an example, we state the physical significance of flux being positive, negative and zero. When $\Phi > 0$, it means that the net flow through the oriented surface Σ in the given direction (i.e. the difference of the volume of the fluid flew in the given direction and that of the opposite direction) is positive, i.e. the vol-

Chapter 10 The Second Type Curve Integral, Surface Integral, and Vector Field

ume flew in the given direction is bigger than the volume flew in the opposite direction; when $\Phi < 0$, it means the volume flew in the given direction is smaller than the volume flew in the opposite direction; when $\Phi = 0$, it means the volume flew in the given direction equals to the volume flew in the opposite direction.

For a closed surface Σ oriented outwards (i. e. the normal vector points out), the flux

$$\Phi = \oiint_\Sigma \boldsymbol{v} \cdot d\boldsymbol{S}$$

When $\Phi > 0$, it means the volume flew out is bigger than the volume flew in, then there are sources that giving out fluid in the solid V enclosed by Σ.

When $\Phi < 0$, it means the volume flew out is smaller than the volume flew in, then there are sinks that absorbing fluid in the solid V enclosed by Σ.

When $\Phi = 0$, it means the volume flew out equals to the volume flew in, then the sources and the sinks are canceled out in V.

If we treat the sinks as negative sources, then every point in the solid V can be thought of as a source. In this moment, the sources have signs, and sometimes strong sometimes weak. In the study of velocity fields, the intensity of the source is no doubt very important.

Example 5 In the electric field generated by a point charge q at the origin O, the electric displacement at the point M is

$$\boldsymbol{D} = \varepsilon \boldsymbol{E} = \frac{q}{4\pi r^2} \boldsymbol{r}^0$$

where $r = |OM|$, $\boldsymbol{r}^0$ is the unit vector starting at O and pointing to M. Suppose Σ is a sphere centered at O, and with radius R, find the electric displacement flux Φ_D getting out of the sphere Σ.

Solution Since on the sphere $r = R$, and $\boldsymbol{r}^0$ is coincident with the external normal vector $\boldsymbol{n}^0$, so

$$\Phi_D = \oiint_\Sigma \boldsymbol{D} \cdot d\boldsymbol{S} = \frac{q}{4\pi R^2} \oiint_\Sigma \boldsymbol{r}^0 \cdot \boldsymbol{n}^0 dS$$

$$= \frac{q}{4\pi R^2} \oiint_\Sigma dS = \frac{q}{4\pi R^2} 4\pi R^2 = q$$

As we can see, the source of the electric displacement flux Φ_D from inside the sphere is just the free point charge q. When q is a positive charge, it is a positive source, and will generate electric displacement lines; when q is a negative charge, it is a negative source, and will absorb electric displacement lines. The size of q will determine the intensity of the source.

In the vector field $F(M)$, the flux Φ through the closed surface Σ is the accumulation of the abilities of generating or absorbing vector lines of the points in the region V enclosed by Σ. Next we will study how to express the abilities of generating or absorbing vector lines of the points in the region V.

Definition 10.4 Suppose M is a point in the vector field $F(M)$, make a small closed surface Σ containing the point M (oriented outwards), use ΔV to denote the solid enclosed by the surface Σ and its volume, let $\lambda = \max\limits_{M_1 \in \Sigma} d(M, M_1)$, if when $\lambda \to 0$, the limit

$$\lim_{\lambda \to 0} \frac{\oiint_\Sigma F \cdot dS}{\Delta V}$$

exists, and is irrelevant to the methods of contraction of Σ, then we call this limit the divergence of the vector field $F(M)$ at the point M, denoted by $\mathrm{div} F(M)$, i.e.

$$\mathrm{div} F(M) = \lim_{\lambda \to 0} \frac{\oiint_\Sigma F \cdot dS}{\Delta V}$$

The divergence $\mathrm{div} F(M)$ is a quantity determined by the vector field, it is the volume density of the flux.

Under the rectangular coordinate system, the divergence of the vector field

$$F = \{P(x,y,z), Q(x,y,z), R(x,y,z)\}$$

at the point $M(x,y,z)$ (suppose the partial derivatives of P, Q, R with respect to x, y, z are continuous) can be calculated by the following formula

$$\mathrm{div} F(M) = \frac{\partial P}{\partial x} + \frac{\partial Q}{\partial y} + \frac{\partial R}{\partial z} = \nabla \cdot F$$

In fact, according to the Gauss formula and the mean value theorem

$$\mathrm{div}\, F(M) = \lim_{\lambda \to 0} \frac{\oiint_\Sigma F \cdot dS}{\Delta V} = \lim_{\lambda \to 0} \frac{1}{\Delta V} \iiint_{\Delta V} \left(\frac{\partial P}{\partial x} + \frac{\partial Q}{\partial y} + \frac{\partial R}{\partial z}\right) dV$$

$$= \lim_{\lambda \to 0} \frac{1}{\Delta V} \left(\frac{\partial P}{\partial x} + \frac{\partial Q}{\partial y} + \frac{\partial R}{\partial z}\right)_{M^*} \Delta V = \left(\frac{\partial P}{\partial x} + \frac{\partial Q}{\partial y} + \frac{\partial R}{\partial z}\right)_M$$

where M^* is a point in the region V enclosed by Σ.

Chapter 10 The Second Type Curve Integral, Surface Integral, and Vector Field

Based on the concept of divergence, the Gauss formula can be expressed as the vector form

$$\oiint_{\Sigma} \boldsymbol{F} \cdot d\boldsymbol{S} = \iiint_{V} \mathrm{div}\boldsymbol{F} dV = \iiint_{V} \nabla \cdot \boldsymbol{F} dV$$

The physical significance: The flux through the oriented closed surface Σ (oriented outwards) equals to the integral of the divergence of the points in the region V enclosed by Σ.

If in a vector field $\boldsymbol{F}(M)$, the divergence at any point is zero, i.e. we always have div $\boldsymbol{F}(M) = 0$, then we call the field a source free vector field. For a simply connected region in a source free vector field, for any closed surface Σ, we always have

$$\oiint_{\Sigma} \boldsymbol{F} \cdot d\boldsymbol{S} = 0$$

At this moment, the second type surface integral on any surface in this region is only relevant to the boundary curve Γ of the surface, and is irrelevant to the shape of the surface, as showed in Fig. 10.26, when div $\boldsymbol{F}(M) = 0$

$$\iint_{\Sigma_1} \boldsymbol{F} d\boldsymbol{S} = \iint_{\Sigma_2} \boldsymbol{F} d\boldsymbol{S}$$

By equation (5), we get the following calculative properties of the divergence:

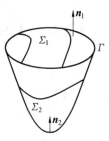

Fig. 10.26

(i) $\mathrm{div}(C\boldsymbol{F}) = C\mathrm{div}\,\boldsymbol{F}$ (C is a constant).
(ii) $\mathrm{div}(\boldsymbol{F}_1 \pm \boldsymbol{F}_2) = \mathrm{div}\,\boldsymbol{F}_1 \pm \mathrm{div}\,\boldsymbol{F}_2$.
(iii) $\mathrm{div}(u\boldsymbol{F}) = u\,\mathrm{div}\,\boldsymbol{F} + \boldsymbol{F} \cdot \mathrm{grad}\,u$ (u is a number valued function).

Proof of (iii) Since $u\boldsymbol{F} = \{uP, uQ, uR\}$, so

$$\mathrm{div}(u\boldsymbol{F}) = \frac{\partial(uP)}{\partial x} + \frac{\partial(uQ)}{\partial y} + \frac{\partial(uR)}{\partial z}$$

$$= u\left(\frac{\partial P}{\partial x} + \frac{\partial P}{\partial y} + \frac{\partial P}{\partial z}\right) + P\frac{\partial u}{\partial x} + Q\frac{\partial u}{\partial y} + R\frac{\partial u}{\partial z}$$

$$= u\,\mathrm{div}\,\boldsymbol{F} + \boldsymbol{F}\,\mathrm{grad}\,u$$

Example 6 In the electrostatic field generated by the point charge q, find the divergence of the electric displacement vector $\boldsymbol{D}$.

Solution Suppose q lies in the origin, then

$$D = \frac{q}{4\pi r^3} r$$

where $r = xi + yj + zk$, $r = |r|$

$$\text{div } D = \frac{q}{4\pi}\left[\frac{\partial}{\partial x}\left(\frac{x}{r^3}\right) + \frac{\partial}{\partial y}\left(\frac{y}{r^3}\right) + \frac{\partial}{\partial z}\left(\frac{z}{r^3}\right)\right]$$

$$= \frac{q}{4\pi}\left[\frac{r^2 - 3x^2}{r^5} + \frac{r^2 - 3y^2}{r^5} + \frac{r^2 - 3z^2}{r^5}\right] = 0 \, (r \neq 0)$$

Except for the point $(0,0,0)$ where the charge lies in, the divergence of the electric displacement at any point are zero. The divergence at the point $(0,0,0)$ does not exist, which is a singular point. By Example 4 and the discussion above we know, the flux of any closed surface containing q is $\oint_\Sigma D \cdot dS = q$.

10.6 The Stokes' Theorem, Circulation and Curl

In this section we will introduce the relation between the space surface integral and the curve integral—the Stokes' theorem. Meanwhile, we will introduce two important concepts of the vector field—circulation and curl.

10.6.1 The Stokes' Theorem

Theorem 10.4 Suppose C is a piecewise smooth oriented closed curve, Σ is an arbitrary piecewise smooth oriented surface with C as the boundary, the orientation of C and the orientation of Σ satisfy the right-hand screw rule. The functions P, Q, R have continuous partial derivatives on some region containing Σ, then we have the Stokes' formula

$$\oint_C P(x,y,z)\,dx + Q(x,y,z)\,dy + R(x,y,z)\,dz$$
$$= \iint_\Sigma \left(\frac{\partial R}{\partial y} - \frac{\partial Q}{\partial z}\right)dydz + \left(\frac{\partial P}{\partial z} - \frac{\partial R}{\partial x}\right)dzdx + \left(\frac{\partial Q}{\partial x} - \frac{\partial P}{\partial y}\right)dxdy \tag{1}$$

i. e.

$$\oint_C P(x,y,z)\,dx + Q(x,y,z)\,dy + R(x,y,z)\,dz$$
$$= \iint_\Sigma \left[\left(\frac{\partial R}{\partial y} - \frac{\partial Q}{\partial z}\right)\cos\alpha + \left(\frac{\partial P}{\partial z} - \frac{\partial R}{\partial x}\right)\cos\beta + \left(\frac{\partial Q}{\partial x} - \frac{\partial P}{\partial y}\right)\cos\gamma\right]dS \tag{2}$$

Chapter 10 The Second Type Curve Integral, Surface Integral, and Vector Field

where $\cos\alpha, \cos\beta, \cos\gamma$ are the directional cosines of the normal vector of Σ in the given direction.

When Σ is a planar region on the Oxy plane, the Stokes' formula is coincident with the Green's formula, so the Stokes' formula is a generalization of the Green's formula to the three dimensional space.

For convenient to memorize, we express the integrant in the equation (2) as

$$\begin{vmatrix} \cos\alpha & \cos\beta & \cos\gamma \\ \dfrac{\partial}{\partial x} & \dfrac{\partial}{\partial y} & \dfrac{\partial}{\partial z} \\ P & Q & R \end{vmatrix}$$

Example 1 Evaluate $I = \oint_\Gamma (y^2 - z^2)\,dx + (z^2 - x^2)\,dy + (x^2 - y^2)\,dz$, where the closed curve Γ is the boundary curve $ABCA$ of the triangle with $A(1,0,0), B(0,1,0), C(0,0,1)$ as vertices (Fig. 10.27).

Solution Let Σ be the triangle $ABC: x + y + z = 1$, $n = \{1,1,1\}$ be the given normal vector, its directional cosines are

$$\cos\alpha = \cos\beta = \cos\gamma = \frac{1}{\sqrt{3}}$$

Then by the Stokes' formula (2), we get

$$I = \iint_\Sigma \begin{vmatrix} \dfrac{1}{\sqrt{3}} & \dfrac{1}{\sqrt{3}} & \dfrac{1}{\sqrt{3}} \\ \dfrac{\partial}{\partial x} & \dfrac{\partial}{\partial y} & \dfrac{\partial}{\partial z} \\ y^2 - z^2 & z^2 - x^2 & x^2 - y^2 \end{vmatrix} dS$$

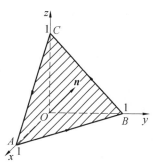

Fig. 10.27

$$= \frac{-4}{\sqrt{3}} \iint_\Sigma (x + y + z)\,dS$$

$$= \frac{-4}{\sqrt{3}} \iint_\Sigma dS = -2$$

10.6.2 The Circulation and Curl of Vector Fields

In the study of vector field, the integral of a vector on an oriented closed curve is very important. Suppose C is an oriented closed curve in the vector field, t^0 is a unit tangent vector homodromous with C, $\mathrm{d}\boldsymbol{s} = \boldsymbol{t}^0 \mathrm{d}s$.

In the force field $\boldsymbol{F}(M)$, the closed curve integral

$$\oint_C \boldsymbol{F} \cdot \mathrm{d}\boldsymbol{S} = \oint_C \boldsymbol{F} \cdot \boldsymbol{t}^0 \mathrm{d}S = \oint_C \mathrm{Prj}_{t^0} \boldsymbol{F} \mathrm{d}s$$

equals to the work done by the force $\boldsymbol{F}$ along the closed curve C.

In the velocity field $\boldsymbol{v}(M)$, the closed curve integral

$$\oint_C \boldsymbol{v} \cdot \mathrm{d}\boldsymbol{S} = \oint_C \boldsymbol{v} \cdot \boldsymbol{t}^0 \mathrm{d}S = \oint_C \mathrm{Prj}_{t^0} \boldsymbol{v} \mathrm{d}s$$

equals to the circumfluence along the closed curve C.

In a magnetic field with intensity $\boldsymbol{H}(M)$, according to the Ampere circuit law, the closed curve integral

$$\oint_C \boldsymbol{H} \cdot \mathrm{d}\boldsymbol{S} = \oint_C \boldsymbol{H} \cdot \boldsymbol{t}^0 \mathrm{d}S = \oint_C \mathrm{Prj}_{t^0} \boldsymbol{H} \mathrm{d}s$$

equals to the algebraic sum of the electric current passing through the surface enclosed by C.

We give the following definition for circulation.

Definition 10.5 In the vector field $\boldsymbol{F}(M)$, suppose C is an oriented closed curve, then we call the curve integral

$$\Gamma = \oint_C \boldsymbol{F} \cdot \mathrm{d}\boldsymbol{S} \tag{3}$$

the circulation of the vector field $\boldsymbol{F}(M)$ along the oriented closed curve C.

Under the rectangular coordinate system, if

$$\boldsymbol{F}(M) = \{P(x,y,z), Q(x,y,z), R(x,y,z)\}$$

then the circulation

$$\Gamma = \oint_C P(x,y,z) \mathrm{d}x + Q(x,y,z) \mathrm{d}y + R(x,y,z) \mathrm{d}z$$

Next, we will study the characteristics of the circulation. On the surface Σ expanded by C, use an arbitrary grid to divide Σ. Since for the second type curve integral, if we reverse the orientation of the integral path then the sign of the integral changes, so the circulation has the additivity property in the following sense (Fig. 10.28): The sum of the circulations of the vector field along the boundary of every small piece of surface (the orientation of the boundary and the orientation of the surface satisfy the right-hand screw rule) equals to the circulation of the vector field along the curve C. If we think of

Chapter 10 The Second Type Curve Integral, Surface Integral, and Vector Field

the circulation along the boundary of each small surface as the mass of the small surface, we give the following concept of the area density of circulation.

Definition 10.6 Suppose M is a point in the vector field $F(M)$, n is a given vector. Make an arbitrary non-closed smooth surface Σ which passed through M, such that n is a normal vector of it at the point M (Fig. 10.29). Suppose C is a close curve on Σ surround the point M, ΔS is the surface enclosed by it, the orientation of C and the orientation of ΔS satisfy the right-hand screw rule, if when C contract to the point M infinitely along Σ (denoted by $\Delta S \to M$), the limit

$$\lim_{\Delta S \to M} \frac{\oint_C F \cdot ds}{\Delta S}$$

exist, and is irrelevant to the choice of Σ and the contracting of C, then we call this limit the area density of the circulation of the vector field $F(M)$ at the point M along the direction of n.

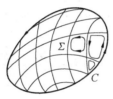

Fig. 10.28   Fig. 10.29

By Definition 10.6 we know that the area density of circulation not only depends on the vector field $F(M)$ but also depends on the surface Σ.

Under the rectangular coordinate system, suppose

$$F = \{P(x,y,z), Q(x,y,z), R(x,y,z)\}$$

then by the Stokes' formula, we have

$$\Gamma = \oint_C F \cdot ds = \oint_C P(x,y,z)\,dx + Q(x,y,z)\,dy + R(x,y,z)\,dz$$

$$= \iint_{\Delta S} \left(\frac{\partial R}{\partial y} - \frac{\partial Q}{\partial z}\right) dydz + \left(\frac{\partial P}{\partial z} - \frac{\partial R}{\partial x}\right) dzdx + \left(\frac{\partial Q}{\partial x} - \frac{\partial P}{\partial y}\right) dxdy$$

$$= \iint_{\Delta S} \left[\left(\frac{\partial R}{\partial y} - \frac{\partial Q}{\partial z}\right)\cos\alpha + \left(\frac{\partial P}{\partial z} - \frac{\partial R}{\partial x}\right)\cos\beta + \left(\frac{\partial Q}{\partial x} - \frac{\partial P}{\partial y}\right)\cos\gamma\right] dS$$

By the mean value theorem for integral, we get

$$\Gamma = \left[\left(\frac{\partial R}{\partial y} - \frac{\partial Q}{\partial z}\right)\cos\alpha + \left(\frac{\partial P}{\partial z} - \frac{\partial R}{\partial x}\right)\cos\beta + \left(\frac{\partial Q}{\partial x} - \frac{\partial P}{\partial y}\right)\cos\gamma\right]_{M^*}\Delta S$$

where M^* is a point on ΔS, when $\Delta S \to M$, $M^* \to M$, so

$$\lim_{\Delta S \to M}\frac{\oint_C \boldsymbol{F}\cdot\mathrm{d}\boldsymbol{s}}{\Delta S} = \left(\frac{\partial R}{\partial y} - \frac{\partial Q}{\partial z}\right)\cos\alpha + \left(\frac{\partial P}{\partial z} - \frac{\partial R}{\partial x}\right)\cos\beta + \left(\frac{\partial Q}{\partial x} - \frac{\partial P}{\partial y}\right)\cos\gamma$$

where $\cos\alpha, \cos\beta, \cos\gamma$ are the directional cosines of the given vector $\boldsymbol{n}$.

The area density of circulation is not only relevant to the position of the point M, but also to the direction of $\boldsymbol{n}$, and it equals to the dot product of the vectors

$$\left(\frac{\partial R}{\partial y} - \frac{\partial Q}{\partial z}\right)\boldsymbol{i} + \left(\frac{\partial P}{\partial z} - \frac{\partial R}{\partial x}\right)\boldsymbol{j} + \left(\frac{\partial Q}{\partial x} - \frac{\partial P}{\partial y}\right)\boldsymbol{k}$$

and

$$\boldsymbol{n}^0 = \cos\alpha\,\boldsymbol{i} + \cos\beta\,\boldsymbol{j} + \cos\gamma\,\boldsymbol{k}$$

Definition 10.7 The curl (or rotation number) of the vector field $\boldsymbol{F}(M)$ at the point M is a vector, whose direction is coincident with the direction of the maximum area density of circulation at the point M, and its size equals to this maximum value, denoted by rot $\boldsymbol{F}(M)$ and

$$\operatorname{rot}\boldsymbol{F}(M) = \left(\frac{\partial R}{\partial y} - \frac{\partial Q}{\partial z}\right)\boldsymbol{i} + \left(\frac{\partial P}{\partial z} - \frac{\partial R}{\partial x}\right)\boldsymbol{j} + \left(\frac{\partial Q}{\partial x} - \frac{\partial P}{\partial y}\right)\boldsymbol{k}$$

i. e.

$$\operatorname{rot}\boldsymbol{F}(M) = \begin{vmatrix} \boldsymbol{i} & \boldsymbol{j} & \boldsymbol{k} \\ \frac{\partial}{\partial x} & \frac{\partial}{\partial y} & \frac{\partial}{\partial z} \\ P & Q & R \end{vmatrix} = \nabla \times \boldsymbol{F} \qquad (4)$$

What worth noticing is that the circulation of the planar velocity field $\boldsymbol{v}(M) = \{P(x,y), Q(x,y)\}$ is

$$\Gamma = \oint_C P(x,y)\,\mathrm{d}x + Q(x,y)\,\mathrm{d}y$$

and its curl is the vector

$$\operatorname{rot}\boldsymbol{v}(M) = \left(\frac{\partial Q}{\partial x} - \frac{\partial P}{\partial y}\right)\boldsymbol{k}$$

The projection of the curl to any direction is exactly the area density of circulation in that direction.

By the definition of curl, the Stokes' formula can be expressed as the vector form

Chapter 10 The Second Type Curve Integral, Surface Integral, and Vector Field

$$\oint_C \mathbf{F} \cdot d\mathbf{s} = \iint_\Sigma \mathrm{rot}\mathbf{F} \cdot d\mathbf{S} = \iint_\Sigma \nabla \times \mathbf{F} \cdot d\mathbf{S}$$

this means that the circulation along the closed curve C equals to the flux of the curls of the points on any oriented surface expanded by C.

When P, Q, R have continuous second order partial derivatives, we can easily get
$$\mathrm{div}(\mathrm{rot}\ \mathbf{F}) = 0$$
this means that the curl field is source free, in other words, the surface Σ in the Stokes' formula could be any surface expanded by the oriented closed curve C.

If any point in a field has zero curl, then this field is called an irrotational field.

The curl has the following properties:
(i) $\mathrm{rot}(C_1\mathbf{F} + C_2\mathbf{G}) = C_1\mathrm{rot}\ \mathbf{F} + C_2\mathrm{rot}\ \mathbf{G}$ (C_1, C_2 are constants);
(ii) $\mathrm{rot}(u\mathbf{F}) = u\mathrm{rot}\ \mathbf{F} + \mathrm{grad}\ u \times \mathbf{F}$ (u is a number valued function);
(iii) $\mathrm{div}(\mathbf{F} \times \mathbf{G}) = \mathbf{G} \cdot \mathrm{rot}\ \mathbf{F} - \mathbf{F} \cdot \mathrm{rot}\ \mathbf{G}$;
(iv) $\mathrm{rot}(\mathrm{grad}\ u) = \mathbf{0}$ (the gradient field is irrotational).

Example 2 Evaluate the surface integral
$$\iint_\Sigma \mathrm{rot}\ \mathbf{F} \cdot d\mathbf{S}$$
where $\mathbf{F} = \{x - z, x^3 + yz, -3xy^2\}$, Σ is the cone $z = 2 - \sqrt{x^2 + y^2}$ ($z \geq 0$), oriented upwards.

Solution Since the curl field is source free, so the surface integral is only relevant to the boundary curve C of the surface, and is irrelevant to the surface Σ. Here the boundary is a planar curve $x^2 + y^2 = 2^2$ on the Oxy plane, so Σ can be replaced by a planar region $D: z = 0$, $x^2 + y^2 \leq 2^2$ expanded by C and oriented upwards, since

$$\mathrm{rot}\ \mathbf{F} = \begin{vmatrix} \mathbf{i} & \mathbf{j} & \mathbf{k} \\ \dfrac{\partial}{\partial x} & \dfrac{\partial}{\partial y} & \dfrac{\partial}{\partial z} \\ x - z & x^3 + yz & -3xy^2 \end{vmatrix}$$
$$= (-6xy - y)\mathbf{i} + (-1 + 3y^2)\mathbf{j} + 3x^2\mathbf{k}$$
$$d\mathbf{S} = dydz\mathbf{i} + dzdx\mathbf{j} + dxdy\mathbf{k} = dxdy\mathbf{k}$$
then
$$\iint_\Sigma \mathrm{rot}\ \mathbf{F} \cdot d\mathbf{S} = \iint_D \mathrm{rot}\ \mathbf{F} \cdot d\mathbf{S} \iint_D 3x^2 dxdy = 12\pi$$

We have the following theorem which is analogue to Theorem 10.2:

Theorem 10.5 Suppose the functions $P(x,y,z), Q(x,y,z), R(x,y,z)$ have continuous partial derivatives on the planar simply connected region G, then the following four statements are equivalent:

(i) For any closed curve C in G, the integral
$$\oint_C P\,dx + Q\,dy + R\,dz = 0$$

(ii) In the region G, the curve integral
$$\int_{AB} P\,dx + Q\,dy + R\,dz$$
is independent of path (but relevant to the initial point A and terminal point B).

(iii) In the region G, the expression $P\,dx + Q\,dy + R\,dz$ is the total differential of some function $u(x,y,z)$, i.e.
$$du = P\,dx + Q\,dy + R\,dz$$

(iv) At any point in G, we always have
$$\frac{\partial R}{\partial y} = \frac{\partial Q}{\partial z}, \frac{\partial P}{\partial z} = \frac{\partial R}{\partial x}, \frac{\partial Q}{\partial x} = \frac{\partial P}{\partial y} \tag{5}$$

This theorem points out: Equation (5) is not only the necessary and sufficient condition for curve integral being independent to path, but also the necessary and sufficient condition for the expression $P\,dx + Q\,dy + R\,dz$ being total differential of some function $u(x,y,z)$, and in this moment the original function
$$u(x,y,z) = \int_{(x_0,y_0,z_0)}^{(x,y,z)} P\,dx + Q\,dy + R\,dz + C$$
since the curve integral is independent of path, we usually take polygonal lines that parallel to the coordinate axes as the integral path if possible, and transform the integral into a definite integral, for example
$$u(x,y,z) = \int_{x_0}^{x} P(x,y_0,z_0)\,dx + \int_{y_0}^{y} Q(x,y,z_0)\,dy + \int_{z_0}^{z} R(x,y,z)\,dz + C \tag{6}$$
where (x_0,y_0,z_0) is any point in the region G.

For a vector field $F = \{P(x,y,z), Q(x,y,z), R(x,y,z)\}$, if it satisfies (i) and (ii) then it is called a conservative field; if it satisfies (iii) the it is called a potential field, in this moment we call $v = -u$ the potential function; if it satisfies (iv) then it is called an irrotational field.

A field is source free and irrotational is called a Harmonic field.

Chapter 10 The Second Type Curve Integral, Surface Integral, and Vector Field

Theorem 10.5 tells us, if a simply connected continuously differentiable vector field is a conservative field then it must be irrotational, and vice versa.

Example 3 Is the expression
$$2xyz^2 dx + [x^2z^2 + z\cos(yz)] dy + [2x^2yz + y\cos(yz)] dz$$
the total differential of some function? If it is, find this function.

Solution Since
$$\frac{\partial R}{\partial y} = 2x^2z + \cos(yz) - yz\sin(yz) = \frac{\partial Q}{\partial z}$$
$$\frac{\partial P}{\partial z} = 4xyz = \frac{\partial R}{\partial x}, \frac{\partial Q}{\partial x} = 2xz^2 = \frac{\partial P}{\partial y}$$
then the expression is the total differential of some function. Let $(x_0, y_0, z_0) = (0,0,0)$, then
$$u = \int_0^z [2x^2yz + y\cos(yz)] dz + C = x^2yz^2 + \sin(yz) + C$$
The function $u = x^2yz^2 + \sin(yz) + C$ is the original function.

10.7 Examples

Example 1 Calculate $\int_L \frac{dx + dy}{|x| + |y|}$, where L is the polygonal line ABC with $A(1, 0), B(0,1), C(-1,0)$.

Solution Divide L into two pieces
$$\overline{AB}: y = 1 - x (0 \leqslant x \leqslant 1, y \geqslant 0)$$
$$\overline{BC}: y = 1 + x (-1 \leqslant x \leqslant 0, y \geqslant 0)$$
thus
$$\int_L \frac{dx + dy}{|x| + |y|} = \int_{AB} \frac{dx + dy}{x + y} + \int_{BC} \frac{dx + dy}{-x + y}$$
$$= \int_1^0 \frac{dx - dx}{1} + \int_0^{-1} \frac{dx + dx}{1} = \int_0^{-1} 2dx = -2$$

Example 2 Calculate $\int_L \sqrt{x^2 + y^2} dx + y[xy + \ln(x + \sqrt{x^2 + y^2})] dy$, where L is a sine curve segment $y = \sin x (\pi \leqslant x \leqslant 2\pi)$ along the direction of increasing x.

Solution It is complicated to directly compute the integral. For convenience of calculation, substitute the integral path by the straight line segment $y = 0$ from the point $A(\pi,0)$ to the point $B(2\pi,0)$ by means of the Green's formula.

Since
$$\frac{\partial Q}{\partial x} = y^2 + \frac{y}{\sqrt{x^2 + y^2}}, \frac{\partial P}{\partial y} = \frac{y}{\sqrt{x^2 + y^2}}$$

we have
$$\int_L \sqrt{x^2 + y^2}\, dx + y[xy + \ln(x + \sqrt{x^2 + y^2})]\, dy$$
$$= \left(\int_{AB} + \oint_{L+BA}\right) \sqrt{x^2 + y^2}\, dx + y[xy + \ln(x + \sqrt{x^2 + y^2})]\, dy$$
$$= \int_\pi^{2\pi} x\, dx + \iint_\sigma y^2\, d\sigma = \frac{3\pi^2}{2} + \int_\pi^{2\pi} dx \int_{\sin x}^0 y^2\, dy$$
$$= \frac{3\pi^2}{2} + \frac{4}{9}$$

Example 3 Evaluate $I = \int_{\widehat{AmB}} (\varphi(y)\cos x - \pi y)\, dx + (\varphi'(y)\sin x - \pi)\, dy$, where $\varphi(y)$ has continuous derivative; $\widehat{AmB}$ is a curve from the point $A(\pi,2)$ to the point $B(3\pi,4)$ and lies below the straight line $\overline{AB}$, the area of the region bounded by $\widehat{AmB}$ and $\overline{AB}$ equals to 2 (as shown in Fig. 10.30).

Solution Here
$$\frac{\partial Q}{\partial x} - \frac{\partial P}{\partial y} = \pi$$

the curve integral is dependent of path. Note that the path of integration is not given specifically in this problem, and the integration expression contains a unknown function $\varphi(y)$, the curve integral cannot be converted into a definite integral. Although the curve integral is dependent of path, we can also use the Green's formula to transform the integral path. By

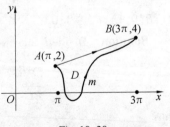

Fig. 10.30

$$\widehat{AmB} = \widehat{AmB} + \overline{BA} - \overline{BA}$$

denote the closed curve $\widehat{AmB} + \overline{BA}$ by C, and $\overline{AB}: x = \pi y - \pi, 2 \leq y \leq 4$, we have

Chapter 10 The Second Type Curve Integral, Surface Integral, and Vector Field

$$I = (\oint_C + \int_{\overline{AB}})(\varphi(y)\cos x - \pi y)dx + (\varphi'(y)\sin x - \pi)dy$$

$$= \iint_D \pi dxdy + \int_2^4 [(\varphi(y)\cos(\pi y - \pi) - \pi y)\pi + (\varphi'(y)\sin(\pi y - \pi) - \pi)]dy$$

$$= 2\pi - \int_2^4 (\pi^2 y + \pi)dy + \int_2^4 [\pi\varphi(y)\cos(\pi y - \pi) + \varphi'(y)\sin(\pi y - \pi)]dy$$

$$= -6\pi^2 + \varphi(y)\sin(\pi y - \pi)\big|_2^4 = -6\pi^2$$

Example 4 Suppose that the function of two variables $Q(x,y)$ has continuous first order partial derivatives in the Oxy plane, that the curve integral $\int_L 2xydx + Q(x,y)dy$ is independent of path, and that for any t the following equality always holds

$$\int_{(0,0)}^{(t,1)} 2xydx + Q(x,y)dy = \int_{(0,0)}^{(1,t)} 2xydx + Q(x,y)dy$$

Find the function $Q(x,y)$.

Solution By the condition that the curve integral is independent of path, we obtain

$$\frac{\partial Q}{\partial x} = \frac{\partial(2xy)}{\partial y} = 2x$$

Hence

$$Q(x,y) = x^2 + C(y)$$

where $C(y)$ is a undetermined function. Thus

$$\int_{(0,0)}^{(t,1)} 2xydx + Q(x,y)dy = \int_0^1 [t^2 + C(y)]dy = t^2 + \int_0^1 C(y)dy$$

$$\int_{(0,0)}^{(1,t)} 2xydx + Q(x,y)dy = \int_0^t [1^2 + C(y)]dy = t + \int_0^t C(y)dy$$

It follows from known conditions that

$$t^2 + \int_0^1 C(y)dy = t + \int_0^t C(y)dy$$

Differentiating both sides with respect to t gives

$$2t = 1 + C(t)$$

So $C(t) = 2t - 1$, i.e. $C(y) = 2y - 1$. Thus

$$Q(x,y) = x^2 + 2y - 1$$

Example 5 A particle moves from the origin to the point $M(\xi,\eta,\zeta)$ along a straight line in the force field $F = \{yz, zx, xy\}$, where the point $M(\xi,\eta,\zeta)$ is located on

Calculus(II)

the first octant region of the ellipsoidal surface $\dfrac{x^2}{a^2} + \dfrac{y^2}{b^2} + \dfrac{z^2}{c^2} = 1$. What are the values of ξ, η, ζ to maximize the work W done by the field force F? What is the maximum work?

Solution Since
$$\begin{vmatrix} i & j & k \\ \dfrac{\partial}{\partial x} & \dfrac{\partial}{\partial y} & \dfrac{\partial}{\partial z} \\ yz & zx & xy \end{vmatrix} = 0$$

F is an irrotational field and the curve integral is independent of path.

The parametric equation of the straight line segment $\overline{OM}$ is
$$x = \xi t, y = \eta t, z = \zeta t, 0 \leqslant t \leqslant 1$$
The work done by the field force in moving the particle from the origin to the point M along the line segment is
$$W = \int_{\overline{OM}} F \cdot ds = \int_{\overline{OM}} yz dx + zx dy + xy dz = \int_0^1 3\xi\eta\zeta t^2 dt = \xi\eta\zeta$$
The problem is converted into the conditional extreme value of $W = \xi\eta\zeta$ with constraint conditions $\dfrac{\xi^2}{a^2} + \dfrac{\eta^2}{b^2} + \dfrac{\zeta^2}{c^2} = 1$. Let
$$F(\xi,\eta,\zeta) = \xi\eta\zeta + \lambda\left(1 - \dfrac{\xi^2}{a^2} - \dfrac{\eta^2}{b^2} - \dfrac{\zeta^2}{c^2}\right)$$
By the system of equations
$$\begin{cases} F'_\xi = \eta\zeta - \dfrac{2\lambda}{a^2}\xi = 0 \\ F'_\eta = \xi\zeta - \dfrac{2\lambda}{b^2}\eta = 0 \\ F'_\zeta = \xi\eta - \dfrac{2\lambda}{c^2}\zeta = 0 \\ \dfrac{\xi^2}{a^2} + \dfrac{\eta^2}{b^2} + \dfrac{\zeta^2}{c^2} - 1 = 0 \end{cases}$$
solve to get
$$\xi = \dfrac{\sqrt{3}}{3}a, \eta = \dfrac{\sqrt{3}}{3}b, \zeta = \dfrac{\sqrt{3}}{3}c$$
Since the maximal value must exist in this actual problem, we obtain

Chapter 10 The Second Type Curve Integral, Surface Integral, and Vector Field

$$W_{max} = \left(\frac{\sqrt{3}}{3}\right)^3 abc = \frac{\sqrt{3}}{9}abc$$

Example 6 Suppose in the first octant for any oriented closed surface Σ the following equality always holds

$$\oiint_\Sigma 2yz\varphi'(x)\,dydz + y^2 z\varphi(x)\,dzdx - yz^2\,e^x dxdy = 0$$

where $\varphi(x) \in C^2, \varphi(0) = \dfrac{1}{2}, \varphi'(0) = 1$. Find $\varphi(x)$.

Solution By the given conditions and the Gauss' formula, for any bounded closed region V, we always have

$$\iiint_V 2yz(\varphi''(x) + \varphi(x) - e^x)\,dV = 0$$

Hence

$$\varphi''(x) + \varphi(x) = e^x$$

It's general solution is

$$\varphi(x) = C_1 \cos x + C_2 \sin x + \frac{1}{2}e^x$$

By the initial conditions $\varphi(0) = \dfrac{1}{2}$ and $\varphi'(0) = 1$, we decide $C_1 = 0, C_2 = \dfrac{1}{2}$, then we get

$$\varphi(x) = \frac{1}{2}(\sin x + e^x)$$

Example 7 Evaluate $J = \iint_\Sigma yx^3\,dydz + xy^3\,dzdx + zdxdy$, where Σ is the lower side of the part of the paraboloid of revolution $z = x^2 + y^2$ that lies within the cylindrical surface $|x| + |y| = 1$.

Solution Since the projection of Σ onto the Oyz plane is symmetric about the z-axis, we have

$$\iint_\Sigma yx^3\,dydz = \iint_{\Sigma(front)} yx^3\,dydz + \iint_{\Sigma(back)} yx^3\,dydz$$

$$= \iint_{\sigma_{yz}} y(\sqrt{z-y^2})^3\,d\sigma - \iint_{\sigma_{yz}} y(-\sqrt{z-y^2})^3\,d\sigma$$

$$= 2\iint_{\sigma_{yz}} y(z-y^2)^{3/2}\,d\sigma = 0$$

The last step is obtained by using the symmetry of double integral (as shown in

Fig. 10.31). And then by the rotation properties of variables x and y, we get

$$\iint_\Sigma xy^3 \, dz\,dx = 0$$

Thus

$$J = 0 + 0 + \iint_\Sigma z\,dx\,dy$$

$$= -\iint_{\sigma_{xy}} (x^2 + y^2)\,dx\,dy$$

$$= -2\iint_{\sigma_{xy}} x^2\,dx\,dy$$

$$= -8\int_0^1 dx \int_0^{1-x} x^2\,dy = -\frac{2}{3}$$

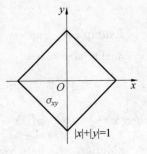

Fig. 10.31

Example 8 Use the Gauss' formula to calculate the triple integral

$$I = \iiint_V (xy + yz + zx)\,dV$$

where V is the first octant part of the solid bounded by the planes $x = 0, y = 0, z = 0, z = 1$ and the cylindrical surface $x^2 + y^2 = 1$.

Solution Taking into account the boundary surface of V (as shown in Fig. 10.32), let

$$P = Q = 0, R = xyz + \frac{1}{2}yz^2 + \frac{1}{2}xz^2$$

then

$$\frac{\partial R}{\partial z} = xy + yz + zx$$

By Gauss' formula

$$I = \iiint_V (xy + yz + zx)\,dV$$

$$= \oiint_{\Sigma\text{outside}} \left[xyz + \frac{1}{2}(x + y) z^2 \right] dx\,dy$$

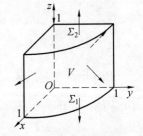

Fig. 10.32

Here Σ consists of three surfaces: the lateral surface of V (i.e. the cylindrical surface whose generatrix is parallel to the z-axis), the bottom side $\Sigma_1 : z = 0$, and the top side $\Sigma_2 : z = 1$.

Hence

Chapter 10 The Second Type Curve Integral, Surface Integral, and Vector Field

$$I = \left(\iint_{\Sigma(\text{upper})} + \iint_{\Sigma(\text{lower})} \right) [xyz + \frac{1}{2}(x+y)z^2] dxdy$$

$$= 0 + \iint_{\sigma_{xy}} [xy + \frac{1}{2}(x+y)] dxdy$$

$$= \int_0^{\pi/2} d\theta \int_0^1 [r^2 \sin\theta\cos\theta + \frac{1}{2}r(\sin\theta + \cos\theta)] r dr = \frac{11}{24}$$

Exercises 10

10.1

1. Evaluate $\oint_l x dy$, where l is the counterclockwise triangle loop enclosed by the coordinate axes and the straight line $\frac{x}{2} + \frac{y}{3} = 1$.

2. Evaluate $\int_l (x^2 - 2xy) dx + (y^2 - 2xy) dy$, where l is the arc of the parabola $y = x^2$ from $x = -1$ to $x = 1$.

3. Evaluate $\int_l (2a - y) dx - (a - y) dy$, where l is one arch of the cycloid $x = a(t - \sin t), y = a(1 - \cos t), 0 \leq t \leq 2\pi$.

4. Evaluate $\oint_l \frac{(x+y)dx - (x-y)dy}{x^2 + y^2}$, where l is the circle $x^2 + y^2 = a^2$ in the clockwise direction.

5. Evaluate $\int_l (x^2 + y^2) dx + (x^2 - y^2) dy$, where l is the segment of the curve $y = 1 - |1 - x|$ as x varies from 0 to 2.

6. Evaluate $\int_\Gamma y dx + z dy + x dz$, where Γ is the segment of the spiral $x = a\cos t, y = a\sin t, z = bt$ from $t = 0$ to $t = 2\pi$.

7. Evaluate $\int_\Gamma x dx + y dy + (x + y - 1) dz$, where Γ is the straight line segment from the point $(1,1,1)$ to the point $(4,7,10)$.

8. Evaluate $\int_l 2xe^{xy} dx + ye^{xy} dy$, where l is the arc segment of the ellipse $x^2 + \frac{y^2}{2} =$

Calculus(Ⅱ)

1 from the point $A(1,0)$ to the point $B(0,\sqrt{2})$ in the counterclockwise direction.

9. Evaluate $\oint_{\Gamma}(y^2+z^2)dx+(z^2+x^2)dy+(x^2+y^2)dz$, where Γ is the curve
$$\begin{cases} x^2+y^2+z^2=4x(z\geq 0) \\ x^2+y^2=2x \end{cases}$$
oriented counterclockwisely if saw from the positive direction of the z-axis.

10. Suppose that the equation of $\widehat{AB}$ in the polar coordinate system is $r=f(\theta)$, where $f(\theta)$ has continuous derivative in $[0,2\pi]$, the point A corresponds to $\theta=\alpha$, and the point $B(0\leq\alpha\leq\beta\leq 2\pi)$ corresponds to $\theta=\beta$. Show that
$$\int_{\widehat{AB}}-ydx+xdy=\int_{\alpha}^{\beta}f^2(\theta)d\theta$$

11. Suppose $\widehat{MEN}$ is the arc segment from the point $M(0,-1)$ via the point $E(1,0)$ to the point $N(0,1)$ along the right half circle $x=\sqrt{1-y^2}$. Evaluate
$$\int_{\widehat{MEN}}|y|dx+y^3dy$$

12. Let $F(M)$ be a force field in the Oxy plane, it's direction point towards the origin, and the size of $F(M)$ equals to the distance between the point M and the origin.

(1) Find the work done by the force field in moving the particle counterclockwise along the ellipse $\frac{x^2}{a^2}+\frac{y^2}{b^2}=1$ from the point $A(a,0)$ to the point $B(0,b)$.

(2) Find the work done by the force field in moving the particle counterclockwise once around the ellipse $\frac{x^2}{a^2}+\frac{y^2}{b^2}=1$.

13. Suppose that Γ is a smooth curve segment with arc length s, that the functions $P(x,y,z)$, $Q(x,y,z)$, $R(x,y,z)$ are continuous in Γ, and that $M=\max_{\Gamma}\{\sqrt{P^2+Q^2+R^2}\}$. Prove that
$$\left|\int_{\Gamma}Pdx+Qdy+Rdz\right|\leq Ms$$

10.2

1. Use the curve integral to compute the area of the figure enclosed by the astroid $x=a\cos^3 t, y=a\sin^3 t$.

Chapter 10 The Second Type Curve Integral, Surface Integral, and Vector Field

2. Evaluate $\oint_C x^2 dx + x e^{y^2} dy$, where C is the boundary with positive direction of the triangular region enclosed by the lines $y = x - 1, y = 1$ and $x = 1$.

3. Let C be a simple closed curve with clockwise direction in the xOy plane and $\oint_C (x - 2y) dx + (4x + 3y) dy = -9$. Find the area of the region enclosed by the curve C.

4. Calculate $\oint_C e^x [(1 - \cos y) dx - (y - \sin y) dy]$, where C is the positively oriented closed boundary of the region $0 < x < \pi, 0 < y < \sin x$.

5. Evaluate $\oint_C (x^3 - x^2 y) dx + (x y^2 - y^3) dy$, where C is the circle $x^2 + y^2 = a^2$ ($a > 0$) in the clockwise direction.

6. Evaluate $\oint_C y(2x - 1) dx - x(x + 1) dy$, where C is the positively oriented ellipse $b^2 x^2 + a^2 y^2 = a^2 b^2$.

7. Calculate $\oint_C \dfrac{y x^2 dx - x y^2 dy}{1 + \sqrt{x^2 + y^2}}$, where C is the closed curve oriented clockwisely consist of the curve $l_1 : y = -\sqrt{1 - x^2}$ and the straight line $l_2 : y = 0 (-1 \leq x \leq 1)$.

8. Evaluate $\int_l (x + y)^2 dx + (x + y^2 \sin y) dy$, where l is the arc segment of a curve from the point $A(1,1)$ to the point $B(-1,1)$.

9. Evaluate $\int_l \sqrt{x^2 + y^2} dx + [x + y \ln(x + \sqrt{x^2 + y^2})] dy$, where l is the arc segment along the upper half circle $y = 1 + \sqrt{1 - (x - 1)^2}$ from the point $B(2,1)$ to the point $A(0,1)$.

10. Calculate $\int_l (3xy + \sin x) dx + (x^2 - y e^y) dy$, where l is the segment along the parabola $y = x^2 - 2x$ from the point $(0,0)$ to the point $(4,8)$.

11. Evaluate the curve integral
$$I = \int_l [u'_x(x,y) + xy] dx + u'_y(x,y) dy$$
where l is the segment of the curve $y = \dfrac{\sin x}{x}$ from the point $A(0,1)$ to the point $B(\pi, 0)$, and $u(x,y)$ has second order continuous partial derivatives with $u(0,1) = 1$,

$u(\pi, 0) = \pi$.

12. Suppose that $u = u(x,y)$, $v = v(x,y)$, $w = w(x,y)$ have continuous first order partial derivatives in the bounded closed region D, and that C is the boundary of the region D. Verify that

$$\iint_D (u\frac{\partial w}{\partial x} + v\frac{\partial w}{\partial y}) dxdy = \oint_{C^+} w(udy - vdx) - \iint_D (\frac{\partial u}{\partial x} + \frac{\partial v}{\partial y}) w dxdy.$$

10.3

1. Prove that the curve integral $\int_l (\cos y dx - \sin y dy)$ only depends on the initial point and the terminal point and is independent of the path. Then calculate

$$\int_{(0,0)}^{(a,b)} e^x(\cos y dx - \sin y dy)$$

2. Prove that the curve integral $\int_l \frac{ydx - xdy}{x^2}$ only depends on the initial point and the terminal point and is independent of the path, where l is an arbitrary curve not intersecting with y-axis. And determine

$$\int_{(2,1)}^{(1,2)} \frac{ydx - xdy}{x^2}$$

3. Evaluate $\int_l \frac{1}{x}\sin\left(xy - \frac{\pi}{4}\right) dx + \frac{1}{y}\sin\left(xy - \frac{\pi}{4}\right) dy$, where l is the straight line segment from the point $A(1,\pi)$ to the point $B(\frac{\pi}{2}, 2)$.

4. Evaluate $\int_l (x^2 + 1 - e^y \sin x) dy - e^y \cos x dx$, where l is the curve segment along $y = x^2$ from the point $O(0,0)$ to the point $A(1,1)$.

5. Suppose $f(x)$ has second order continuous derivative with $f(0) = 0, f'(0) = 1$, and the curve integral

$$\int_l [f'(x) + 6f(x) + 4e^{-x}] y dx + f'(x) dy$$

is independent of path. Compute

$$\int_{(0,0)}^{(1,1)} [f'(x) + 6f(x) + 4e^{-x}] y dx + f'(x) dy$$

6. Let the function $f(x)$ be differentiable with $f(1) = 1$. Determine $f(x)$ so that the curve integral

Chapter 10 The Second Type Curve Integral, Surface Integral, and Vector Field

$$\int_{\widehat{AB}}[\sin x - f(x)]\frac{y}{x}dx + [f(x) - x^2]dy$$

is independent of path. (Here $\widehat{AB}$ is not intersecting the y-axis), and determine the integral value from the point $A\left(-\frac{3\pi}{2},\pi\right)$ to the point $B\left(-\frac{\pi}{2},0\right)$.

7. Suppose that the curve integral $\int_l F(x,y)(ydx + xdy)$ is independent of path, that $F(x,y)$ is differentiable, and that the figure of the implicit function determined by the equation $F(x,y)=0$ passes through the point $(1,2)$. Find the function $y=f(x)$ defined by $F(x,y)=0$.

8. Let $f(x),g(x) \in C(-\infty, +\infty)$, and the curve integral

$$\int_l g(x)ydx + f(x)dy$$

is independent of path. Try to prove that

$$f(x) = f(0) + \int_0^x g(t)dt$$

9. Evaluate the closed curve integral $\oint_C \frac{-ydx + xdy}{x^2+y^2}$, where C is the ellipse $\frac{x^2}{a^2} + \frac{y^2}{b^2} = 1$ in the counterclockwise direction.

10. Let C be any closed plane non self-intersection curve. Find the value of the constant a such that the curve integral

$$\oint_C \frac{xdx - aydy}{x^2+y^2} = 0$$

where C is a closed curve not passing through the origin $(0,0)$.

11. Let $\boldsymbol{F} = (2xy^3 - y^2\cos x)\boldsymbol{i} + (1 - 2y\sin x + 3x^2y^2)\boldsymbol{j}$ be a planar force field. Find the work done by $\boldsymbol{F}$ in moving a particle from the point $O(0,0)$ to the point $A\left(\frac{\pi}{2},1\right)$ along the curve $l:2x = \pi y^2$.

12. Suppose that the gravitational force F between the particle A and the particle M is $\frac{k}{r^2}$ (k is a constant), and that r is the distance between A and M. Now let A be fixed at the point $(0,1)$. Find the work done by the gravitational force F in moving M along the curve $y = \sqrt{2x - x^2}$ from the point $(0,0)$ to the point $(2,0)$.

13. Check that the expression
$$\frac{y\,dx}{3x^2 - 2xy + 3y^2} - \frac{x\,dy}{3x^2 - 2xy + 3y^2}$$
is the total differential of a function $u(x,y)$ in an arbitrary simple connected region not containing the origin. Find the function $u(x,y)$ in the region $x > 0$.

14. Check that the expression $(2x\cos y - y^2\sin x)\,dx + (2y\cos x - x^2\sin y)\,dy$ is the total differential of a two-variable function $u(x,y)$. Find the function $u(x,y)$ and compute the curve integral
$$\int_{(0,0)}^{(\frac{\pi}{2},\pi)} (2x\cos y - y^2\sin x)\,dx + (2y\cos x - x^2\sin y)\,dy$$

15. For what value of a is the expression
$$\frac{(x + ay)\,dx + y\,dy}{(x + y)^2}$$
a total differential of some function?

16. Determine the constant λ such that the vector
$$\mathbf{F}(x,y) = 2xy\,(x^4 + y^2)^\lambda \mathbf{i} - x^2\,(x^4 + y^2)^\lambda \mathbf{j}$$
in the right half plane $x > 0$ is the gradient of the function $u(x,y)$ of two variables. And find the function $u(x,y)$.

17. Suppose that two partial increments of the function $z = f(x,y)$ at an arbitrary point (x,y) are as follows
$$\Delta_x z = (2 + 3x^2 y^2)\Delta x + 3xy^2 \Delta x y^2 + y^2 \Delta x^3$$
$$\Delta_y z = 2x^2 y \Delta y + x^3 \Delta y^2$$
and $f(0,0) = 1$. Find the function $f(x,y)$.

18. Check that the following equations are total differential equations, and find the general solution.
   (1) $(3x^2 + 6y^2 x)\,dx + (6x^2 y + 4y^3)\,dy = 0$;
   (2) $[\cos(x + y^2) + 3y]\,dx + [2y\cos(x + y^2) + 3x]\,dy = 0$;
   (3) $(x\cos y + \cos x)y' - y\sin x + \sin y = 0$.

19. Suppose that $f(x)$ has second order continuous derivative with $f(0) = 0$, $f'(0) = 1$, and that
$$[xy(x + y) - f(x)y]\,dx + [f'(x) + x^2 y]\,dy = 0$$
is a total differential equation. Find $f(x)$, and find the general solution of the total dif-

Chapter 10 The Second Type Curve Integral, Surface Integral, and Vector Field

ferential equation.

20. Prove that the method of separation of variables for solving first order differential equations, essentially is a transformation from an equation to a total differential equation by multiplying the integral factor on both sides of the equation.

21. Let $F = \{2x\cos y - y^2\sin x, 2y\cos x - x^2\sin y\}$ be a planar vector field.

(1) Prove that F is a conservative field;

(2) Find the potential function;

(3) Evaluate the curve integral from the point $(-\pi, \pi)$ to the point $\left(3\pi, \dfrac{\pi}{2}\right)$.

10.4

1. Evaluate the surface integral $\iint_\Sigma (z-1)\,dxdy$, where Σ is the inside of the sphere $x^2 + y^2 + z^2 = 1$ located in the first octant.

2. Evaluate $\iint_S xyz\,dxdy$, where S is the outside of the part of the sphere $x^2 + y^2 + z^2 = 1$ that lies in $x \geqslant 0, y \geqslant 0$.

3. Evaluate $\iint_S x\,dydz + y\,dzdx + z\,dxdy$, where S is the upper side of the part of the paraboloid of revolution $z = x^2 + y^2$ contained in $z \leqslant 1$.

4. Evaluate $\oiint_S (x+y+z)\,dxdy - (y-z)\,dydz$, where S is the outside of the surface of the cube enclosed by the three coordinate plane and the planes $x = 1, y = 1, z = 1$.

5. Let $v = xi + yj + zk$ be a flow velocity field.

(1) Find the net flux I_1 across the conical surface $\Sigma_1 : x^2 + y^2 = z^2 (0 \leqslant z \leqslant h)$ to the lower side.

(2) Find the net flux I_2 across the plane $\Sigma_2 : z = h(x^2 + y^2 \leqslant h)$ to the upper side.

6. Evaluate $\oiint_S \dfrac{x\,dydz + z^2\,dxdy}{x^2 + y^2 + z^2}$, where S is the outside of the boundary surface of the solid enclosed by the cylindrical surface $x^2 + y^2 = R^2$ and two planes $z = R, z = -R$ $(R > 0)$.

7. Evaluate $\iint_S F \cdot dS$, where $F = \dfrac{xi + yj + zk}{\sqrt{x^2 + y^2 + z^2}}$, and S is the lower side of the upper hemisphere $z = \sqrt{R^2 - x^2 - y^2}$.

Calculus(II)

8. Suppose σ_{xy} is the projection of the surface S onto the Oxy plane. Determine whether
$$\iint_S f(x,y)\,dxdy = \iint_{\sigma_{xy}} f(x,y)\,dxdy$$
holds and state the reason.

10.5

1. Show that the volume of the solid enclosed by a closed surface S is given by
$$V = \frac{1}{3}\oiint_S [x\cos\alpha + y\cos\beta + z\cos\gamma]\,dS$$
where $\cos\alpha, \cos\beta, \cos\gamma$ are the directional cosines of the outward normal vector to the surface S.

2. Evaluate $\oiint_\Sigma xz^2\,dydz + yx^2\,dzdx + zy^2\,dzdy$, where Σ is the outside of the sphere $x^2 + y^2 + z^2 = a^2$.

3. Evaluate $\oiint_\Sigma xz\,dydz + x^2y\,dzdx + y^2z\,dxdy$, where Σ is the outside of the surface of the first octant part of the solid enclosed by the paraboloid of revolution $z = x^2 + y^2$, the cylindrical surface $x^2 + y^2 = 1$ and the coordinate planes.

4. Suppose that the function $f(u)$ has continuous derivative. Compute the surface integral
$$\oiint_\Sigma \frac{x}{y}f(\frac{x}{y})\,dydz + f(\frac{x}{y})\,dydx + [z - \frac{z}{y}f(\frac{x}{y})]\,dxdy$$
where Σ is the outside of the surface of the solid enclosed by $y = x^2 + z^2 + 1$ and $y = 9 - x^2 - z^2$.

5. Evaluate $\iint_S x^3\,dydz + y^3\,dzdx + z^3\,dxdy$, where S is the lower side of the surface $z = \sqrt{x^2 + y^2}\,(0 \leq z \leq h)$.

6. Evaluate $\iint_S (8y + 1)x\,dydz + 2(1 - y)^2\,dzdx - 4yz\,dxdy$, where S is the lateral surface of the solid of revolution generated by rotating the curve $\begin{cases} z = \sqrt{y - 1} \\ x = 0 \end{cases} (1 \leq y \leq 3)$ about the y-axis, and the included angle between its normal vector and the positive direction of y-axis is always greater than $\frac{\pi}{2}$.

Chapter 10 The Second Type Curve Integral, Surface Integral, and Vector Field

7. Evaluate $\iint_S x^2 dydz + y^2 dzdx + 2czdxdy$, where S is the lower side of the upper hemisphere $z = \sqrt{R^2 - (x-a)^2 - (y-b)^2}$.

8. Suppose that $V = \{(x,y,z) \mid -\sqrt{2ax - x^2 - y^2} \leq z \leq 0\}$ and S is the outside of the boundary surface of V. Evaluate

$$\oiint_S \frac{axdydz + 2(x+a)ydzdx}{\sqrt{(x-a)^2 + y^2 + z^2}}$$

9. Suppose the space region Ω is enclosed by the surface $z = a^2 - x^2 - y^2$ and the plane $z = 0$, that S is the outside of the boundary surface of Ω, and that V is the volume of Ω. Show that

$$\oiint_S x^2 y z^2 dydz - x y^2 z^2 dzdx + z(1 + xyz) dxdy = V$$

10. Suppose the bounded closed space region V is symmetric about the plane $x = 0$ and the plane $y = x$, and S is the outside of the boundary surface of V. Prove that the volume of V is given by

$$\oiint_S f(x) yzdydz - xf(y) z^2 dzdx + z(1 + xyf(z)) dxdy$$

11. Let $F = \dfrac{2^y}{\sqrt{x^2 + z^2}} \boldsymbol{j}$. Compute

$$\oiint_S \boldsymbol{F} \cdot d\boldsymbol{S}$$

where S is the outside of the boundary surface of the solid enclosed by the surface $y = \sqrt{x^2 + z^2}$ and the planes $y = 1, y = 2$.

12. Suppose Σ is the outside of the boundary surface of the solid bounded by the surface $z = x^2 + y^2$ and the plane $z = 1$. Find the outward flux Φ of the vector field $\boldsymbol{A} = x^2\boldsymbol{i} + y^2\boldsymbol{j} + z^2\boldsymbol{k}$ across Σ.

13. Suppose there is a vector field

$$\boldsymbol{F} = \frac{1}{\sqrt{x^2 + y^2 + 4z^2 + 3}} (xy^2\boldsymbol{i} + yz^2\boldsymbol{j} + zx^2\boldsymbol{k})$$

Find the outward flux Φ of $\boldsymbol{F}$ across the ellipsoid $x^2 + y^2 + 4z^2 = 1$.

14. Evaluate $\iint_S \dfrac{\boldsymbol{r} \cdot d\boldsymbol{S}}{r^3}$, where $\boldsymbol{r} = x\boldsymbol{i} + y\boldsymbol{j} + z\boldsymbol{k}, r = |\boldsymbol{r}|$.

Calculus(Ⅱ)

(1) S is the outside of an arbitrary simple closed surface neither passing through nor enclosing the origin;

(2) S is the outside of an arbitrary simple closed surface containing the origin.

15. Find the divergence of the vector field A at the given point M.

(1) $A = x^3 i + y^3 j + z^3 k, M(1,0,-1)$;

(2) $A = 4xi + 2xyj - 2k, M(7,3,0)$;

(3) $A = xyzr, r = xi + yj + zk, M(1,2,3)$.

16. Let $r = xi + yj + zk, r = |r|$.

(1) Find $f(r)$ such that $\text{div}[f(r)r] = 0$

(2) Find $f(r)$ such that $\text{div}[\text{grad} f(r)] = 0$.

10.6

1. Evaluate the integral on the closed space curve C

$$\oint_C (y-z)dx + (z-x)dy + (x-y)dz$$

where C is the curve of intersection of the cylindrical surface $x^2 + y^2 = a^2$ and the plane $\dfrac{x}{a} + \dfrac{z}{h} = 1 (a > 0, h > 0)$, and the direction of C is counterclockwise looking from the positive direction of the z-axis.

2. Suppose that C is the curve of intersection of the sphere $x^2 + y^2 + z^2 = 2Rx$ and the cylindrical surface $x^2 + y^2 = 2rx (0 < r < R, z \geq 0)$, and the direction of C is clockwise looking from the positive direction of the z-axis. Evaluate

$$\oint_C (y^2 + z^2)dx + (z^2 + x^2)dy + (x^2 + y^2)dz$$

3. Prove that the curve integral

$$\int_\Gamma yzdx + zxdy + xydz$$

is independent of path (only depends on the initial point and the terminal point). Evaluate the integral from the point $A(1,1,0)$ to the point $B(1,1,1)$.

4. Compute the curve integral

$$\int_{\widehat{AmB}} (x^2 - yz)dx + (y^2 - xz)dy + (z^2 - xy)dz$$

where $\widehat{AmB}$ is the curve segment of the spiral $x = a\cos\theta, y = a\sin\theta, z = \dfrac{h}{2\pi}\theta$ from the

Chapter 10 The Second Type Curve Integral, Surface Integral, and Vector Field

point $A(a,0,0)$ to the point $B(a,0,h)$.

5. Suppose that the components of the vector $A(M)$ have continuous second-order partial derivatives. Show that in the vector field A for any piecewise smooth closed surface Σ, we have

$$\oint_{\Sigma} \text{rot } A \cdot dS = 0$$

6. Let Σ be the upper side of the upper half sphere $x^2 + y^2 + z^2 = 9$, C be the boundary of Σ, and $A = \{2y, 3x, -z^2\}$. Try to compute $\iint_{\Sigma} \text{rot } A \cdot dS$ by the following given methods:

(1) by means of the first type surface integral;
(2) by means of the second type surface integral;
(3) by means of the Gauss formula;
(4) by means of the Stokes' formula.

7. Suppose that there is a planar flow velocity field

$$v(x,y) = [e^x(y^3 - 2y) - y^2]i + [e^x(3y^2 - 2) - x]j$$

(1) Find the curl at each point in the field;
(2) Find the circulation along the ellipse $C: 4(x-3)^2 + 9y^2 = 36$ in the counterclockwise direction.

8. Find the circulation of the vector field $A = -yi + xj + ak$ (a is a constant) along the closed curve C.

(1) C is the circle $x^2 + y^2 = 1, z = 0$ in the counterclockwise direction;
(2) C is the circle $(x+2)^2 + y^2 = 1, z = 0$ in the counterclockwise direction.

9. Evaluate the curl of each of the following vector fields:

(1) $A = yi + zj + xk$;   (2) $A = x^2i + y^2j + z^2k$;
(3) $A = yzi + zxj + xyk$;   (4) $A = (y^2 + z^2)i + (z^2 + x^2)j + (x^2 + y^2)k$;
(5) $A = xyz(i + j + k)$;   (6) $A = P(x)i + Q(y)j + R(z)k$.

10. Prove that the vector field $A = (y\cos(xy))i + (x\cos(xy))j + (\sin z)k$ is a conservative field, and find the potential function.

11. Suppose that the function $Q(x,y,z)$ has continuous first-order partial derivatives with $Q(0,y,0) = 0$, and that the expression

$$axzdx + Q(x,y,z)dy + (x^2 + 2y^2z - 1)dz$$

Calculus(II)

is the total differential of a function $u(x,y,z)$. Determine the constant a, and the functions Q and u.

10.7

1. Find a curve L among the curve family $y = a\sin x (a > 0)$ passing through two points $O(0,0)$ and $A(\pi,0)$ such that the integral $\int_L (1 + y^3) dx + (2x + y) dy$ along L from the point O to the point A is minimized.

2. A particle M moves from the point $A(1,2)$ to the point $B(3,4)$ along a right half circle with diameter AB. During the process, a varying force $\boldsymbol{F}$ is acting on M. The size of $\boldsymbol{F}$ equals the distance between the point M and the origin O, and its direction is perpendicular to the line segment OM. Moreover, the included angle between the force $\boldsymbol{F}$ and the positive direction of y-axis is less than $\pi/2$. Find the work done by $\boldsymbol{F}$ on the particle M.

3. Compute the plane curve integral
$$\int_l \frac{(x-y)dx + (x+y)dy}{x^2 + y^2}$$
where l is the arc segment of the cycloid $x = t - \sin t - \pi, y = 1 - \cos t$ from $t = 0$ to $t = 2\pi$.

4. Determine the value of the parameter t such that the curve integral
$$I = \int_l \frac{x(x^2+y^2)^t}{y} dx - \frac{x^2(x^2+y^2)^t}{y^2} dy$$
is independent of path in the region not containing the straight line $y = 0$, and evaluate the integral I from the point $A(1,1)$ to the point $B(0,2)$.

5. Suppose that the curve integral
$$\int_l f(x,y)(ydx + xdy)$$
is independent of path in the first quadrant, where $f(x,y)$ has continuous first-order partial derivative, $f'_y(x,y) \neq 0$, and $f(1,2) = 0$. Find the function $y = y(x)$ determined by the equation $f(x,y) = 0$.

6. Evaluate the surface integral
$$\iint_\Sigma (z^2 x + y e^z) dydz + x^2 y dzdx + (\sin^3 x + y^2 z) dxdy$$
where Σ is the upper side of the lower hemisphere $z = -\sqrt{R^2 - x^2 - y^2}$.

Chapter 10 The Second Type Curve Integral, Surface Integral, and Vector Field

7. Evaluate the surface integral
$$\oiint_S (2x - 2x^3 - e^{-\pi})\,dydz + (zy^2 + 6x^2y + z^2x)\,dzdx - z^2y\,dxdy$$
where S is the outside of the boundary surface of the solid enclosed by the paraboloid $z = 4 - x^2 - y^2$, coordinate planes Oxz, Oyz and the planes $z = \frac{1}{2}y, x = 1, y = 1$.

8. Try to transform the surface integral
$$\oiint_S \frac{x\cos\alpha + y\cos\beta + z\cos\gamma}{\sqrt{x^2 + y^2 + z^2}}\,dS$$
into a triple integral, where $\cos\alpha, \cos\beta, \cos\gamma$ are the directional cosines of the inner normal vector to the surface S.

9. Let $\boldsymbol{A} = (x^3 - y^2)\boldsymbol{i} + (y^3 - z^2)\boldsymbol{j} + (z^3 - x^2)\boldsymbol{k}$ be a vector field. Evaluate its divergence, curl, and find the outward flux Φ of $\boldsymbol{A}$ across the surface S, where S is the closed surface enclosed by the hemisphere $y = R + \sqrt{R^2 - x^2 - z^2}$ ($R > 0$) and the conical surface $y = \sqrt{x^2 + z^2}$. And Γ is the circulation of $\boldsymbol{A}$ along the curve C. Here C is the curve of intersection between the cylindrical surface $x^2 + y^2 = Rx$ and the sphere $z = \sqrt{R^2 - x^2 - y^2}$, oriented counterclockwisely if saw from the positive direction of the z-axis.

10. Suppose that $u = u(x,y), v = v(x,y)$ have continuous partial derivatives, C is the positively oriented boundary of the plane region C. Show that the following double integral satisfies the formula of integration by parts
$$\iint_D u\frac{\partial v}{\partial x}\,dxdy = \oint_C uv\cos(\boldsymbol{n},x)\,ds - \iint_D v\frac{\partial u}{\partial x}\,dxdy$$
where $\boldsymbol{n}$ is the outer normal vector to the curve C.

11. Suppose that $u = u(x,y,z)$ has continuous second order partial derivatives. Show that
$$\oiint_S \frac{\partial u}{\partial n}\,dS = \iiint_V (u''_{xx} + u''_{yy} + u''_{zz})\,dV$$
where S is the boundary surface of V, and $\boldsymbol{n}$ is the outer normal vector to S.

12. Suppose that S is a smooth simple closed surface, that $u = u(x,y,z)$ has continuous first order partial derivatives in the space closed region V enclosed by S, and that $v = v(x,y,z)$ has continuous second order partial derivatives and satisfies the Laplace e-

quation

$$\frac{\partial^2 v}{\partial x^2} + \frac{\partial^2 v}{\partial y^2} + \frac{\partial^2 v}{\partial z^2} = 0$$

Where n is the outer normal vector to the surface S at the point (x,y,z). Prove that

$$\oiint_S u \frac{\partial v}{\partial n} \mathrm{d}S = \iiint_v (\mathrm{grad}\ u \cdot \mathrm{grad}\ v)\,\mathrm{d}x\mathrm{d}y\mathrm{d}z$$

Chapter 11 Infinite Series

The infinite series theory is an important part of the higher mathematics. Their importance in calculus stems from Newton's idea of representing functions as sums of infinite series. For instance, in finding areas he often integrated a function by first expressing it as a series and then integrating each term of the series. Many of the functions that arise in mathematical physics and chemistry, such as Bessel functions, are defined as sums of series, so it is important to be familiar with the basic concepts of convergence of infinite sequences and series.

Physicists also use series in another way, as we will see in Section 11.8. In studying fields as diverse as optics, special relativity, and electromagnetism, they analyze phenomena by replacing a function by the first few terms in the series that represents it.

Definition 11.1 The expression of the sum of the terms of an infinite sequence $u_1, u_2, \cdots, u_n, \cdots$

$$u_1 + u_2 + \cdots + u_n + \cdots$$

is called an infinite series (series for short), denoted by $\sum_{n=1}^{\infty} u_n$, that is

$$\sum_{n=1}^{\infty} u_n = u_1 + u_2 + \cdots + u_n + \cdots \tag{1}$$

where u_n is called the general term.

We call the series whose terms are all constants numerical series, such as

$$\frac{3}{10} + \frac{3}{100} + \cdots + \frac{3}{10^n} + \cdots$$

$$1 - \frac{1}{2} + \frac{1}{3} - \frac{1}{4} + \cdots + (-1)^{n-1} \frac{1}{n} + \cdots$$

$$1 - 1 + 1 - 1 + \cdots + (-1)^{n-1} + \cdots$$

The series whose terms are functions are called functional series, such as

$$1 + x + x^2 + \cdots + x^n + \cdots$$

$$x - \frac{x^3}{3!} + \frac{x^5}{5!} - \cdots + (-1)^{n-1}\frac{x^{2n-1}}{(2n-1)!} + \cdots$$

$$\sin x + \frac{1}{3}\sin 3x + \cdots + \frac{1}{2n-1}\sin(2n-1)x + \cdots$$

11.1 Convergence and Divergence of Infinite Series

11.1.1 Concepts of The Convergence and Divergence of Infinite Series

What is the meaning of the definition formula (1) of the infinite series? How to calculate the sum? If we follow the ordinary addition, add one term to another, we can't finish it forever. Then how to calculate it?

We call the first n terms of the infinite series (1)

$$S_n = u_1 + u_2 + \cdots + u_n$$

the partial sum (the first n terms) of the series (1). So there is a partial sum sequence of the series (1)

$$S_1, S_2, \cdots, S_n, \cdots \qquad (2)$$

Definition 11.2 If the partial sum sequence (2) of the series (1) has a limit

$$\lim_{n \to +\infty} S_n = S$$

then the series (1) is called convergent, and its limit S is called the sum of the series (1), denoted by

$$S = \sum_{n=1}^{\infty} u_n = u_1 + u_2 + \cdots + u_n + \cdots$$

Otherwise the series is called divergent.

The convergence and divergence of a series is a basic problem, and it is equivalent to the question asking if the partial sum sequence has a limit.

For the convergent series (1), the difference

$$r_n = S - S_n = u_{n+1} + u_{n+2} + \cdots$$

is called the remainder sum of the series (1). It is obvious that $\lim_{n \to +\infty} r_n = 0$, so when n is sufficiently large, S can be substituted by S_n approximately, and the error is $|r_n|$.

Example 1 Study the convergence and divergence of the series

Chapter 11　Infinite Series

$$\sum_{n=2}^{+\infty} \ln\left(1 - \frac{1}{n^2}\right)$$

If it is convergent, compute the sum.

Solution　Because

$$\ln\left(1 - \frac{1}{n^2}\right) = \ln\frac{n^2 - 1}{n^2} = \ln(n^2 - 1) - \ln n^2$$

$$= \ln(n+1) - 2\ln n + \ln(n-1)$$

So, the partial sum

$$S_n = \sum_{k=2}^{n} \ln\left(1 - \frac{1}{k^2}\right)$$

$$= (\ln 3 - 2\ln 2 + \ln 1) + (\ln 4 - 2\ln 3 + \ln 2) + \cdots +$$
$$(\ln(n+1) - 2\ln n + \ln(n-1))$$

$$= -\ln 2 + \ln(n+1) - \ln n = \ln\left(1 + \frac{1}{n}\right) - \ln 2$$

so

$$\lim_{n \to +\infty} S_n = -\ln 2$$

Then the series discussed is convergent, the sum is $-\ln 2$.

Example 2　Prove that the geometric series

$$\sum_{n=1}^{\infty} ar^{n-1} = a + ar + ar^2 + \cdots + ar^{n-1} + \cdots (a \neq 0) \qquad (3)$$

is convergent when $|r| < 1$ and is divergent when $|r| \geq 1$.

Solution　When the common ratio $r \neq 1$, the partial sum

$$S_n = a + ar + ar^2 + \cdots + ar^{n-1} = \frac{a - ar^n}{1 - r} = \frac{a}{1 - r} - \frac{ar^n}{1 - r}$$

1° If $|r| < 1$, since $\lim_{n \to +\infty} r^n = 0$, so

$$\lim_{n \to +\infty} S_n = \lim_{n \to +\infty}\left(\frac{a}{1-r} - \frac{ar^n}{1-r}\right) = \frac{a}{1-r}$$

So when $|r| < 1$, the geometric series (3) is convergent, the sum is $\frac{a}{1-r}$.

2° If $|r| > 1$, because $\lim_{n \to +\infty} r^n = \infty$, so S_n has no limit, and the geometric series (3) is divergent.

When the common ratio $r = 1$, $S_n = na$; When the common ratio $r = -1$

$$S_n = \begin{cases} a, & \text{when } n \text{ is odd} \\ 0, & \text{when } n \text{ is even} \end{cases}$$

It can be concluded when $n \to +\infty$, S_n has no limits. So when $|r| = 1$, the geometric series (3) is also divergent.

Example 3 Prove that the series $\sum_{n=1}^{\infty} \dfrac{n}{2^n}$ is convergent, and compute its sum.

Solution Because

$$S_n = \frac{1}{2} + \frac{2}{2^2} + \frac{3}{2^3} + \cdots + \frac{n}{2^n}$$

$$2S_n = 1 + \frac{2}{2} + \frac{3}{2^2} + \cdots + \frac{n}{2^{n-1}}$$

The later formula minus the former one, we get

$$S_n = 1 + \left(\frac{2}{2} - \frac{1}{2}\right) + \left(\frac{3}{2^2} - \frac{2}{2^2}\right) + \cdots + \left(\frac{n}{2^{n-1}} - \frac{n-1}{2^{n-1}}\right) - \frac{n}{2^n}$$

$$= 1 + \frac{1}{2} + \frac{1}{2^2} + \cdots + \frac{1}{2^{n-1}} - \frac{n}{2^n}$$

$$= \frac{1 - \dfrac{1}{2^n}}{1 - \dfrac{1}{2}} - \frac{n}{2^n} = 2 - \frac{1}{2^{n-1}} - \frac{n}{2^n}.$$

So

$$S = \lim_{n \to +\infty} S_n = \lim_{n \to +\infty} \left(2 - \frac{1}{2^{n-1}} - \frac{n}{2^n}\right) = 2$$

Example 4 Prove that the harmonic series

$$\sum_{n=1}^{\infty} \frac{1}{n} = 1 + \frac{1}{2} + \frac{1}{3} + \cdots + \frac{1}{n} + \cdots \tag{4}$$

is divergent.

Solution Using the inequality $x > \ln(1+x) \ (x > 0)$, we get

$$S_n > \ln(1+1) + \ln\left(1+\frac{1}{2}\right) + \cdots + \ln\left(1+\frac{1}{n}\right)$$

$$= \ln 2 + \ln 3 - \ln 2 + \cdots + \ln(n+1) - \ln n$$

$$= \ln(n+1)$$

and $\lim\limits_{n \to +\infty} \ln(n+1) = +\infty$, so $\lim\limits_{n \to +\infty} S_n = +\infty$, we know that the harmonic series (4) is divergent.

Chapter 11　Infinite Series

Determining the convergence or divergence of the infinite series by computing the limit of the sequence of partial sums is the most basic method, but it is often very difficult. So we should look for a better method which is the main task in the following sections.

11.1.2　Several Basic Properties of Infinite Series

By the concepts of convergence and divergence of the infinite series and the properties of limit, it is not hard to get the following properties of the infinite series:

Property 1　When k is a nonzero constant, the series $\sum_{n=1}^{\infty} ku_n$ and the series $\sum_{n=1}^{\infty} u_n$ have the same convergence or divergence. When they are both convergent, we have

$$\sum_{n=1}^{\infty} ku_n = k \sum_{n=1}^{\infty} u_n$$

Proof　By the partial sum of the series

$$\sum_{i=1}^{n} ku_i = k \sum_{i=1}^{n} u_i$$

and the properties of the limits, we have

$$\lim_{n \to +\infty} \sum_{i=1}^{n} ku_i = k \lim_{n \to +\infty} \sum_{i=1}^{n} u_i$$

it is easy to get the conclusions.

Property 2　If the series $\sum_{n=1}^{\infty} u_n$ and $\sum_{n=1}^{\infty} v_n$ are both convergent, then the series $\sum_{n=1}^{\infty} (u_n \pm v_n)$ are also convergent. And

$$\sum_{n=1}^{\infty} u_n \pm \sum_{n=1}^{\infty} v_n = \sum_{n=1}^{\infty} (u_n \pm v_n)$$

Proof　By the partial sum of the series

$$\sum_{i=1}^{n} (u_i \pm v_i) = \sum_{i=1}^{n} u_i \pm \sum_{n=1}^{n} v_i$$

and the properties of the limits

$$\lim_{n \to +\infty} \sum_{i=1}^{n} (u_i \pm v_i) = \lim_{n \to +\infty} \sum_{i=1}^{n} u_i \pm \lim_{n \to +\infty} \sum_{i=1}^{n} v_i$$

Property 2 is established.

By the above property we know that if one of two series is convergent and the other one is divergent, then the series obtained by adding (or subtracting) the corresponding items must be divergent. But if both of the series are divergent, then the series obtained by adding (or subtracting) the corresponding items may not be divergent. For example, the series

$$\sum_{n=1}^{\infty} (-1)^n, \sum_{n=1}^{\infty} (-1)^{n-1}$$

are both divergent, but

$$\sum_{n=1}^{\infty} [(-1)^n + (-1)^{n-1}] = \sum_{n=1}^{\infty} 0 = 0$$

is convergent.

Property 3 For a series, deleting, adding or changing finitely many items at random won't affect the convergence or divergence of the series. But for the convergent series, the sum may be changed.

Proof Suppose in the series $\sum_{n=1}^{\infty} u_n$, getting rid of $u_{i_1}, u_{i_2}, \cdots, u_{i_l}$, we get the new series

$$\sum_{n=1}^{\infty} \hat{u}_n$$

Suppose $u_{i_1} + u_{i_2} + \cdots + u_{i_l} = a$, S_n and $\hat{S}_n$ are the sum of the first n terms of the two series, respectively. Obviously, when $n + l > i_l$, we have

$$\hat{S}_n = S_{n+l} - a$$

So we can see that the series $\sum_{n=1}^{\infty} \hat{u}_n$ and the series $\sum_{n=1}^{\infty} u_n$ have the same convergence or divergence. But when $\sum_{n=1}^{\infty} u_n = S$, $\sum_{n=1}^{\infty} \hat{u}_n = S - a$.

Similarly to the above, adding finitely many items at random, the convergence or divergence of the series is also not changed but the sum of the series may be changed.

Changing finitely many items is same as to deleting some items and then adding items at the original places, so the conclusion is true.

Property 4 The terms of a convergent series can be grouped in any way (provided that the orders of the terms are maintained), and the new series is still convergent to

Chapter 11 Infinite Series

the same sum as the original series. For example, if we group the items as following

$$(u_1 + u_2 + \cdots + u_{k_1}) + (u_{k_1+1} + u_{k_1+2} + \cdots + u_{k_2}) + \cdots + (u_{k_{n-1}+1} + u_{k_{n-1}+2} + \cdots + u_{k_n}) + \cdots \tag{5}$$

the resulting series is still convergent to the same sum as the original series.

Proof Since the partial sum of the new series (5) $\{\hat{S}_n\} = \{S_{k_n}\}$, and

$$\lim_{n \to +\infty} S_n = S,$$

so

$$\lim_{n \to +\infty} \hat{S}_n = \lim_{n \to +\infty} S_{k_n} = S$$

By Property 4, the divergent series is still divergent if the brackets are deleted.

However, the converse of Property 4 is not true. For example

$$(1 - 1) + (1 - 1) + \cdots + (1 - 1) + \cdots$$

is convergent, the sum is zero, but the series

$$1 - 1 + 1 - 1 + \cdots + (-1)^{n-1} + \cdots$$

is divergent.

Property 5 (Necessary Condition for the Convergence of Series) If the series $\sum_{n=1}^{\infty} u_n$ is convergent, then

$$\lim_{n \to +\infty} u_n = 0$$

In other words, the limit of the general term is zero (an infinitesimal).

Proof Suppose $S = \sum_{n=1}^{\infty} u_n$, then

$$\lim_{n \to +\infty} u_n = \lim_{n \to +\infty} (S_n - S_{n-1}) = S - S = 0$$

By Property 5, if the limit of the general term of some series is not zero, it can be concluded that the series is divergent. For example

$$\sum_{n=1}^{\infty} \frac{n!}{a^n} (a > 0)$$

because $\lim_{n \to +\infty} \frac{n!}{a^n} = \infty$, so this series is divergent.

The general term being infinitesimal is only the necessary condition of the convergence of series, not the sufficient condition. For example, the general term $\frac{1}{n}$ of the harmonic series $\sum_{n=1}^{\infty} \frac{1}{n}$ is infinitesimal, but the harmonic series is divergent.

Example 5 Is

$$\sum_{n=1}^{+\infty} \frac{n^{n+\frac{1}{n}}}{\left(n+\frac{1}{n}\right)^n}$$

convergent or divergent?

Solution Because

$$\lim_{n \to +\infty} u_n = \lim_{n \to \infty} \frac{n^{n+\frac{1}{n}}}{\left(n+\frac{1}{n}\right)^n} = \lim_{n \to \infty} \frac{n^n \cdot \sqrt[n]{n}}{n^n \left(\left(1+\frac{1}{n^2}\right)^{n^2}\right)^{\frac{1}{n}}} = 1$$

The limit of the general term of the original series is not zero, so the original series is divergent.

Example 6 Is $\sum_{n=1}^{\infty} \left(\frac{1}{3n} - \frac{\ln^n 3}{3^n}\right)$ convergent or divergent?

Solution Because the harmonic series $\sum_{n=1}^{\infty} \frac{1}{n}$ is divergent, by Property 1, the series $\sum_{n=1}^{\infty} \frac{1}{3n}$ is divergent, and the common ratio of the series

$$\sum_{n=1}^{\infty} \left(\frac{\ln 3}{3}\right)^n$$

is $r = \frac{\ln 3}{3}$, $|r| = \frac{\ln 3}{3} < 1$, so this geometric series is convergent. By Property 2, the series

$$\sum_{n=1}^{\infty} \left(\frac{1}{3n} - \frac{\ln^n 3}{3^n}\right)$$

is divergent.

Example 7 Prove that

$$\lim_{n \to \infty} \frac{a_n}{(1+a_1)(1+a_2)\cdots(1+a_n)} = 0$$

where $a_i > 0 (i = 1, 2, \cdots)$.

Solution Because the partial sum of the series

$$S_n = \sum_{n=1}^{\infty} \frac{a_n}{(1+a_1)(1+a_2)\cdots(1+a_n)} \qquad (6)$$

Chapter 11　Infinite Series

is monotonically increasing, and

$$S_n = \frac{a_1(1+a_2)\cdots(1+a_n) + a_2(1+a_3)\cdots(1+a_n) + \cdots + a_{n-1}(1+a_n) + a_n}{(1+a_1)(1+a_2)\cdots(1+a_n)}$$

$$= \frac{a_1(1+a_2)\cdots(1+a_n) + a_2(1+a_3)\cdots(1+a_n) + \cdots + a_{n-1}(1+a_n) + (1+a_n) - 1}{(1+a_1)(1+a_2)\cdots(1+a_n)}$$

$$= \frac{(1+a_1)(1+a_2)\cdots(1+a_n) - 1}{(1+a_1)(1+a_2)\cdots(1+a_n)} = 1 - \frac{1}{(1+a_1)(1+a_2)\cdots(1+a_n)} < 1$$

so, $\{S_n\}$ is monotonically increasing and has upper bound, so $\{S_n\}$ has a limit, that is the series (6) is convergent, and by Property 5, the limit of the general term of the series (6) is zero.

11.2　The Discriminances for Convergence and Divergence of Infinite Series with Positive Terms

If every term of a series $\sum_{n=1}^{\infty} u_n$ is non-negative, then it is called a series with positive terms (or positive series). Because the partial sum sequence $\{S_n\}$ is non-decreasing, that is

$$S_1 \leqslant S_2 \leqslant \cdots \leqslant S_n \leqslant \cdots$$

then if $\{S_n\}$ has an upper bound, it has limit, and the series $\sum_{n=1}^{\infty} u_n$ is convergent; If $\{S_n\}$ has no upper limit, then $\lim_{n \to +\infty} S_n = +\infty$, the series $\sum_{n=1}^{\infty} u_n$ is divergent. In all, we have:

Theorem 11.1　A series with positive terms converges if and only if its partial sums are bounded from above.

It is easy to conclude that the terms of a series with positive terms can be grouped in any way, its convergence doesn't change, and for the convergent series, the sum doesn't change. If the sum of a convergent series with positive terms is S then $S \geqslant S_n$. By Theorem 11.1, it is easy to get:

Theorem 11.2 (Ordinary Comparison Test)　Suppose $\sum_{n=1}^{\infty} u_n$, $\sum_{n=1}^{\infty} v_n$ are both

series with positive terms, and
$$u_n \leq v_n (n = 1, 2, \cdots)$$
then if $\sum_{n=1}^{\infty} v_n$ converges, so does $\sum_{n=1}^{\infty} u_n$, and if $\sum_{n=1}^{\infty} u_n$ diverges, so does $\sum_{n=1}^{\infty} v_n$.

Proof Suppose the partial sums of $\sum_{n=1}^{\infty} u_n$ and $\sum_{n=1}^{\infty} v_n$ are S_n and σ_n in turn, because $u_n \leq v_n$, so $S_n \leq \sigma_n$. When $\sum_{n=1}^{\infty} v_n$ converges, σ_n has an upper bound, and so S_n also has an upper bound, by Theorem 11.1, $\sum_{n=1}^{\infty} u_n$ converges.

When $\sum_{n=1}^{\infty} u_n$ diverges, $S_n \to +\infty$, and so $\sigma_n \to +\infty$ $(n \to +\infty)$, it is easy to get that $\sum_{n=1}^{\infty} v_n$ diverges.

Corollary If for the two series with positive terms $\sum_{n=1}^{\infty} u_n$ and $\sum_{n=1}^{\infty} v_n$, there exists a constant $C > 0$ and a nonnegative integer N, such that when $n \geq N$
$$u_n \leq C v_n$$
then when $\sum_{n=1}^{\infty} v_n$ converges, so does $\sum_{n=1}^{\infty} u_n$, and when $\sum_{n=1}^{\infty} u_n$ diverges, so does $\sum_{n=1}^{\infty} v_n$.

Example 1 Prove that the P-Series
$$\sum_{n=1}^{\infty} \frac{1}{n^P} = 1 + \frac{1}{2^P} + \frac{1}{3^P} + \cdots + \frac{1}{n^P} + \cdots$$
diverges when $P \leq 1$, and converges when $P > 1$.

Proof When $p \leq 1$, we have
$$\frac{1}{n^P} \geq \frac{1}{n} (n = 1, 2, \cdots)$$
and the harmonic series diverges, so by the Ordinary Comparison Test we know when $P \leq 1$, the P-Series $\sum_{n=1}^{\infty} \frac{1}{n^P}$ diverges.

When $P > 1$, because the terms of the series with positive terms can be grouped at random, its convergence doesn't change, so the P-Series is grouped as following
$$1 + \left(\frac{1}{2^P} + \frac{1}{3^P}\right) + \left(\frac{1}{4^P} + \frac{1}{5^P} + \frac{1}{6^P} + \frac{1}{7^P}\right) + \left(\frac{1}{8^P} + \cdots + \frac{1}{15^P}\right) + \cdots$$

Chapter 11 Infinite Series

its every item is not greater than the following series with positive terms

$$1 + \left(\frac{1}{2^P} + \frac{1}{2^P}\right) + \left(\frac{1}{4^P} + \frac{1}{4^P} + \frac{1}{4^P} + \frac{1}{4^P}\right) + \left(\frac{1}{8^P} + \cdots + \frac{1}{8^P}\right) + \cdots$$

That is the according terms of

$$1 + \frac{1}{2^{P-1}} + \frac{1}{4^{P-1}} + \frac{1}{8^{P-1}} + \cdots$$

The last series is a convergent geometric series, the common ratio $r = \frac{1}{2^{P-1}} < 1$. So by the Ordinary Comparison Test we know that when $P > 1$, the P-Series $\sum_{n=1}^{\infty} \frac{1}{n^P}$ converges.

When using the Ordinary Comparison Test for series with positive terms, it is necessary to know the convergence and divergence of some series as the standard of comparison. The geometric series $\sum_{n=1}^{\infty} ar^n$ and the P-Series $\sum_{n=1}^{\infty} \frac{1}{n^P}$ are often be used as the standard.

Example 2 Suppose $a_n \geqslant 0 (n = 1,2,\cdots)$ and $\sum_{n=1}^{+\infty} a_n$ are convergent, determine the convergence or divergence of $\sum_{n=1}^{+\infty} \frac{1}{n}\sqrt{a_n}$.

Solution Because

$$0 \leqslant \frac{1}{n}\sqrt{a_n} \leqslant \frac{1}{2}\left(\frac{1}{n^2} + a_n\right)$$

$\sum_{n=1}^{+\infty} \frac{1}{n^2}$ and $\sum_{n=1}^{+\infty} a_n$ are convergent, so $\sum_{n=1}^{+\infty} \frac{1}{n}\sqrt{a_n}$ converges.

Example 3 Discuss the convergence and divergence of the following series with positive terms:

1° $\sum_{n=1}^{\infty} 2^n \sin \frac{\pi}{3^n}$; 2° $\sum_{n=1}^{\infty} \frac{1}{\sqrt[3]{n(n+1)}}$; 3° $\sum_{n=1}^{\infty} \int_0^{1/n} \frac{\sqrt{x}}{1+x^2} dx$.

Solution 1° Because

$$0 < u_n = 2^n \sin \frac{\pi}{3^n} < 2^n \cdot \frac{\pi}{3^n} = \pi \left(\frac{2}{3}\right)^n$$

the geometric series $\sum_{n=1}^{\infty} \pi \left(\frac{2}{3}\right)^n$ converges, so by the Ordinary Comparison Test

$$\sum_{n=1}^{\infty} 2^n \sin \frac{\pi}{3^n}$$

converges.

2° Because

$$u_n = \frac{1}{\sqrt[3]{n(n+1)}} > \frac{1}{(n+1)^{2/3}}$$

and $\sum_{n=1}^{\infty} \frac{1}{(n+1)^{2/3}} = \sum_{n=2}^{\infty} \frac{1}{n^{2/3}}$ is a divergent P-Series ($P = \frac{2}{3} < 1$), so by the Ordinary Comparison Test

$$\sum_{n=1}^{\infty} \frac{1}{\sqrt[3]{n(n+1)}}$$

diverges.

3° Because

$$0 < u_n < \int_0^{1/n} \frac{\sqrt{x}}{1+x^2} dx < \int_0^{1/n} \sqrt{x}\, dx = \frac{2}{3} \frac{1}{n^{3/2}}$$

and the P-Series $\sum_{n=1}^{\infty} \frac{1}{n^{3/2}}$ converges $x \neq 0$, so

$$\sum_{n=1}^{\infty} \int_0^{1/n} \frac{\sqrt{x}}{1+x^2} dx$$

converges.

Theorem 11.3 (Limit Comparison Test) Suppose $\sum_{n=1}^{\infty} u_n$ and $\sum_{n=1}^{\infty} v_n$ are series with positive terms, if

$$\lim_{n \to +\infty} \frac{u_n}{v_n} = C$$

then

(i) When $0 < C < +\infty$, the two series have the same convergence or divergence;

(ii) When $C = 0$, if $\sum_{n=1}^{\infty} v_n$ converges, so does $\sum_{n=1}^{\infty} u_n$;

(iii) When $C = \infty$, if $\sum_{n=1}^{\infty} v_n$ diverges, so does $\sum_{n=1}^{\infty} u_n$.

Proof We only prove (i), when $0 < C < +\infty$, choose $\varepsilon = \frac{C}{2}$, $\exists N$, when $n >$

Chapter 11 Infinite Series

N, $\left|\dfrac{u_n}{v_n} - C\right| < \dfrac{C}{2}$, that is

$$\dfrac{C}{2}v_n < u_n < \dfrac{3C}{2}v_n, \forall n > N$$

By the corollary of the Ordinary Comparison Test, the conclusion (i) is true.

The proofs of (ii), (iii) are left to readers.

In the following, we give other methods to determine the convergence and divergence of the series.

Theorem 11.4 (Ratio Comparison Test or D'Alembert Test) For the positive series $\sum_{n=1}^{\infty} u_n$, if

$$\lim_{n \to +\infty} \dfrac{u_{n+1}}{u_n} = \rho$$

then when $\rho < 1$, the series converges; when $\rho > 1$ (or $\rho = +\infty$), the series diverges.

Proof By the definition of the limit of sequences, $\forall \varepsilon > 0$, $\exists N$, when $n \geq N$

$$\left|\dfrac{u_{n+1}}{u_n} - \rho\right| < \varepsilon$$

$$\rho - \varepsilon < \dfrac{u_{n+1}}{u_n} < \rho + \varepsilon, \forall n \geq N \qquad (1)$$

(i) When $\rho < 1$, choose ε appropriately small such that $\rho + \varepsilon = r < 1$. So by (1)

$$u_{n+1} < r u_n, \forall n \geq N$$

and

$$u_{N+k} < r u_{N+k-1} < \cdots < r^k u_N (k = 1, 2, \cdots)$$

Because $0 < r < 1$, $\sum_{k=1}^{\infty} u_N r^k$ converges, by the ordinary Comparison Test, $\sum_{k=1}^{\infty} u_{N+k} = \sum_{n=N+1}^{\infty} u_n$ converges, so $\sum_{n=1}^{\infty} u_n$ converges.

(ii) When $\rho > 1$, choose ε appropriately small, such that $\rho - \varepsilon > 1$, so by (1)

$$u_{n+1} > u_n, \forall n \geq N$$

Noticing that at this time u_n is non-decreasing, so $\lim_{n \to +\infty} u_n \neq 0$, and the series $\sum_{n=1}^{\infty} u_n$ diverges.

Theorem 11.5 (Root Test or Cauchy Test) For the positive series $\sum_{n=1}^{\infty} u_n$, if

Calculus(II)

$$\lim_{n\to+\infty} \sqrt[n]{u_n} = \rho$$

when $\rho < 1$, the series converges; when $\rho > 1$ (or $\rho = +\infty$), the series diverges.

The proof is similar to the previous theorem and is left to the readers.

It should be pointed out that

1° When we use the Ratio Comparison Test or the Root Test to determine the series to be divergent ($\rho > 1$), the general term of the series doesn't converges to zero. Later we will use this point.

2° When $\rho = 1$, the Ratio Comparison Test and the Root Test don't work. For example, for the P-Series $\sum_{n=1}^{\infty} \frac{1}{n^p}$, we have

$$\lim_{n\to+\infty} \frac{u_{n+1}}{u_n} = \lim_{n\to+\infty} \left(\frac{n}{n+1}\right)^p = 1$$

so the Ratio Comparison Test can't determine the convergence or divergence of the P-Series.

Example 4 Determine the convergence and divergence of $\sum_{n=1}^{\infty} \frac{n}{2^n} \cos^2 \frac{n\pi}{3}$.

Solution Because $0 \leq \cos^2 \frac{n\pi}{3} \leq 1$, so

$$0 \leq \frac{n}{2^n} \cos^2 \frac{n\pi}{3} \leq \frac{n}{2^n} (n = 1, 2, \cdots)$$

and because

$$\lim_{n\to+\infty} \left(\frac{n+1}{2^{n+1}} \Big/ \frac{n}{2^n}\right) = \lim_{n\to+\infty} \frac{n+1}{2n} = \frac{1}{2} < 1$$

so the series $\sum_{n=1}^{\infty} \frac{n}{2^n}$ converges, and by the Ordinary Comparison Test, $\sum_{n=1}^{\infty} \frac{n}{2^n} \cos^2 \frac{n\pi}{3}$ also converges.

Example 5 Discuss the convergence and divergence of $\sum_{n=1}^{\infty} \left(\frac{n}{2n+1}\right)^{an}$.

Solution Because

$$\lim_{n\to+\infty} \sqrt[n]{u_n} = \lim_{n\to+\infty} \sqrt[n]{\left(\frac{n}{2n+1}\right)^{an}} = \lim_{n\to+\infty} \left(\frac{n}{2n+1}\right)^a = \left(\frac{1}{2}\right)^a$$

so, when $a > 0$, $\left(\frac{1}{2}\right)^a < 1$, the series converges; when $a < 0$, $\left(\frac{1}{2}\right)^a > 1$, the series di-

Chapter 11 Infinite Series

verges; when $a = 0$, the Ratio Test doesn't work, but at this moment the series is $\sum\limits_{n=1}^{\infty} 1$, and the series diverges.

Two supplements:

1° The general form of the Ratio Comparison Test or the Root Test is: When $n \geqslant N$, if $\dfrac{u_{n+1}}{u_n} \leqslant r < 1$ (or $\sqrt[n]{u_n} \leqslant r < 1$), then the positive series $\sum\limits_{n=1}^{\infty} u_n$ converges, if $\dfrac{u_{n+1}}{u_n} \geqslant 1$ (or $\sqrt[n]{u_n} \geqslant 1$), then the positive series $\sum\limits_{n=1}^{\infty} n_n$ diverges.

2° If S is substituted by the partial sum S_n, when using the Ordinary Comparison Test, the error r_n is evaluated by

$$r_n = u_{n+1} + u_{n+2} + \cdots < ru_n + r^2 u_n + \cdots = \frac{ru_n}{1-r} \tag{2}$$

When using the Ratio Test, then

$$r_n = u_{n+1} + u_{n+2} + \cdots < r^{n+1} + r^{n+2} + \cdots = \frac{r^{n+1}}{1-r} \tag{3}$$

Example 6 Prove that the series $\sum\limits_{n=2}^{\infty} \dfrac{1}{(n-1)!}$ converges, and evaluate the error.

Solution Because

$$\lim_{n \to +\infty} \frac{u_{n+1}}{u_n} = \lim_{n \to +\infty} \frac{(n-1)!}{n!} = \lim_{n \to +\infty} \frac{1}{n} = 0 < 1$$

so, the series converges. The error when S is substituted by the partial sum S_n is

$$r_n = \frac{1}{n!} + \frac{1}{(n+1)!} + \cdots = \frac{1}{n!}\left[1 + \frac{1}{n+1} + \frac{1}{(n+1)(n+2)} + \cdots\right]$$

$$< \frac{1}{n!}\left(1 + \frac{1}{n} + \frac{1}{n^2} + \cdots\right) = \frac{1}{(n-1)!(n-1)}$$

Example 7 Using the convergence of series, prove $\lim\limits_{n \to +\infty} \dfrac{n^n}{(n!)^2} = 0$.

Solution Study the series $\sum\limits_{n=1}^{\infty} \dfrac{n^n}{(n!)^2}$, because

$$\lim_{n \to +\infty} \frac{u_{n+1}}{u_n} = \lim_{n \to +\infty} \left\{ \frac{(n+1)^{n+1}}{[(n+1)!]^2} \cdot \frac{(n!)^2}{n^n} \right\}$$

211

$$= \lim_{n \to +\infty} \left[\frac{1}{n+1} \left(1 + \frac{1}{n}\right)^n \right] = 0 < 1$$

so the series $\sum_{n=1}^{\infty} \frac{n^n}{(n!)^2}$ converges. By the necessary condition of the convergence of the series we know that

$$\lim_{n \to +\infty} \frac{n^n}{(n!)^2} = 0$$

Theorem 11.6 (The Integral Test) Suppose $\sum_{n=1}^{\infty} u_n$ is a positive series, the function $f(x)$ is nonnegative, continuous, monotonically increasing in $[a, +\infty)(a > 0)$, and

$$f(n) = u_n \ (n \geqslant N)$$

then the series $\sum_{n=1}^{\infty} u_n$ has the same convergence or divergence with the improper integral $\int_a^{+\infty} f(x) \, dx$.

Proof For simplicity, suppose $a = 1, N = 1$, see Fig. 11.1. By the known conditions, for any positive integer k, we have

$$u_{k+1} = f(k+1) \leqslant \int_k^{k+1} f(x) \, dx \leqslant f(k) = u_k$$

and so

$$S_{n+1} - u_1 \leqslant \int_1^{n+1} f(x) \, dx \leqslant S_n$$

$$(n = 1, 2, \cdots)$$

Because $f(x) > 0$, so $\int_1^b f(x) \, dx$ is a monotonically increasing function of b.

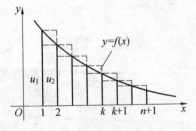

Fig. 11.1

Since S_n is also monotonically increasing, if the improper integral $\int_1^{+\infty} f(x) \, dx$ converges to I, then $\int_1^{n+1} f(x) \, dx < I$, so $S_{n+1} \leqslant I + u_1$, that is $\{S_n\}$ is bounded. By Theorem 11.1, the series $\sum_{n=1}^{\infty} u_n$ is convergent. If the improper integral $\int_1^{+\infty} f(x) \, dx$ diverges, then $\lim_{n \to +\infty} \int_1^{n+1} f(x) \, dx = +\infty$, and $\lim_{n \to +\infty} S_n = +\infty$, so

Chapter 11　Infinite Series

$\sum_{n=1}^{\infty} u_n$ diverges.

By Theorem 11.6, it is easy to see that the P-series $\sum_{n=1}^{\infty} \dfrac{1}{n^P}$ and the improper integral $\int_{1}^{+\infty} \dfrac{1}{x^P} dx$ have the same convergence.

Example 9　Prove that the series $\sum_{n=2}^{\infty} \dfrac{1}{n(\ln n)^P}$ is convergent when $P > 1$; divergent when $0 < P \leqslant 1$.

Solution　Suppose
$$f(x) = \dfrac{1}{x(\ln x)^P}, x \geqslant 2$$
then the function $f(x) > 0$ is continuous and monotonically increasing in the interval $[2, +\infty)$, $f(n) = \dfrac{1}{n(\ln n)^P}$, when $P = 1$
$$\int_{2}^{+\infty} \dfrac{dx}{x \ln x} = \ln \ln x \Big|_{2}^{\infty} = +\infty$$
the improper integral diverges. When $P \neq 1$
$$\int_{2}^{+\infty} \dfrac{dx}{x(\ln x)^P} = \dfrac{1}{1-P}(\ln x)^{1-P}\Big|_{2}^{+\infty} = \begin{cases} +\infty, & \text{when } P < 1 \\ \dfrac{1}{P-1}(\ln 2)^{1-P}, & \text{when } P > 1 \end{cases}$$
So by the Integral Test: when $P > 1$, the series discussed converges; when $0 < P \leqslant 1$, the series discussed diverges.

11.3　Series With Arbitrary Terms, Absolute Convergence

The series which has both positive and negative terms is called a series with arbitrary term (or arbitrary series). If there are finitely many positive terms or negative terms, then the rest terms have the same sign, then the convergence or divergence can be determined by the convergence or divergence of the positive series. If the series has infinitely many positive terms and negative terms, then how to solve the convergence problem of the series? Suppose

$$\sum_{n=1}^{\infty} u_n = u_1 + u_2 + \cdots + u_n + \cdots \qquad (1)$$

is an arbitrary series. By taking absolute values of every term we get a positive series

$$\sum_{n=1}^{\infty} |u_n| = |u_1| + |u_2| + \cdots + |u_n| + \cdots \qquad (2)$$

Definition 11.3 If the series (2) converges, then the series (1) is called absolute convergent. If the series (2) diverges, and the series (1) converges, then the series (1) is called conditional convergent.

Theorem 11.7 The series (1) is absolute convergent if and only if the series formed by the positive terms and the negative terms of the series (1), respectively

$$\sum_{n=1}^{\infty} \frac{|u_n| + u_n}{2}, \sum_{n=1}^{\infty} \frac{u_n - |u_n|}{2} \qquad (3)$$

are both convergent.

Proof Noticing that $\sum_{n=1}^{\infty} \frac{|u_n| + u_n}{2}, \sum_{n=1}^{\infty} \left(-\frac{u_n - |u_n|}{2} \right)$ are two positive series,

and

$$0 \leqslant \frac{|u_n| + u_n}{2} \leqslant |u_n|, 0 \leqslant \frac{|u_n| - u_n}{2} \leqslant |u_n|$$

If the series (1) is absolute convergent, then by the comparison test of the positive series it is known that

$$\sum_{n=1}^{\infty} \frac{|u_n| + u_n}{2}, \sum_{n=1}^{\infty} \frac{|u_n| - u_n}{2}$$

both converges.

On the contrary, if the two series of (3) both converges, then by the properties of the series and

$$|u_n| = \frac{|u_n| + u_n}{2} + \frac{|u_n| - u_n}{2}$$

the series (1) is absolute convergent.

Theorem 11.8 If the series (1) is absolute convergent, then the series (1) is convergent.

Proof Because

$$u_n = \frac{|u_n| + u_n}{2} - \frac{|u_n| - u_n}{2}$$

Chapter 11 Infinite Series

using Theorem 11.7 and the property of the series, the series (1) converges.

Noticing, the conditional convergence and the absolute convergence are both sufficient conditions for the series being convergent, but not the necessary condition.

In all, If the two series of (3) both converges, then the series (1) is absolute convergent; If one of the two series in (3) converges and the other diverges, then the series (1) must be divergent; If the two series of (3) both diverge, then the series (1) may be conditional convergent and may be divergent.

If the series is alternated by positive terms and negative terms, the series is called an alternating series. Suppose $u_n > 0, n = 1, 2, \cdots$, then the alternating series is

$$\sum_{n=1}^{\infty} (-1)^{n-1} u_n = u_1 - u_2 + u_3 - \cdots + (-1)^{n-1} u_n + \cdots \qquad (4)$$

or

$$\sum_{n=1}^{\infty} (-1)^n u_n = -u_1 + u_2 - u_3 + \cdots + (-1)^n u_n + \cdots \qquad (5)$$

Theorem 11.9 (Leibniz Test) It the alternating series (4) satisfies

(i) $\lim\limits_{n \to +\infty} u_n = 0$;

(ii) $u_n \geq u_{n+1}, n = 1, 2, \cdots$,

then the series (4) converges, and the sum $S \leq u_1$, the remainder $r_n = S - S_n$ and its absolute value $|r_n| \leq u_{n+1}$.

Proof The sum of the first $2m$ terms of the series (4) is

$$S_{2m} = (u_1 - u_2) + (u_3 - u_4) + \cdots + (u_{2m-1} - u_{2m})$$
$$S_{2m} = u_1 - (u_2 - u_3) - \cdots - (u_{2m-2} - u_{2m-1}) - u_{2m}$$

By the condition (ii), $\{S_{2m}\}$ is monotonically increasing and has the upper bound $S_{2m} \leq u_1$, so

$$\lim_{m \to +\infty} S_{2m} = S \leq u_1$$

On the other hand, by condition (i), we have

$$\lim_{m \to +\infty} S_{2m+1} = \lim_{m \to +\infty} (S_{2m} + u_{2m+1}) = S$$

In all, no matter n is odd or even, we have

$$\lim_{n \to +\infty} S_n = S$$

so the series (4) converges, and $S \leq u_1$.

The absolute value of the remainder terms

Calculus(II)

$$|r_n| \leq u_{n+1}$$

Example 1 Determine the convergence or divergence of the series, if it is convergent, show if it is conditional convergent or absolute convergent.

$$1° \sum_{n=1}^{\infty} (-1)^{n-1} \frac{k+n}{n^2}; \quad 2° \sum_{n=1}^{\infty} \sin(\pi\sqrt{n^2+1}).$$

Solution 1° Since $\lim_{n\to\infty} \dfrac{\frac{k+n}{n^2}}{\frac{1}{n}} = 1$, and the harmonic series $\sum_{n=1}^{\infty} \dfrac{1}{n}$ is divergent, so the series 1° is not absolute convergent.

Because

$$\lim_{n\to+\infty} u_n = \lim_{n\to+\infty} \frac{n+k}{n^2} = 0, u_n = \frac{1}{n} + \frac{k}{n^2} > \frac{1}{n+1} + \frac{k}{(n+1)^2} = u_{n+1}(n=1,2,\cdots)$$

so by the Leibniz Test, the series 1° converges. In all, the series 1° is conditional convergent.

2° Because

$$\sin(\pi\sqrt{n^2+1}) = (-1)^n \sin\frac{\pi}{\sqrt{n^2+1}+n}$$

so

$$\sum_{n=1}^{\infty} \sin(\pi\sqrt{n^2+1}) = \sum_{n=1}^{\infty} (-1)^n \sin\frac{\pi}{\sqrt{n^2+1}+n}$$

is an alternating series. Because

$$\lim_{n\to+\infty} \frac{u_n}{\frac{1}{n}} = \lim_{n\to+\infty} \frac{\sin\frac{\pi}{\sqrt{n^2+1}+n}}{\frac{1}{n}} = \lim_{n\to+\infty} \frac{\frac{\pi}{\sqrt{n^2+1}+n}}{\frac{1}{n}} = \frac{\pi}{2}$$

By the limit form of the comparison Test, the series

$$\sum_{n=1}^{\infty} |\sin(\pi\sqrt{n^2+1})|$$

diverges, i.e. the original series 2° is not absolute convergent. But because u_n is an infinitesimal, and

$$u_n = \sin\frac{\pi}{\sqrt{n^2+1}+n} > \sin\frac{\pi}{\sqrt{(n+1)^2+1}+(n+1)} = u_{n+1}$$

So the series 2° is convergent. In all, the series 2° is conditional convergent.

Example 2 Determine the convergence and divergence of the series, if it is convergent, show if it is conditional convergent or absolute convergent.

1° $\sum_{n=1}^{\infty} (-1)^{\frac{n(n+1)}{2}} \frac{1}{2^n}$;   2° $\sum_{n=1}^{\infty} \frac{(-n)^n}{n!}$.

Solution   1° Because

$$\sum_{n=1}^{\infty} \left|(-1)^{\frac{n(n+1)}{2}} \frac{1}{2^n}\right| = \sum_{n=1}^{\infty} \frac{1}{2^n}$$

And the geometric series $\sum_{n=1}^{\infty} \frac{1}{2^n}$ is convergent, so the series 1° is absolute convergent.

2° Because

$$\sum_{n=1}^{\infty} \left|\frac{(-n)^n}{n!}\right| = \sum_{n=1}^{\infty} \frac{n^n}{n!}$$

and

$$\lim_{n \to +\infty} \left[\frac{(n+1)^{n+1}}{(n+1)!} \Big/ \frac{n^n}{n!}\right] = \lim_{n \to +\infty} \left(\frac{n+1}{n}\right)^n = e > 1$$

then by the Ratio Test of the positive series, the series $\sum_{n=1}^{\infty} \frac{n^n}{n!}$ diverges, and the series 2° is not absolute convergent. Because $\frac{u_{n+1}}{u_n} > 1$, so the series 2° is divergent.

Example 3 Suppose the constant $\lambda \geq 0$, and the series $\sum_{n=1}^{\infty} a_n^2$ converges, determine if the series $\sum_{n=1}^{\infty} (-1)^n \frac{|a_n|}{\sqrt{n^2 + \lambda}}$ is convergent or divergent, if it is convergent, explain if it is conditional convergent or absolute convergent.

Solution   Using $a^2 + b^2 \geq 2ab$, we get

$$|a_n| \frac{1}{\sqrt{n^2 + \lambda}} \leq \frac{1}{2}\left(|a_n|^2 + \frac{1}{n^2 + \lambda}\right)$$

Because the positive series

$$\sum_{n=1}^{\infty} |a_n|^2 = \sum_{n=1}^{\infty} a_n^2, \sum_{n=1}^{\infty} \frac{1}{n^2 + \lambda}$$

are both convergent, so the series

$$\sum_{n=1}^{\infty} \frac{1}{2}\left(|a_n|^2 + \frac{1}{n^2 + \lambda}\right)$$

is convergent. So by the Comparison Test of the positive series, the series
$$\sum_{n=1}^{\infty} \frac{|a_n|}{\sqrt{n^2+\lambda}}$$
converges, so $\sum_{n=1}^{\infty} (-1)^n \frac{|a_n|}{\sqrt{n^2+\lambda}}$ is absolute convergent.

We give two properties of the absolute convergent series in the following, but the proofs will not be given.

Property 1 If the series $\sum_{n=1}^{\infty} u_n$ is not absolute convergent, and the sum is S, then the new series obtained by exchanging any terms at random $\sum_{n=1}^{\infty} u_n^*$ is also absolute convergent, and the sum is still S.

The conditional convergent series doesn't have this property.

Property 2 If the series $\sum_{n=1}^{\infty} u_n$, $\sum_{n=1}^{\infty} v_n$ are both absolute convergent, the sum of them are S and σ respectively, then their Cauchy Product
$$\left(\sum_{n=1}^{\infty} u_n\right)\left(\sum_{n=1}^{\infty} v_n\right) = (u_1 v_1) + (u_1 v_2 + u_2 v_1) + \cdots + (u_1 v_n + u_2 v_{n-2} + \cdots + u_n v_1) + \cdots$$
is also absolute convergent, and the sum is $S\sigma$.

11.4* The Discriminances for Convergence of Improper Integral, Γ Function

In the following, we generalize the convergence tests to improper integrals. We just give the tests, not to prove them.

Principal of Comparison Suppose $f(x), g(x) \in C[a, +\infty)$ and $0 \leqslant f(x) \leqslant g(x)$, then when $\int_a^{+\infty} g(x)\,dx$ converges, $\int_a^{+\infty} f(x)\,dx$ also converges; when $\int_a^{+\infty} f(x)\,dx$ diverges, then $\int_a^{+\infty} g(x)\,dx$ also diverges.

Because
$$\int_a^{+\infty} \frac{1}{x^p}\,dx \,(a > 0)$$

Chapter 11 Infinite Series

is convergent when $P > 1$ and divergent when $P \leq 1$, so, if take $\dfrac{M}{x^P}(M > 0,\text{ constant})$ as the comparison function, we will get the following tests.

Comparison Test I Suppose $f(x) \in C[a, +\infty)(a > 0), f(x) \geq 0$.

(i) If there exist $M > 0, P > 1$, such that $f(x) \leq \dfrac{M}{x^P}, x \in [a, +\infty)$, then $\displaystyle\int_a^{+\infty} f(x)\,dx$ converges;

(ii) If there exist $M > 0, P \leq 1$, such that $f(x) \geq \dfrac{M}{x^P}, x \in [a, +\infty)$, then $\displaystyle\int_a^{+\infty} f(x)\,dx$ diverges.

Example 1 Determine the convergence and divergence of $\displaystyle\int_1^{+\infty} \dfrac{1}{\sqrt[3]{x^5+1}}dx$.

Solution Because
$$0 < \dfrac{1}{\sqrt[3]{x^5+1}} < \dfrac{1}{x^{5/3}}, x \in [1, +\infty)$$
by the Comparison Test I, this improper integral converges.

Infinitesimal Test I Suppose $f(x) \in C[a, +\infty)(a > 0), f(x) \geq 0$. If there exists a constant $P > 0$ such that $\lim\limits_{x \to +\infty} x^P f(x) = C$, then

(i) When $P > 1$ and $0 \leq C < +\infty$, $\displaystyle\int_a^{+\infty} f(x)\,dx$ converges;

(ii) When $P \leq 1$ and $0 < C \leq +\infty$, $\displaystyle\int_a^{+\infty} f(x)\,dx$ diverges.

Example 2 Determine the convergence or divergence of the improper integral

$1°\ \displaystyle\int_1^{+\infty} \dfrac{1}{x\sqrt{1+x^2}}dx$; $2°\ \displaystyle\int_1^{+\infty} \dfrac{\arctan x}{x}dx$.

Solution $1°$ Because
$$\lim_{x \to +\infty} x^2 \dfrac{1}{x\sqrt{1+x^2}} = \lim_{x \to +\infty} \dfrac{1}{\sqrt{\dfrac{1}{x^2}+1}} = 1$$

$P = 2$, so $1°$ converges.

$2°$ Because

Calculus(II)

$$\lim_{x \to +\infty} x \frac{\arctan x}{x} = \lim_{x \to +\infty} \arctan x = \frac{\pi}{2}$$

$P = 1$, $2°$ diverges.

Comparison Test II Suppose $f(x) \in C(a,b]$ and $f(x) \geq 0$, $\lim\limits_{x \to a^+} f(x) = +\infty$.

(i) If there exist constants $M > 0$ and $q < 1$ such that $f(x) \leq \dfrac{M}{(x-a)^q} (a < x \leq b)$, then the improper integral $\int_a^b f(x)\,dx$ converges;

(ii) If there exist constants $M > 0$ and $q \geq 1$ such that $f(x) \geq \dfrac{M}{(x-a)^q} (a < x \leq b)$, then the improper integral $\int_a^b f(x)\,dx$ diverges.

Infinite Test II Suppose $f(x) \in C(a,b]$, and $f(x) \geq 0$, $\lim\limits_{x \to a^+} f(x) = +\infty$. If there exists $q > 0$, such that

$$\lim_{x \to a^+} (x-a)^q f(x) = l$$

then

(i) When $0 < q < 1$, and $0 \leq l < +\infty$, the improper integral $\int_a^b f(x)\,dx$ converges;

(ii) When $q \geq 1$, and $0 < l \leq +\infty$, the improper integral $\int_a^b f(x)\,dx$ diverges.

Example 3 Determine the convergence and divergence of the following improper integral:

$1°$ $\int_1^3 \dfrac{1}{\ln x}\,dx$; $2°$ $\int_0^1 \dfrac{1}{\sqrt{(1-x^2)(1-k^2x^2)}}\,dx \ (k^2 < 1)$.

Solution $1°$ By L'Hospital's Rule

$$\lim_{x \to 1^+} (x-1) \frac{1}{\ln x} = \lim_{x \to 1^+} x = 1 > 0$$

$q = 1$, so the improper integral $1°$ diverges.

$2°$ Here the singular point is $x = 1$, since

$$\lim_{x \to 1^-} (1-x)^{\frac{1}{2}} \frac{1}{\sqrt{(1-x^2)(1-k^2x^2)}} = \lim_{x \to 1^-} \frac{1}{\sqrt{(1+x)(1-k^2x^2)}}$$

$$= \frac{1}{\sqrt{2(1-k^2)}}$$

Chapter 11 Infinite Series

$q = \dfrac{1}{2}$, then the improper integral 2° diverges.

If the integrant takes both positive and negative values, we have

Theorem of absolute convergence If $\int_a^{+\infty} |f(x)| \, dx$ converges, then so is $\int_a^{+\infty} f(x) \, dx$, and we call the improper integral $\int_a^{+\infty} f(x) \, dx$ absolute converges (or absolute integrable).

Example 4 Discuss if the improper integral $\int_0^{+\infty} e^{-ax} \sin(bx) \, dx$ ($a, b > 0$ are constants) converges or diverges.

Solution Since
$$|e^{-ax}\sin(bx)| \leqslant e^{-ax}$$
and $\int_0^{+\infty} e^{-ax} dx$ converges, then by the principal of comparison we know $\int_0^{+\infty} |e^{-ax}\sin(bx)| \, dx$ converges, then $\int_0^{+\infty} e^{-ax}\sin(bx) \, dx$ absolute converges.

Γ Function We will study the improper integral $\int_0^{+\infty} e^{-t} t^{x-1} dt$ in this subsection, since
$$\int_0^{+\infty} e^{-t} t^{x-1} dt = \int_0^1 e^{-t} t^{x-1} dt + \int_1^{+\infty} e^{-t} t^{x-1} dt$$
And the integral $\int_0^1 e^{-t} t^{x-1} dt$, is a definite integral when $x \geqslant 1$; and is an improper integral with singular point $t = 0$ when $x < 1$, since
$$\lim_{t \to 0^+} t^{1-x}(e^{-t} t^{x-1}) = \lim_{t \to 0^+} e^{-t} = 1$$
then by the infinite test II we know, when $0 < x < 1$, the improper integral $\int_0^1 e^{-t} t^{x-1} dt$ converges. When $x \leqslant 0$, since
$$e^{-t} t^{x-1} \geqslant e^{-1} t^{x-1}, t \in (0, 1]$$
and when $x \leqslant 0$, the improper integral $\int_0^1 e^{-1} t^{x-1} dt$ diverges, so the improper integral $\int_0^1 e^{-t} t^{x-1} dt$ diverges when $x \leqslant 0$. For the improper integral $\int_1^{+\infty} e^{-t} t^{x-1} dt$, since when $x >$ 0

$$\lim_{t\to+\infty} t^2(e^{-t}t^{x-1}) = \lim_{t\to+\infty} e^{-t}t^{x+1} = 0$$

so the improper integral $\int_1^{+\infty} e^{-t}t^{x-1}dt$ converges. In all, when $x > 0$, the improper integral $\int_0^{+\infty} e^{-t}t^{x-1}dt$ converges, and its value is relevant to x, we call it Γ function, denoted by $\Gamma(x)$, i.e.

$$\Gamma(x) = \int_0^{+\infty} e^{-t}t^{x-1}dt \quad (x > 0) \tag{1}$$

Γ function has the following important properties:

Property 1 $\Gamma(1) = 1$.

By the definition equation (1) we know

$$\Gamma(1) = \int_0^{+\infty} e^{-t}dt = 1$$

Property 2 $\Gamma(x+1) = x\Gamma(x) \ (x > 0)$.

According to integration by parts

$$\Gamma(x+1) = \int_0^{+\infty} e^{-t}t^x dt$$

$$= -e^{-t}t^x \Big|_0^{+\infty} + x\int_0^{+\infty} e^{-t}t^{x-1}dt$$

$$= x\int_0^{+\infty} e^{-t}t^{x-1}dt = x\Gamma(x)$$

Specially, we have

$$\Gamma(n+1) = n\Gamma(n)$$
$$= n(n-1)\Gamma(n-1)$$
$$= \cdots = n(n-1)\cdots\Gamma(1) = n!$$

By property 2, we can expand the Γ function to the negative axis, when $x < 0$, define

$$\Gamma(x) = \frac{\Gamma(x+1)}{x}$$

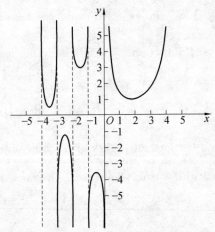

Fig. 11.2

Thus, Γ function can be defined as (its graph is showed in Fig. 11.2)

Chapter 11 Infinite Series

$$\Gamma(x) = \begin{cases} \int_0^{+\infty} e^{-t} t^{x-1} dt, \text{when } x > 0 \\ \dfrac{\Gamma(x+1)}{x}, \text{when } x < 0, x \neq -1, -2, \cdots \end{cases}$$

11.5 Series with Function Terms, Uniform Convergence

11.5.1 Series with Function Terms

Suppose the functions $u_n(x)$ $(n = 1, 2, \cdots)$ are defined on the set X, for the functional series

$$\sum_{n=1}^{\infty} u_n(x) = u_1(x) + u_2(x) + \cdots + u_n(x) + \cdots \tag{1}$$

when $x_0 \in X$, if the series with number terms

$$\sum_{n=1}^{\infty} u_n(x_0) = u_1(x_0) + u_2(x_0) + \cdots + u_n(x_0) + \cdots \tag{2}$$

converges, then x_0 is called a convergent point of the functional series (1), otherwise it is called a divergent point of the series (1). The set which is formed by all the convergent points is called the convergent set of the functional series (1), the set of all the divergent points is called the divergent set of (1).

Suppose J is the convergent set of the functional series (1), $\forall x \in J$, the series (1) always has its sum. The sum is a function on J, denoted by $S(x)$, it is called the sum function of the series (1).

For example, the convergent set of the geometric series

$$\sum_{n=0}^{\infty} x^n = 1 + x + x^2 + \cdots + x^n + \cdots$$

is $|x| < 1$, and its divergent set is $|x| \geqslant 1$, the sum function is $\dfrac{1}{1-x}$, i.e.

$$\sum_{n=0}^{\infty} x^n = \dfrac{1}{1-x}, \forall x \in (-1, 1)$$

Suppose $S_n(x)$ is the partial sum of the functional series (1), then when $x \in J$

$$\lim_{n \to +\infty} S_n(x) = S(x) \tag{3}$$

The term $r_n(x) = S(x) - S_n(x)$ is called the remainder terms (remainder sum) of the

functional series, it is obvious that
$$\lim_{n \to +\infty} r_n(x) = 0, \forall x \in J \tag{4}$$

Example 1 Find the convergent set of the functional series $\sum_{n=1}^{\infty} x^n (1-x)^n$.

Solution Because the series is the geometric series with

$$|r| = \left|\frac{u_{n+1}}{u_n}\right| = \left|\frac{x^{n+1}(1-x)^{n+1}}{x^n(1-x)^n}\right| = |x(1-x)|$$

when $|x(1-x)| < 1$ that is $\frac{1-\sqrt{5}}{2} < x < \frac{1+\sqrt{5}}{2}$, the series discussed is convergent;

When $x \geq \frac{1+\sqrt{5}}{2}$ or $x \leq \frac{1-\sqrt{5}}{2}$, the series diverges, the convergent set of the discussed series is the interval $(\frac{1-\sqrt{5}}{2}, \frac{1+\sqrt{5}}{2})$.

Example 2 Find the convergent set of the functional series $\sum_{n=1}^{\infty} (-1)^{n-1} \frac{x^{3n}}{n}$.

Solution Because

$$\lim_{n \to +\infty} \left|\frac{u_{n+1}}{u_n}\right| = \lim_{n \to +\infty} \frac{\frac{|x|^{3n+3}}{n+1}}{\frac{|x|^{3n}}{n}} = \lim_{n \to +\infty} \frac{n}{n+1} |x|^3 = |x|^3$$

by the Ratio Test of the positive series, when $|x| < 1$, the discussed series is absolute convergent; when $|x| > 1$, the series diverges.

When $x = 1$, the series is $\sum_{n=1}^{\infty} (-1)^{n-1} \frac{1}{n}$, and is conditional convergent; When $x = -1$, the series is $\sum_{n=1}^{\infty} \frac{-1}{n}$, and is divergent.

In all, the convergent set of the discussed series is $(-1, 1]$.

To determine the convergence or divergence of the functional series is to determine that of the number series in which the variable x is considered as a constant.

11.5.2 Uniform Convergence*

The functional series (1) introduce previously converges to its sum function $S(x)$ on its convergent set J pointwisely. We give it a more detailed definition as follows using

Chapter 11　Infinite Series

the "$\varepsilon - N$" language:
$$\forall \varepsilon > 0, \forall x \in J, \exists N = N(\varepsilon, x), \text{such that when } n > N, \text{we always have}$$
$$|r_n(x)| = |S_n(x) - S(x)| < \varepsilon$$

In this definition, different values of x in J correspond to different N, there may not be a universal N for all the values of x. Next, we will introduce a stronger convergence, it requires that for a series on some interval I there exists a N only related to ε, i.e. there is a universal N for any x in I.

Definition 11.4　If for $\forall \varepsilon > 0, \exists N = N(\varepsilon)$, such that when $n > N$, we always have
$$|r_n(x)| = |S_n(x) - S(x)| < \varepsilon, \forall x \in I$$
then we say the functional series $\sum_{n=1}^{\infty} u_n(x)$ is uniformly convergent on the interval I.

Example 3　Discuss the uniform convergence of the series
$$\frac{1}{x+1} - \frac{1}{(x+1)(x+2)} - \frac{1}{(x+2)(x+3)} - \cdots - \frac{1}{(x+n-1)(x+n)} - \cdots$$
on the interval $0 \leq x < \infty$.

Solution　Since
$$\frac{1}{(x+n-1)(x+n)} = \frac{1}{x+n-1} - \frac{1}{x+n}$$
so, $S_n(x) = \dfrac{1}{x+n}$, then
$$S(x) = \lim_{n \to +\infty} S_n(x) = \lim_{n \to +\infty} \frac{1}{x+n} = 0$$
i.e. when $x \geq 0$, the series convergent to the sum function as $n \to \infty$. Also because
$$|r_n(x)| = |S_n(x) - S(x)| = \frac{1}{x+n} \leq \frac{1}{n}, \quad \text{when } 0 \leq x < \infty$$
$\forall \varepsilon > 0$, take $N = \left[\dfrac{1}{\varepsilon}\right] + 1$, then when $n > N$, always have
$$|r_n(x)| < \varepsilon, \forall x \in [0, +\infty)$$
so the series is uniformly convergent on the interval $[0, +\infty)$.

Example 4　Determine if the functional series
$$x + (x^2 - x) + (x^3 - x^2) + \cdots + (x^n - x^{n-1}) + \cdots$$
is uniformly convergent on the interval $[0,1]$ or not.

Solution Since
$$S(x) = \lim_{n\to\infty} S_n(x) = \lim_{n\to\infty} x^n = \begin{cases} 0, & \text{when } 0 \le x \le 1 \\ 1, & \text{when } x = 1 \end{cases}$$

So, when $0 < x < 1$
$$|r_n(x)| = |S_n(x) - S(x)| = x^n$$
then for $\forall \varepsilon > 0$, if we want $|r_n(x)| < \varepsilon$, then must have $n\ln x < \ln\varepsilon$, i.e.
$$n > \frac{\ln \varepsilon}{\ln x} (0 < x < 1)$$

When $x \to 1^-$, since $\frac{\ln \varepsilon}{\ln x} \to +\infty$, so we cannot find the universal N when x is in the interval $(0,1)$, thus, the series is not uniformly convergent on the interval $(0,1)$ (Fig. 11.3).

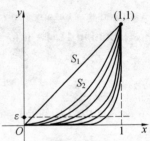

Fig. 11.3

But, for any positive number r less than 1, the series is uniformly convergent on the interval $[0,r]$, because we can take $N = \frac{\ln \varepsilon}{\ln r}$.

Theorem 11.10 (Weierstrass M-test) Suppose $\sum_{n=1}^{\infty} M_n$ is a convergent positive series, such that when $x \in I$
$$|u_n(x)| \le M_n (n = 1, 2, \cdots)$$
then the functional series $\sum_{n=1}^{\infty} u_n(x)$ is uniformly convergent on the interval I.

Proof By the assumptions and the comparison test for positive series we know, the series $\sum_{n=1}^{\infty} u_n(x)$ is absolute convergent for any $x \in I$.

Since the positive series $\sum_{n=1}^{\infty} M_n$ is convergent, $\forall \varepsilon > 0, \exists N = N(\varepsilon)$, such that when $n > N$, we have
$$r_N = M_{N+1} + M_{N+2} + \cdots + M_{N+P} + \cdots < \varepsilon$$
then for any $x \in I$ and any positive integer P, we always have
$$|u_{N+1}(x) + u_{N+2}(x) + \cdots + u_{N+P}(x)|$$
$$\le |u_{N+1}(x)| + |u_{N+2}(x)| + \cdots + |u_{N+P}(x)|$$

Chapter 11 Infinite Series

$$\leq M_{N+1} + M_{N+2} + \cdots + M_{N+P} \leq r_N < \varepsilon$$

Let $P \to +\infty$, and notice that $\sum_{n=N+1}^{\infty} u_n(x)$ is absolute convergent on the interval I, then when $n > N$, we always have

$$|u_{N+1}(x) + u_{N+2}(x) + \cdots| \leq r_N < \varepsilon, \forall x \in I$$

This proved that the series $\sum_{n=1}^{\infty} u_n(x)$ is uniformly convergent on the interval I.

The series $\sum_{n=1}^{\infty} M_n$ in Theorem 11.10 is called the control series (or dominate series).

Example 5 Determine if the series $\sum_{n=1}^{\infty} \dfrac{x}{1+n^4 x^2}$ is uniformly convergent on the interval $x \geq 0$.

Solution Since $1 + n^4 x^2 \geq 2n^2 x, x \geq 0$, then

$$\frac{x}{1+n^4 x^2} \leq \frac{1}{2n^2}$$

and the positive series $\sum_{n=1}^{\infty} \dfrac{1}{n^2}$ is a convergent P-series, so the series studied is uniformly convergent on the interval $[0, +\infty)$.

Uniformly convergent series have a lot of important analytic properties:

Property 1 (Continuity of the sum function of functional series) If the functional series $\sum_{n=1}^{\infty} u_n(x)$ is uniformly convergent on the interval $[a,b]$, and every term $u_n(x)$ of the series is continuous on the interval $[a,b]$, then the sum function $S(x)$ is also continuous on the interval $[a,b]$.

Proof For any $x, x_0 \in [a,b]$, since the series $\sum_{n=1}^{\infty} u_n(x)$ is uniformly convergent on interval $[a,b]$, so for any given $\varepsilon > 0$, there exists a number n_1 only relevant to ε, such that

$$|S(x) - S_{n_1}(x)| < \frac{\varepsilon}{3}$$

$$|S_{n_1}(x_0) - S(x_0)| < \frac{\varepsilon}{3}$$

also because every term $u_n(x)$ is continuous, so the partial sum of the first n_1 terms, $S_{n_1}(x)$, is continuous at x_0, then there exists $\delta > 0$, such that when $|x - x_0| < \delta$, we have

$$|S_{n_1}(x) - S_{n_1}(x_0)| < \frac{\varepsilon}{3}$$

All in all, for any given $\varepsilon > 0$, there exists $\delta > 0$, such that when $|x - x_0| < \delta$, we always have

$$|S(x) - S(x_0)|$$
$$\leq |S(x) - S_{n_1}(x)| + |S_{n_1}(x) - S_{n_1}(x_0)| + |S_{n_1}(x_0) - S(x_0)|$$
$$\leq \frac{\varepsilon}{3} + \frac{\varepsilon}{3} + \frac{\varepsilon}{3} = \varepsilon$$

This proved that the sum function $S(x)$ is continuous at x_0. Since x_0 is taken arbitrarily, the sum function $S(x)$ is continuous on the interval $[a,b]$.

For the functional series in Example 4

$$x + (x^2 - x) + (x^3 - x^2) + \cdots + (x^n - x^{n-1}) + \cdots$$

although every term $(x^n - x^{n-1})$ is continuous on the interval $[0,1]$, but its sum function

$$S(x) = \begin{cases} 0, 0 \leq x < 1 \\ 1, x = 1 \end{cases}$$

is not continuous at $x = 1$, so this series is not uniformly convergent on the interval $[0, 1]$.

Similarly we get: For a uniformly convergent series, if every term has limit, then the limit of the sum function equals to the sum of the limits of the terms, i. e.

$$\lim_{x \to x_0} S(x) = \sum_{n=1}^{\infty} [\lim_{x \to x_0} u_n(x)]$$

in other words

$$\lim_{x \to x_0} \sum_{n=1}^{\infty} u_n(x) = \sum_{n=1}^{\infty} \lim_{x \to x_0} u_n(x)$$

Property 2 (Integrate functional series term by term) If the functional series $\sum_{n=1}^{\infty} u_n(x)$ is uniformly convergent on the interval $[a,b]$, and every term of the series $u_n(x)$ is continuous on the interval $[a,b]$, then the sum function $S(x)$ is integrable, and can be integrated term by term, i. e.

Chapter 11 Infinite Series

$$\int_{x_0}^{x} S(x)\,dx = \int_{x_0}^{x} u_1(x)\,dx + \int_{x_0}^{x} u_2(x)\,dx + \cdots + \int_{x_0}^{x} u_n(x)\,dx + \cdots$$

where x_0, x are two arbitrary points in the interval $[a,b]$. The new series after termwise integration is also uniformly convergent on the interval $[a,b]$.

Proof By Property 1 we know, $S(x)$ is continuous on $[a,b]$, then the integral $\int_{x_0}^{x} S(x)\,dx$ exists.

Since the functional series $\sum_{n=1}^{\infty} u_n(x)$ is uniformly convergent on the interval $[a,b]$, then for any given $\varepsilon > 0$, there exists a number N only determined by ε, such that when $n > N$, we have

$$|r_n(x)| = |S(x) - S_n(x)| < \varepsilon, \forall x \in [a,b]$$

so, when $n > N$

$$\left|\int_{x_0}^{x} S_n(x)\,dx - \int_{x_0}^{x} S(x)\,dx\right| = \left|\int_{x_0}^{x} [S_n(x) - S(x)]\,dx\right|$$
$$\leq \left|\int_{x_0}^{x} |r_n(x)|\,dx\right| < |x - x_0|\varepsilon < (b-a)\varepsilon, \forall x \in [a,b]$$

and

$$\int_{x_0}^{x} S_n(x)\,dx = \int_{x_0}^{x} u_1(x)\,dx + \int_{x_0}^{x} u_2(x)\,dx + \cdots + \int_{x_0}^{x} u_n(x)\,dx$$

By the definition of uniform convergence of functional series, the series

$$\int_{x_0}^{x} u_1(x)\,dx + \int_{x_0}^{x} u_2(x)\,dx + \cdots + \int_{x_0}^{x} u_n(x)\,dx + \cdots$$

uniformly converges to $\int_{x_0}^{x} S(x)\,dx$ on the interval $[a,b]$, then

$$\int_{x_0}^{x} S(x)\,dx = \int_{x_0}^{x} u_1(x)\,dx + \int_{x_0}^{x} u_2(x)\,dx + \cdots + \int_{x_0}^{x} u_n(x)\,dx + \cdots$$

Property 2 can usually be written as

$$\int_{x_0}^{x} \left[\sum_{n=1}^{\infty} u_n(x)\right] dx = \sum_{n=1}^{\infty} \left[\int_{x_0}^{x} u_n(x)\,dx\right], x_0, x \in [a,b]$$

Property 3 (Differentiate functional series term by term) Suppose the series $\sum_{n=1}^{\infty} u_n(x)$ is convergent on the interval $[a,b]$. If every term $u_n(x)$ has continuous derivative on the interval $[a,b]$, i.e. $u'_n(x) \in C[a,b](n = 1,2,\cdots)$, and the series

$$\sum_{n=1}^{\infty} u'_n(x) = u'_1(x) + u'_2(x) + \cdots + u'_n(x) + \cdots$$

is uniformly convergent on the interval $[a,b]$, then the series $\sum_{n=1}^{\infty} u_n(x)$ is also uniformly convergent on this interval, and the sum function $S(x)$ has continuous derivative on $[a,b]$, and can be differentiated term by term, i.e.

$$S'(x) = u'_1(x) + u'_2(x) + \cdots + u'_n(x) + \cdots$$

Proof Let

$$S^*(x) = \sum_{n=1}^{\infty} u'_n(x), x \in [a,b]$$

by Property 2, we have

$$\int_{x_0}^{x} S^*(x) dx = \sum_{n=1}^{\infty} \int_{x_0}^{x} u'_n(x) dx = \sum_{n=1}^{\infty} [u_n(x) - u_n(x_0)]$$

$$= \sum_{n=1}^{\infty} u_n(x) - \sum_{n=1}^{\infty} u_n(x_0) = S(x) - S(x_0)$$

By property 1 we know, the series $S^*(x)$ is continuous, so $\int_{x_0}^{x} S^*(x) dx$ is differentiable. By differentiating both sides of the above equation, we get

$$S^*(x) = S'(x), x \in [a,b]$$

This tells us that the series $\sum_{n=1}^{\infty} u'_n(x)$ uniformly converges to $S'(x)$ on the interval $[a, b]$. From the above proof, the series

$$\sum_{n=1}^{\infty} u_n(x) = \sum_{n=1}^{\infty} \int_{x_0}^{x} u'_n(x) dx + \sum_{n=1}^{\infty} u_n(x_0)$$

where $\sum_{n=1}^{\infty} u_n(x_0)$ is a convergent series with number terms, and the series $\sum_{n=1}^{\infty} \int_{x_0}^{x} u'(x) dx$ is obtained from the uniformly convergent functional series $\sum_{n=1}^{\infty} u'_n(x)$ by termwise integration. By Property 2 we know, $\sum_{n=1}^{\infty} \int_{x_0}^{x} u'_n(x) dx$ is uniformly convergent on $[a,b]$. Easy to prove, the sum of a uniformly convergent functional series and a convergent number series is still uniformly convergent, so $\sum_{n=1}^{\infty} u_n(x)$ is uniformly convergent on $[a,b]$.

Chapter 11 Infinite Series

In property 3, the uniform convergence of the series obtained from termswise differentiation cannot be replaced by the uniform convergence of the original series, for example, the series

$$\frac{\sin x}{1^2} + \frac{\sin(2^2 x)}{2^2} + \cdots + \frac{\sin(n^2 x)}{n^2} + \cdots$$

Since $\left|\dfrac{\sin(n^2 x)}{n^2}\right| \leq \dfrac{1}{n^2}$, $\sum\limits_{n=1}^{\infty} \dfrac{1}{n^2}$ convergent, then by Weierstrass M test, the series $\sum\limits_{n=1}^{\infty} \dfrac{\sin(n^2 x)}{n^2}$ is uniformly convergent on any interval, but the general term of the series obtained by termwise differentiation

$$\cos x + \cos(2^2 x) + \cdots + \cos(n^2 x) + \cdots$$

not approaches zero, so the convergent set is the empty set, then the original series cannot be differentiate term by term.

Property 3 can usually be written as

$$\left[\sum_{n=1}^{\infty} u_n(x)\right]' = \sum_{n=1}^{\infty} u'_n(x)$$

i. e. under the condition that $\sum\limits_{n=1}^{\infty} u'_n(x)$ is uniformly convergent, the functional series $\sum\limits_{n=1}^{\infty} u_n(x)$ can be differentiated term by term.

11.6 Power Series

A power series about x is a series of the form

$$\sum_{n=0}^{\infty} a_n x^n = a_0 + a_1 x + a_2 x^2 + \cdots + a_n x^n + \cdots \qquad (1)$$

where the constants $a_n (n = 0, 1, 2, \cdots)$ are called the coefficients of the power series. More generally, a functional series of the form

$$\sum_{n=0}^{\infty} a_n (x - x_0)^n = a_0 + a_1(x - x_0) + a_2(x - x_0)^2 + \cdots + a_n(x - x_0)^n + \cdots$$

$$(2)$$

is called the power series about $(x - x_0)$, where x_0 is a fixed value.

Obviously, by the transform $t = x - x_0$, the series (2) is transformed into the form of the series(1), so we will discuss the power series(1) mainly in the following.

11.6.1 The Radius and Interval of Convergence of Power Series

Abel Lemma If the power series (1) is convergent at the point $x = x_0 (x_0 \neq 0)$, then for any point x of $(-|x_0|, |x_0|)$, the power series(1) converges absolutely; If the power series (1) is divergent at $x = x_0$, then when $x > |x_0|$ or $x < -|x_0|$, the power series (1) is divergent.

Proof (i) Suppose the power series(1) is convergent at $x = x_0 (x_0 \neq 0)$, i.e. the series with number terms

$$a_0 + a_1 x_0 + a_2 x_0^2 + \cdots + a_n x_0^n + \cdots$$

converges. By the necessary condition of the convergence it is known

$$\lim_{n \to +\infty} a_n x_0^n = 0$$

and so the sequence $\{a_n x_0^n\}$ is bounded, that is there is a constant $M > 0$, such that

$$|a_n x_0^n| \leq M (n = 0, 1, 2, \cdots)$$

so

$$|a_n x^n| = |a_n x_0^n| \left|\frac{x}{x_0}\right|^n \leq M \left|\frac{x}{x_0}\right|^n (n = 0, 1, 2, \cdots)$$

When $|x| < |x_0|$, $\left|\frac{x}{x_0}\right| < 1$, the geometric series $\sum_{n=0}^{\infty} M \left|\frac{x}{x_0}\right|^n$ converges. By the Comparison Test, it is known that the series $\sum_{n=0}^{\infty} |a_n x^n|$ converges, that is the power series (1) is absolute convergent at any point of the open interval $(-|x_0|, |x_0|)$.

(ii) Suppose the power series(1) is divergent at the point $x = x_0$, and there is a point x_1, satisfying $|x_1| > |x_0|$, and making the power series convergent at the point x_1. Then by the proof of(i), the power series(1) is convergent at x_0, it is contradictory with the conditions.

Because every term of the power series has definition on $(-\infty, +\infty)$., so for every real number x, the power series (1) may diverge or converge. But any power series (1) converges at the original point, so by the Abel Lemma, we get the following corollary:

Corollary There are three convergent types for the power series(1):

(i) There is a constant $R > 0$, such that when $|x| < R$, the power series (1) is

Chapter 11 Infinite Series

absolute convergent, when $|x| > R$, the power series (1) diverges;

(ii) Except $x = 0$, the power series (1) diverges everywhere, i.e. $R = 0$;

(iii) The power series (1) converges at any point x, at this moment, $R = \infty$.

R is called the convergent radius of the power series (1). The open interval $(-R, R)$ is called the convergent interval of the power series (1). The power series is absolute convergent in the convergent interval.

Except $R = 0$, in general, the convergent set is an interval with radius R and centered at origin. For the condition (i), we also need to discuss the following two number series, corresponding to $x = \pm R$

$$\sum_{n=0}^{\infty} a_n(-R)^n, \quad \sum_{n=0}^{\infty} a_n R^n$$

Their convergence will help to determine the convergent set finally.

In the following, we will discuss how to compute the convergent radius R and the convergent set.

Theorem 11.11 For the power series (1), if

$$\lim_{n \to \infty} \left| \frac{a_{n+1}}{a_n} \right| = b \quad \text{or} \quad \lim_{n \to \infty} \sqrt[n]{|a_n|} = b$$

then the convergent radius of the power series (1)

$$R = \begin{cases} \dfrac{1}{b}, & \text{when } 0 < b < +\infty \\ +\infty, & \text{when } b = 0 \\ 0, & \text{when } b = +\infty \end{cases}$$

Proof We only discuss $\lim\limits_{n \to \infty} \left| \dfrac{a_{n+1}}{a_n} \right| = b$, because

$$\sum_{n=0}^{\infty} |a_n x^n| = |a_0| + |a_1 x| + |a_2 x^2| + \cdots + |a_n x^n| + \cdots \quad (3)$$

and the limit

$$\lim_{n \to \infty} \frac{|a_{n+1} x^{x+1}|}{|a_n x^n|} = \lim_{n \to \infty} \left| \frac{a_{n+1}}{a_n} \right| |x| = b|x|$$

so by the Ratio Test

(i) When $0 < b < +\infty$, if $|x| < \dfrac{1}{b}$, that is $b|x| < 1$, then the series (3) converges, and so the series(1) is absolute convergent; If $|x| > \dfrac{1}{b}$, that is $b|x| > 1$, then the series(3) diverges, and we can say the series(1) diverges (Because we use the ratio test here, so the series(1) is not absolutely convergent). In all, at this moment the convergent radius $R = \dfrac{1}{b}$.

(ii) When $b = 0$, we have $b|x| = 0$, so the series(3) converges everywhere, that is the series(1) converges everywhere, $R = +\infty$.

(iii) When $b = +\infty$, except for $x = 0$, $b|x| = +\infty$, the series (3) diverges everywhere except for $x = 0$, so the series(1) diverges at $x \neq 0$, $R = 0$.

In the conditions of Theorem 11.11, the convergent radius can be computed by the following formula

$$R = \lim_{n \to \infty} \left| \frac{a_n}{a_{n+1}} \right| \quad \text{or} \quad R = \lim_{n \to \infty} \frac{1}{\sqrt[n]{|a_n|}}$$

Example 1 Compute the convergent radius and the convergent set of the following power series:

1° $\displaystyle\sum_{n=1}^{\infty} \frac{x^n}{2^n \cdot n}$; 2° $\displaystyle\sum_{n=1}^{\infty} \frac{(n!)^2}{(2n)!} x^n$;

3° $\displaystyle\sum_{n=0}^{\infty} \frac{x^n}{(2n)!!}$; 4° $\displaystyle\sum_{n=1}^{\infty} n^n x^n$.

Solution 1° The convergent radius

$$R = \lim_{n \to \infty} \left| \frac{a_n}{a_{n+1}} \right| = \lim_{n \to \infty} \left[\frac{1}{2^n \cdot n} \Big/ \frac{1}{2^{n+1}(n+1)} \right] = \lim_{n \to \infty} \frac{2(n+1)}{n} = 2$$

so the convergent interval is $(-2, 2)$.

When $x = -2$, the series 1° is $\displaystyle\sum_{n=1}^{\infty} (-1)^n \frac{1}{n}$, which is a convergent alternating series, when $x = 2$, the series 1° becomes $\displaystyle\sum_{n=1}^{\infty} \frac{1}{n}$, which is a harmonic series, and is divergent.

So the convergent set of the series 1° is $[-2, 2]$.

2° The convergent radius

$$R = \lim_{n \to \infty} \left| \frac{a_n}{a_{n+1}} \right| = \lim_{n \to \infty} \left\{ \frac{(n!)^2}{(2n)!} \Big/ \frac{[(n+1)!]^2}{[2(n+1)]!} \right\} = \lim_{n \to \infty} \frac{2(2n+1)}{(n+1)} = 4$$

Chapter 11 Infinite Series

So the convergent interval is $(-4, 4)$.

When $x = 4$, the series $2°$ is a positive series
$$\sum_{n=1}^{\infty} \frac{(n!)^2}{(2n)!} 4^n$$

because
$$\frac{u_{n+1}}{u_n} = \frac{2n+2}{2n+1} > 1$$

so, $u_n \not\to 0$ (when $n \to \infty$), so $\sum_{n=1}^{\infty} \frac{(n!)^2}{(2n)!} 4^n$ diverges. When $x = -4$, the according number series of the series $2°$ also diverges. So, the convergent set of the power series $2°$ is $(-4, 4)$.

$3°$ Because the convergent radius
$$R = \lim_{n \to \infty} \left| \frac{a_n}{a_{n+1}} \right| = \lim_{n \to \infty} \left[\frac{1}{(2n)!!} \bigg/ \frac{1}{(2n+2)!!} \right] = \lim_{n \to \infty} (2n+2) = \infty$$

so, the convergent set of the power series $3°$ is $(-\infty, +\infty)$.

$4°$ Because the convergent radius
$$R = \lim_{n \to \infty} \frac{1}{\sqrt[n]{|a_n|}} = \lim_{n \to \infty} \frac{1}{\sqrt[n]{n^n}} = \lim_{n \to \infty} \frac{1}{n} = 0$$

so, the power series $4°$ is convergent only at the point $x = 0$.

Example 2 If the convergent set of the series $\sum_{n=0}^{+\infty} a^{n^2} x^n (a > 0)$ is $(-\infty, +\infty)$, discuss the range of a.

Solution Because $\left| \frac{a_n}{a_{n+1}} \right| = \frac{a^{n^2}}{a^{(n+1)^2}} = \frac{1}{a^{2n+1}}$, so
$$\lim_{n \to \infty} \left| \frac{a_n}{a_{n+1}} \right| = +\infty$$

if and only if $a < 1$. So $a < 1$.

Example 3 Find the convergent set of the power series $\sum_{n=1}^{+\infty} \frac{(x-2)^{2n}}{n 4^n}$.

Solution Set $t = (x-2)^2$ and the series becomes a power series of t, that is $\sum_{n=1}^{+\infty} \frac{t^n}{n 4^n}$. The convergent radius of this power series is

$$R_t = \lim_{n\to\infty} \frac{(n+1)4^{n+1}}{n4^n} = 4$$

When $x - 2 = \pm 2$, the original series is $\sum_{n=1}^{+\infty} \frac{1}{n}$, which is divergent.

Then the convergent set is $-2 < x - 2 < 2$, that is $0 < x < 4$.

In general, the convergent interval of the power series of $(x - x_0)$ is symmetric about the point x_0, so we don't need to make transformation. Firstly, we compute the convergent radius R, and then discuss the convergence of the corresponding number series at the two endpoints of the interval $(x_0 - R, x_0 + R)$. Finally we determine the convergent set.

Example 4 Compute the convergent set of the functional series
$$\ln x + \sum_{n=0}^{\infty} (-1)^n \frac{x^{2n+1}}{(2n+1)!}$$

Solution If we delete the first term, then the original series becomes a power series absent of the even degree terms. Since
$$\lim_{n\to\infty} \left| \frac{u_{n+1}}{u_n} \right| = \lim_{n\to\infty} \left[\frac{|x|^{2n+3}}{(2n+3)!} \cdot \frac{(2n+1)!}{|x|^{2n+1}} \right] = \lim_{n\to\infty} \frac{|x|^2}{(2n+2)(2n+3)} = 0$$
so getting rid of the first term, the series is absolute convergent everywhere. Because the domain of the first term is $x > 0$, so the convergent set of the whole series is $(0, +\infty)$.

Example 5 Discuss the convergent set of the functional series
$$\sum_{n=0}^{\infty} \frac{1}{2n+1} \left(\frac{1 - \sin x}{2} \right)^n$$

Solution Making transformation, setting $y = \frac{1 - \sin x}{2} \geq 0$, the series becomes the power series $\sum_{n=0}^{\infty} \frac{y^n}{2n+1}$, because
$$R_y = \lim_{n\to\infty} \frac{1}{2n+1}(2n+3) = 1$$
when $y = 1$, the series is $\sum_{n=0}^{\infty} \frac{1}{2n+1}$, which is divergent; and because $0 \leq y < 1$, the convergent set of the original functional series is $x \neq 2k\pi - \frac{\pi}{2}, k = 0, \pm 1, \cdots$

11.6.2 Computation of the Power Series

Suppose there are two power series

$$\sum_{n=0}^{\infty} a_n x^n = f(x), x \in (-A, A)$$

$$\sum_{n=0}^{\infty} b_n x^n = g(x), x \in (-B, B)$$

Let $R = \min\{A, B\}$. Obviously, the two power series are both absolute convergent in the interval $(-R, R)$, by the properties of the absolute convergent series, we have:

1. Addition and subtraction

$$\left(\sum_{n=0}^{\infty} a_n x^n\right) \pm \left(\sum_{n=0}^{\infty} b_n x^n\right) = \sum_{n=0}^{\infty} (a_n \pm b_n) x^n = f(x) \pm g(x), x \in (-R, R)$$

and $\sum_{n=0}^{\infty} (a_n \pm b_n) x^n$ is absolute convergent in $(-R, R)$.

2. Multiplication

$$\left(\sum_{n=0}^{\infty} a_n x^n\right) \left(\sum_{n=0}^{\infty} b_n x^n\right)$$
$$= a_0 b_0 + (a_0 b_1 + a_1 b_0) x + \cdots + (a_0 b_n + a_1 b_{n-1} + \cdots + a_n b_0) x^n + \cdots$$
$$= \sum_{n=0}^{\infty} \left(\sum_{i=0}^{n} a_i b_{n-i}\right) x^n = f(x) g(x), x \in (-R, R)$$

and $\sum_{n=0}^{\infty} \left(\sum_{i=0}^{n} a_i b_{n-i}\right) x^n$ is absolute convergent in $(-R, R)$.

3. Division

Because division is the inverse operation of multiplication, when $b_0 \neq 0$, define the quotient of the two power series as a power series

$$\frac{\sum_{n=0}^{\infty} a_n x^n}{\sum_{n=0}^{\infty} b_n x^n} = \frac{a_0 + a_1 x + a_2 x^2 + \cdots + a_n x^n + \cdots}{b_0 + b_1 x + b_2 x^2 + \cdots + b_n x^n + \cdots}$$

$$= c_0 + c_1 x + c_2 x^2 + \cdots + c_n x^n + \cdots = \sum_{n=0}^{\infty} c_n x^n$$

where $\sum_{n=0}^{\infty} c_n x^n$ satisfies

$$\left(\sum_{n=0}^{\infty} b_n x^n\right)\left(\sum_{n=0}^{\infty} c_n x^n\right) = \sum_{n=0}^{\infty} a_n x^n$$

Because the coefficients of the corresponding terms of the series on the two sides of the equation are the same, we can determine $c_n (n = 1, 2, \cdots)$, for example

$$c_0 = a_0/b_0, c_1 = (a_1 b_0 - a_0 b_1)/b_0^2$$
$$c_2 = (a_2 b_0^2 - a_1 b_0 b_1 - a_0 b_0 b_2 + a_0 b_1^2)/b_0^3, \cdots$$

The convergent radius of the series $\sum_{n=0}^{\infty} c_n x^n = \dfrac{f(x)}{g(x)}$ is smaller than the above R sometimes.

For example, the convergent radius of the power series 1 and $1 - x$ are both $+\infty$, but the quotient of them is

$$\frac{1}{1-x} = 1 + x + x^2 + \cdots + x^n + \cdots$$

and the convergent radius $R = 1$.

For the addition (subtraction) and the multiplication, we can only ensure the absolute convergence in the interval $(-R, R)$. When $A \neq B$, the convergent radius is just R; When $A = B$, the convergent radius is not smaller than this R.

Example 6 Find the convergent set of the power series $\sum_{n=1}^{\infty} (2^n + \sqrt{n})(x+1)^n$.

Solution Divide the original series into two series

$$\sum_{n=1}^{\infty} 2^n (x+1)^n, \sum_{n=1}^{\infty} \sqrt{n}(x+1)^n$$

The convergent radius of the former one

$$R_1 = \lim_{n \to \infty} \frac{2^n}{2^{n+1}} = \frac{1}{2}$$

The convergent radius of the later one

$$R_2 = \lim_{n \to \infty} \frac{\sqrt{n}}{\sqrt{n+1}} = 1$$

So, the convergent radius of the original power series

$$R = \min\left\{\frac{1}{2}, 1\right\} = \frac{1}{2}$$

i. e. the convergent interval is

$$-\frac{3}{2} < x < -\frac{1}{2}$$

When $x = -\frac{3}{2}$ and $x = -\frac{1}{2}$, the corresponding series is as follows in turn

$$\sum_{n=1}^{\infty}(-1)^n \frac{2^n + \sqrt{n}}{2^n}, \quad \sum_{n=1}^{\infty} \frac{2^n + \sqrt{n}}{2^n}$$

Because $\lim_{n\to\infty} \frac{2^n + \sqrt{n}}{2^n} = 1 \neq 0$, so the two series both diverges, and so the convergent set of the original power series is $\left(-\frac{3}{2}, -\frac{1}{2}\right)$.

In the following we discuss the analytical properties of the power series. Firstly we introduce a theorem about the uniform convergence of the power series.

Theorem 11.12 The power series(1) is convergent uniformly in every closed subinterval of $[-R_1, R_1]$ (Where $0 < R_1 < R$).

Proof Suppose $R_1 \in (-R, R)$, so the series (1) is absolute convergent at the point R_1, that is $\sum_{n=0}^{\infty} |a_n R_1^n|$ converges. For any point x in the interval $[-R_1, R_1]$, we have

$$|a_n x^n| \leq |a_n R_1^n|, n = 1, 2, \cdots$$

So, by Theorem 11.10, the series(1) is uniformly convergent in $[-R_1, R_1]$.

The power series have the following analytic properties:

4. In the convergent set, the sum function $f(x)$ of the power series $\sum_{n=0}^{\infty} a_n x^n$ is continuous.

5. In the convergent set, the power series can be integrated term by term, and the convergent radius doesn't change. That is

$$\int_0^x f(x)\,dx = \sum_{n=0}^{\infty}\left(a_n \int_0^x x^n\,dx\right) = \sum_{n=0}^{\infty} \frac{a_n}{n+1} x^{n+1}$$

6. In the convergent set, the power series can be differentiated term by term, and the convergent radius doesn't change. That is

$$f'(x) = \sum_{n=0}^{\infty}(a_n x^n)' = \sum_{n=1}^{\infty} na_n x^{n-1}$$

For the power series, after differentiation and integration term by term, although

Calculus(II)

the convergent radius and interval don't change, the convergent set may change. For example, for the series

$$\frac{1}{1+x} = 1 - x + x^2 - \cdots + (-1)^n x^n + \cdots$$

the convergent set is $(-1,1)$, but the power series after integration term by term

$$\ln(1+x) = x - \frac{x^2}{2} + \frac{x^3}{3} - \cdots + (-1)^{n-1}\frac{x^n}{n} + \cdots$$

has convergent set $(-1,1]$. The reason is when $x = 1$, the left function (sum function) has definition and is continuous, and the right series converges.

By Property 6, the function expressed by a power series is a "good" function and is infinitely continuously differentiable. Sometimes the convergent interval of the power series is not the same as the domain of the sum function. In the convergent set, the computation of the power series is similar to the computation of the polynomials. Sometimes the sum function of a power series is not an elementary function which makes it useful in the integrating computation and solving differential equations.

11.7 Expanding Functions as Power Series

As we shown, the sum of a power series is a continuous and infinitely differentiable function in the interval of convergence. In this section we shall discuss the reverse problem:

Given a function $f(x)$, what conditions guarantee that it can be represented as a power series?

The properties of power series indicate that if a function $f(x)$ is the sum of a power series it must be infinitely differentiable. But, as what will be shown, this condition is not sufficient.

11.7.1 Direct Expanding Method, Taylor Series

Reviewing the Taylor Formula in Chapter 4: If the function $f(x)$ has $n + 1$-th order derivatives, then $f(x)$ can be expressed as

$$f(x) = f(x_0) + \frac{f'(x_0)}{1!}(x - x_0) + \frac{f''(x_0)}{2!}(x - x_0)^2 + \cdots +$$

Chapter 11 Infinite Series

$$\frac{f^{(n)}(x_0)}{n!}(x-x_0)^n + R_n(x) \tag{1}$$

where

$$R_n(x) = \frac{f^{(n+1)}(\xi)}{(n+1)!}(x-x_0)^{n+1}, \xi \text{ is between } x_0 \text{ and } x$$

Equation (1) is just the Taylor Formula of the function $f(x)$ expanding at the point x_0, $R_n(x)$ is the Lagrange remainder.

If $f(x)$ is continuous and infinitely differentiable in some neighborhood $U(x_0)$ of the point x_0, denoted $f(x) \in C^\infty(U(x_0))$. It is natural to ask if the function $f(x)$ can be expanded as the following power series

$$f(x_0) + \frac{f'(x_0)}{1!}(x-x_0) + \frac{f''(x_0)}{2!}(x-x_0)^2 + \cdots + \frac{f^{(n)}(x_0)}{n!}(x-x_0)^n + \cdots \tag{2}$$

To study this problem, we firstly give the following definition:

The power series (2) is called the Taylor Series of the function at the point x_0. Especially, when $x_0 = 0$, the power series

$$f(0) + \frac{f'(0)}{1!}x + \frac{f''(0)}{2!}x^2 + \cdots + \frac{f^{(n)}(0)}{n!}x^n + \cdots \tag{3}$$

is the Maclaurin Series of $f(x)$.

Obviously, the range of the Taylor series (2) converging to the function $f(x)$ is determined by the range $R_n(x) \to 0$.

Theorem 11.13 Suppose $f(x) \in C^\infty(U(x_0))$, then the necessary and sufficient condition for its Taylor series

$$\sum_{n=0}^{\infty} \frac{f^{(n)}(x_0)}{n!}(x-x_0)^n$$

being convergent to $f(x)$ in $U(x_0)$ is

$$\lim_{n\to\infty} R_n(x) = 0, \forall x \in U(x_0) \tag{4}$$

Proof Let $S_n(x)$ be the sum of the first $n+1$ terms of the Taylor series (2), by the Taylor formula (1)

$$R_n(x) = f(x) - S_n(x), S_n(x) = f(x) - R_n(x)$$

(**Necessity**) Suppose the Taylor series (2) converges to $f(x)$ in $U(x_0)$, then $\forall x \in U(x_0)$, $\lim_{n\to\infty} S_n(x) = f(x)$, and so

$$\lim_{n\to\infty} R_n(x) = \lim_{n\to\infty}[f(x) - S_n(x)] = 0, \forall x \in U(x_0)$$

(**Sufficiency**) Suppose (4) is right, then
$$\lim_{n\to\infty} S_n(x) = \lim_{n\to\infty}[f(x) - R_n(x)] = f(x), \forall x \in U(x_0)$$
i.e. in $U(x_0)$, the Taylor Series (2) converges to $f(x)$.

When condition (4) is not satisfied, although the Taylor series (2) of the function converges, it doesn't converge to $f(x)$. Such as, the function
$$f(x) = \begin{cases} e^{-1/x^2}, & \text{when } x \neq 0 \\ 0, & \text{when } x = 0 \end{cases}$$
see Fig. 11.4. Because
$$f(0) = f'(0) = f''(0) = \cdots = 0$$
so, all coefficients of the Maclaurin series of the function $f(x)$ are zero, it is obvious that it converges to zero in the whole axis. Except for $x = 0$, it doesn't converge to the original function $f(x)$ at any point x.

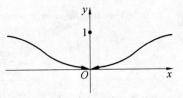

Fig. 11.4

This is because the equation (4) doesn't exist except for the original point.

Theorem 11.14 (Uniqueness Theorem) If the function $f(x)$ can be expanded as a power series in some neighborhood of the point x_0
$$f(x) = \sum_{n=0}^{\infty} a_n (x - x_0)^n \tag{5}$$
then its coefficients
$$a_n = \frac{f^{(n)}(x_0)}{n!} (n = 0, 1, 2, \cdots)$$
here $0! = 1, f^{(0)}(x_0) = f(x_0)$.

Proof Because the power series can be differentiated term by term in the convergent interval, so
$$f(x) = a_0 + a_1(x - x_0) + a_2 (x - x_0)^2 + \cdots + a_n (x - x_0)^n + \cdots$$
$$f'(x) = a_1 + 2a_2(x - x_0) + 3a_3 (x - x_0)^2 + \cdots + na_n (x - x_0)^{n-1} + \cdots$$
$$f''(x) = 2! \, a_2 + 3 \cdot 2a_3(x - x_0) + \cdots + n(n-1)a_n (x - x_0)^{n-2} + \cdots$$
$$\vdots$$
$$f^{(n)}(x) = n! \, a_n + \cdots$$
$$\vdots$$

Chapter 11 Infinite Series

Let $x = x_0$, we get

$$a_0 = f(x_0), a_1 = \frac{f'(x_0)}{1!}, a_2 = \frac{f''(x_0)}{2!}, \cdots, a_n = \frac{f^{(n)}(x_0)}{n!}, \cdots$$

These two theorems tell us: In some neighborhood of x_0, if the function $f(x)$ has derivatives of any order, and the remainder $R_n(x)$ of its Taylor formula converges to zero (when $n \to \infty$), then $f(x)$ can be expanded as a power series, and its expansion is unique, which is just the Taylor series of $f(x)$. So, to expand the function as a power series near x_0, we have the direct expanding method as following:

The first step, compute derivatives of any order of $f(x)$, that is

$$f'(x), f''(x), \cdots, f^{(n)}(x), \cdots$$

The second step, compute

$$f(x_0), f'(x_0), f''(x_0), \cdots, f^{(n)}(x_0), \cdots$$

The third step, writing the Taylor series

$$f(x_0) + \frac{f'(x_0)}{1!}(x - x_0) + \frac{f''(x_0)}{2!}(x - x_0)^2 + \cdots + \frac{f^{(n)}(x_0)}{n!}(x - x_0)^n + \cdots$$

and determine the convergent radius and convergent set;

The fourth step, in the convergent set, compute the interval such that $\lim\limits_{n \to \infty} R_n(x) = 0$, this interval is the expanding interval of the power series of the function.

Example 1 Expand $f(x) = e^x$ as a power series of x.

Solution By $f^{(n)}(x) = e^x (n = 0,1,2,\cdots)$, $f^{(n)}(0) = 1 (n = 0,1,2,\cdots)$, so the Taylor series of e^x is

$$1 + x + \frac{x^2}{2!} + \frac{x^3}{3!} + \cdots + \frac{x^n}{n!} + \cdots$$

Its convergent radius is

$$R = \lim_{n \to \infty} \left| \frac{a_n}{a_{n+1}} \right| = \lim_{n \to \infty} \frac{1}{n!}(n+1)! = +\infty$$

The remainder of the Taylor formula is

$$R_n(x) = \frac{e^\xi}{(n+1)!} x^{n+1}, \xi \text{ is between } 0, x$$

It satisfies

$$|R_n(x)| = \left| \frac{e^\xi}{(n+1)!} x^{n+1} \right| \leq e^{|x|} \frac{|x|^{n+1}}{(n+1)!}$$

For any fixed $x \in (-\infty, +\infty)$, $e^{|x|}$ is a given number, but $\dfrac{|x|^{n+1}}{(n+1)!}$ is the general term of the power series $\sum_{n=0}^{\infty} \dfrac{|x|^n}{n!}$ which is convergent everywhere, so on the interval $(-\infty, +\infty)$, we have

$$\lim_{n \to \infty} R_n(x) = 0$$

so

$$e^x = \sum_{n=0}^{\infty} \frac{x^n}{n!} = 1 + x + \frac{x^2}{2!} + \cdots + \frac{x^n}{n!} + \cdots, x \in (-\infty, +\infty) \tag{6}$$

Example 2 Expanding $f(x) = \sin x$ as a power series of x.

Solution By the Taylor formula of $\sin x$ we know its Taylor series is

$$x - \frac{x^3}{3!} + \frac{x^5}{5!} - \cdots + (-1)^n \frac{x^{2n+1}}{(2n+1)!} + \cdots$$

The convergent radius is $R = \infty$.

For any point x in the convergent interval, there exists a point $\xi \in (0, x)$, such that

$$|R_{2n+2}(x)| = \left| \frac{f^{(2n+3)}(\xi)}{(2n+3)!} x^{2n+3} \right|$$

$$= \frac{|x|^{2n+3}}{(2n+3)!} \left| \sin\left[\xi + (2n+3)\frac{\pi}{2}\right] \right|$$

$$\leq \frac{|x|^{2n+3}}{(2n+3)!} \to 0, \text{ as } n \to \infty$$

so

$$\sin x = \sum_{n=0}^{\infty} (-1)^n \frac{x^{2n+1}}{(2n+1)!} = x - \frac{x^3}{3!} + \frac{x^5}{5!} - \cdots +$$

$$(-1)^n \frac{x^{2n+1}}{(2n+1)!} + \cdots, x \in (-\infty, +\infty) \tag{7}$$

Example 3 Expanding $f(x) = (1+x)^\alpha$ as a power series of x where α is any real constant.

Solution By the Taylor formula of $(1+x)^\alpha$ we know its Taylor series is

$$1 + \alpha x + \frac{\alpha(\alpha-1)}{2!} x^2 + \cdots + \frac{\alpha(\alpha-1)\cdots(\alpha-n+1)}{n!} x^n + \cdots$$

and the convergent radius

Chapter 11 Infinite Series

$$R = \lim_{n\to\infty}\left|\frac{a_n}{a_{n+1}}\right| = \lim_{n\to\infty}\left|\frac{n+1}{\alpha-n}\right| = 1$$

so the convergent interval of the Taylor series of $(1+x)^\alpha$ is $(-1,1)$. At $x = \pm 1$, for different α, the convergence is different.

To avoid discussing the limit of the remainder, we suppose the sum function of the Taylor series of $(1+x)^\alpha$ on the interval $(-1,1)$ is $F(x)$, that is

$$F(x) = 1 + \alpha x + \frac{\alpha(\alpha-1)}{2!}x^2 + \cdots + \frac{\alpha(\alpha-1)\cdots(\alpha-n+1)}{n!}x^n + \cdots, x \in (-1,1)$$

In the following, we prove $F(x) = (1+x)^\alpha, x \in (-1,1)$. By differentiating term by term

$$F'(x) = \alpha\left[1 + \frac{\alpha-1}{1!}x + \cdots + \frac{(\alpha-1)\cdots(\alpha-n+1)}{(n-1)!}x^{n-1} + \cdots\right]$$

After multiplying $(1+x)$ on both sides, the coefficients of x^n in the right square brackets are

$$\frac{(\alpha-1)\cdots(\alpha-n+1)}{(n-1)!} + \frac{(\alpha-1)\cdots(\alpha-n)}{n!} = \frac{\alpha(\alpha-1)\cdots(\alpha-n+1)}{n!}$$

So, the following differential equation is obtained

$$(1+x)F'(x) = \alpha\left[1 + \alpha x + \frac{\alpha(\alpha-1)}{2!}x^2 + \cdots + \frac{\alpha(\alpha-1)\cdots(\alpha-n+1)}{n!}x^n + \cdots\right]$$

$$= \alpha F(x), x \in (-1,1)$$

It satisfies the condition $F(0) = 1$. By the method of separation of variables

$$F(x) = (1+x)^\alpha, x \in (-1,1)$$

so

$$(1+x)^\alpha = 1 + \alpha x + \frac{\alpha(\alpha-1)}{2!}x^2 + \cdots + \frac{\alpha(\alpha-1)\cdots(\alpha-n+1)}{n!}x^n + \cdots$$

$$x \in (-1,1) \tag{8}$$

Equation(8) is called the Newton binomial expansion. When α is a positive integer, (8) is the binomial formula.

When $\alpha = \frac{1}{2}, -\frac{1}{2}$, here are the expansions in turn

$$\sqrt{1+x} = 1 + \frac{1}{2}x - \frac{1}{2\cdot4}x^2 + \frac{1\cdot3}{2\cdot4\cdot6}x^3 - \cdots +$$

$$(-1)^{n-1}\frac{1\cdot3\cdot5\cdot\cdots\cdot(2n-3)}{2\cdot4\cdot6\cdot\cdots\cdot(2n)}x^n + \cdots, x \in [-1,1] \tag{9}$$

Calculus(Ⅱ)

$$\frac{1}{\sqrt{1+x}} = 1 - \frac{1}{2}x + \frac{1\cdot 3}{2\cdot 4}x^2 - \frac{1\cdot 3\cdot 5}{2\cdot 4\cdot 6}x^3 + \cdots +$$
$$(-1)^n \frac{1\cdot 3\cdot 5\cdot \cdots \cdot (2n-1)}{2\cdot 4\cdot 6\cdot \cdots \cdot (2n)} x^n + \cdots, x \in (-1,1] \qquad (10)$$

11.7.2 Indirect Expanding Method

Sometimes the computation are very huge in the direct expanding method. In the following, we will introduce the indirect expanding method. It uses the known expansions of functions, for example the formulas (6),(7),(8), and by the variable substitution, computation of the power series and so on, we get the method of expanding power series of functions. By the uniqueness of expansion, it gets the same result as the direct expanding.

Example 4 Expand $f(x) = \cos x$ as a power series of x.

Solution Differentiating the two sides term by term of the expansion (7) of $\sin x$, we get the expansion of $\cos x$

$$\cos x = \sum_{n=0}^{\infty} (-1)^n \frac{x^{2n}}{(2n)!} = 1 - \frac{x^2}{2!} + \frac{x^4}{4!} - \cdots + (-1)^n \frac{x^{2n}}{(2n)!} + \cdots$$
$$x \in (-\infty, +\infty) \qquad (11)$$

Example 5 Expand the function $f(x) = \ln(1+x)$ as the Maclaurin Series.

Solution Because $[\ln(1+x)]' = \dfrac{1}{1+x}$, and

$$\frac{1}{1+x} = 1 - x + x^2 - \cdots + (-1)^n x^n + \cdots, x \in (-1,1) \qquad (12)$$

Integrating term by term from 0 to x, we get

$$\ln(1+x) = x - \frac{x^2}{2} + \frac{x^3}{3} - \cdots + (-1)^{n-1} \frac{x^n}{n} + \cdots, x \in (-1,1] \qquad (13)$$

The equation (13) is also established for $x = 1$, it is because when $x = 1$, the right series converges, the left function is continuous at $x = 1$, we have

$$\ln 2 = 1 - \frac{1}{2} + \frac{1}{3} - \cdots + (-1)^{n-1} \frac{1}{n} + \cdots$$

Example 6 Expand the function $f(x) = \dfrac{1}{2}\ln\dfrac{1+x}{1-x}$ as a Maclaurin Series.

Solution Make the substitution for (13), substitute x by $-x$, getting the expan-

Chapter 11 Infinite Series

sion

$$\ln(1-x) = -x - \frac{x^2}{2} - \frac{x^3}{3} - \cdots - \frac{x^n}{n} - \cdots, x \in [-1, 1) \quad (14)$$

The equation (13) minus (14), then divided by 2, we get

$$\frac{1}{2}\ln\frac{1+x}{1-x} = x + \frac{x^3}{3} + \frac{x^5}{5} + \cdots + \frac{x^{2n-1}}{2n-1} + \cdots, x \in (-1, 1) \quad (15)$$

Example 7 Evaluate $1 - \frac{1}{3} + \frac{1}{5} - \frac{1}{7} + \cdots$

Solution Integrating

$$\frac{1}{1+x^2} = 1 - x^2 + x^4 - x^6 + \cdots + (-1)^n x^{2n} + \cdots, x \in (-1, 1)$$

from 0 to x, and term by term, we get

$$\arctan x = x - \frac{x^3}{3} + \frac{x^5}{5} - \cdots + (-1)^n \frac{x^{2n+1}}{2n+1} + \cdots, x \in [-1, 1] \quad (16)$$

The equation (16) is also true for $x = \pm 1$, so

$$\frac{\pi}{4} = \arctan 1 = 1 - \frac{1}{3} + \frac{1}{5} - \frac{1}{7} + \cdots$$

Notice the convergence of the two endpoints of the interval when using the indirect expanding method.

Example 8 Expand the function $\sin x$ as a power series of $\left(x - \frac{\pi}{4}\right)$.

Solution Making the transformation, letting $t = x - \frac{\pi}{4}$, then $x = t + \frac{\pi}{4}$, so

$$\sin x = \sin\left(t + \frac{\pi}{4}\right) = \sin\frac{\pi}{4}\cos t + \cos\frac{\pi}{4}\sin t = \frac{\sqrt{2}}{2}(\cos t + \sin t)$$

$$= \frac{\sqrt{2}}{2}\left[\sum_{n=0}^{\infty}(-1)^n \frac{t^{2n}}{(2n)!} + \sum_{n=0}^{\infty}(-1)^n \frac{t^{2n+1}}{(2n+1)!}\right]$$

$$= \frac{\sqrt{2}}{2}\sum_{n=0}^{\infty}(-1)^n \left[\frac{t^{2n}}{(2n)!} + \frac{t^{2n+1}}{(2n+1)!}\right]$$

$$= \frac{\sqrt{2}}{2}\left[1 + \left(x - \frac{\pi}{4}\right) - \frac{\left(x - \frac{\pi}{4}\right)^2}{2!} - \frac{\left(x - \frac{\pi}{4}\right)^3}{3!} + \frac{\left(1 - \frac{\pi}{4}\right)^4}{4!} + \frac{\left(x - \frac{\pi}{4}\right)^5}{5!} - \cdots\right]$$

$x \in (-\infty, +\infty)$

Example 9 Evaluate the Taylor series of $f(x) = \dfrac{1}{x^2 + x}$ at the point $x_0 = -2$.

Solution Making the transformation, letting $t = x + 2$, then $x = t - 2$

$$\frac{1}{x^2 + x} = \frac{1}{x} - \frac{1}{x+1} = \frac{1}{t-2} - \frac{1}{t-1} = \frac{1}{1-t} - \frac{\frac{1}{2}}{1-\frac{t}{2}}$$

Because

$$\frac{1}{1-t} = \sum_{n=0}^{\infty} t^n, t \in (-1,1)$$

$$\frac{\frac{1}{2}}{1-\frac{t}{2}} = \sum_{n=0}^{\infty} \frac{1}{2}\left(\frac{t}{2}\right)^n, t \in (-2,2)$$

so

$$\frac{1}{1-t} - \frac{\frac{1}{2}}{1-\frac{t}{2}} = \sum_{n=0}^{\infty} \frac{2^{n+1} - 1}{2^{n+1}} t^n, t \in (-1,1)$$

so

$$\frac{1}{x^2 + x} = \sum_{n=0}^{\infty} \frac{2^{n+1} - 1}{2^{n+1}} (x+2)^n, x \in (-3, -1)$$

Example 10 Expand $f(x) = \dfrac{e^x}{1-x}$ as a power series of x, and compute $f'''(0)$.

Solution By

$$\frac{1}{1-x} = 1 + x + x^2 + \cdots + x^n + \cdots, |x| < 1$$

$$e^x = 1 + \frac{x}{1!} + \frac{x^2}{2!} + \cdots + \frac{x^n}{n!} + \cdots, |x| < +\infty$$

Multiply them together to get

$$\frac{e^x}{1-x} = 1 + \left(1 + \frac{1}{1!}\right)x + \left(1 + \frac{1}{1!} + \frac{1}{2!}\right)x^2 + \cdots + \left(1 + \frac{1}{1!} + \frac{1}{2!} + \cdots + \frac{1}{n!}\right)x^n + \cdots, |x| < 1$$

Because $f^{(n)}(0) = n!\, a_n$, so

$$f'''(0) = 3! \left(1 + \frac{1}{1!} + \frac{1}{2!} + \frac{1}{3!}\right) = 16$$

Example 11 Suppose $f(x) = \begin{cases} \dfrac{\sin x}{x}, & x \neq 0 \\ 1, & x = 0 \end{cases}$ try to write out the Maclaurin series of the function $\ln f(x)$ to the term x^4.

Solution Because

$$f(x) = 1 - \frac{x^2}{3!} + \frac{x^4}{5!} - \cdots, x \in (-\infty, \infty)$$

$$\ln(1+t) = t - \frac{t^2}{2} + \frac{t^3}{3} - \cdots, x \in (-1, 1]$$

so

$$\ln f(x) = \left(-\frac{x^2}{3!} + \frac{x^4}{5!} - \cdots\right) - \frac{1}{2}\left(-\frac{x^2}{3!} + \frac{x^4}{5!} - \cdots\right)^2 + \cdots$$

$$= -\frac{x^2}{3!} + \frac{x^4}{5!} - \frac{x^4}{2(3!)^2} + \cdots$$

$$= -\frac{x^2}{6} - \frac{x^4}{180} - \cdots, x \in (-\pi, \pi)$$

Sometimes the infinite series are used to popularize the new concepts, for example, we can use the expansion of e^x to define the exponential matrix: Suppose A is a $n \times n$ square matrix, define

$$e^A = E + A + \frac{A^2}{2!} + \cdots + \frac{A^n}{n!} + \cdots \tag{17}$$

where E is the nth order unit matrix.

And for example, substitute x (where $i = \sqrt{-1}$) as ix, define

$$e^{ix} = 1 + (ix) + \frac{(ix)^2}{2!} + \cdots + \frac{(ix)^n}{n!} + \cdots \tag{18}$$

It is easy to conclude

$$e^{ix} = \left(1 - \frac{x^2}{2!} + \frac{x^4}{4!} - \cdots\right) + i\left(x - \frac{x^3}{3!} + \frac{x^5}{5!} - \cdots\right)$$

$$= \cos x + i\sin x \tag{19}$$

Substitute the x of (19) by $-x$, we get

$$e^{-ix} = \cos x - i\sin x \tag{20}$$

Adding (19), (20) together and divide by 2, or subtracting and divide by 2i, we get

Calculus(II)

$$\cos x = \frac{e^{ix} + e^{-ix}}{2}, \sin x = \frac{e^{ix} - e^{-ix}}{2i} \tag{21}$$

The equations (19) ~ (21) are called the Euler Formulas, they indicate the relations between the trigonometric functions and the exponential functions.

11.7.3 Computing the Sum Function of the Power Series

The question discussed in this paragraph is the inverse question of the above paragraph—computing the sum function of the power series. Because not all the sum functions are elementary functions, so we should not compute the sum of any power series. But now, we can use the sum formula of the geometric series, the expansion of e^x, $\sin x$ or more functions, by the variables transforming, the computation of the power series and so on, to compute the sum function of some power series.

Example 12 Compute the sum function of the power series $\sum_{n=1}^{\infty} \frac{x^{4n+1}}{4n+1}$.

Solutions Transform the series to

$$x \sum_{n=1}^{\infty} \frac{x^{4n}}{4n+1} = x \sum_{n=1}^{\infty} \frac{t^n}{4n+1} (t = x^4 \geq 0)$$

then

$$R_t = \lim_{n \to \infty} \left(\frac{1}{4n+1} \bigg/ \frac{1}{4n+5} \right) = 1$$

And when $t=1$, $\sum_{n=1}^{\infty} \frac{1}{4n+1}$ diverges, so the convergent set of the series $\sum_{n=1}^{\infty} \frac{t^n}{4n+1}$ is $[0,1]$, so the convergent set of the discussed series is $(-1,1)$. Suppose the sum function is $S(x)$, that is to suppose

$$S(x) = \sum_{n=1}^{\infty} \frac{x^{4n+1}}{4n+1}, x \in (-1,1)$$

then $S(0) = 0$. Differentiating term by term, we get

$$S'(x) = \sum_{n=1}^{\infty} x^{4n} = \frac{x^4}{1-x^4}$$

Integrating from 0 to x, the sum function is

$$S(x) = S(x) - S(0) = \int_0^x \frac{t^4}{1-t^4} dt = \int_0^x \left(\frac{1/2}{1-t^2} + \frac{1/2}{1+t^2} - 1 \right) dt$$

$$= \frac{1}{4} \ln \left| \frac{1+x}{1-x} \right| + \frac{1}{2} \arctan x - x, x \in (-1,1)$$

Chapter 11 Infinite Series

Example 13 Compute the sum function of $\sum_{n=1}^{+\infty} n(n+2)x^n$.

Solution The convergent radius

$$R = \lim_{n\to\infty} \left|\frac{a_n}{a_{n+1}}\right| = \lim_{n\to\infty} \frac{n(n+2)}{(n+1)(n+3)} = 1$$

And when $x = \pm 1$, the series $\sum_{n=1}^{\infty} n(n+1)$, $\sum_{n=1}^{\infty} (-1)^{n-1} n(n-1)$ both diverge. In all, the convergent set of the power series is $(-1, 1)$.

$$\sum_{n=1}^{+\infty} n(n+2)x^n = \sum_{n=1}^{+\infty} (n+1)nx^n + \sum_{n=1}^{+\infty} nx^n$$

$$= x \frac{d^2}{dx^2} \left(\sum_{n=1}^{+\infty} x^{n+1} \right) + x \frac{d}{dx} \left(\sum_{n=1}^{+\infty} x^n \right)$$

$$= x \frac{d^2}{dx^2} \left(\frac{1}{1-x} - 1 - x \right) + x \frac{d}{dx} \left(\frac{1}{1-x} - 1 \right)$$

$$= \frac{2x}{(1-x)^3} + \frac{x}{(1-x)^2} = \frac{x(3-x)}{(1-x)^3}, \quad -1 < x < 1$$

Example 14 Compute the sum function of the power function $\sum_{n=1}^{\infty} \frac{1}{n2^n} x^{n-1}$.

Solution The convergent radius

$$R = \lim_{n\to\infty} \left|\frac{a_n}{a_{n+1}}\right| = \lim_{n\to\infty} \frac{\frac{1}{n2^n}}{\frac{1}{(n+1)2^{n+1}}} = 2$$

When $x = 2$, the series is $\sum_{n=1}^{\infty} \frac{1}{2n}$, diverges, when $x = -2$, the series is $\sum_{n=1}^{\infty} (-1)^{n-1} \frac{1}{2n}$, converges, so the convergent set of the original set is $[-2, 2)$.

Suppose the sum function to be computed is $S(x)$, that is

$$S(x) = \sum_{n=1}^{\infty} \frac{1}{n2^n} x^{n-1}, x \in [-2, 2)$$

and then

$$xS(x) = \sum_{n=1}^{\infty} \frac{1}{n} \left(\frac{x}{2}\right)^n$$

$$[xS(x)]' = \sum_{n=1}^{\infty} \frac{1}{2}\left(\frac{x}{2}\right)^{n-1} = \frac{\frac{1}{2}}{1-\frac{x}{2}} = \frac{1}{2-x}$$

Integrating from 0 to x for the two sides of the above equation, we get

$$xS(x) = \int_0^x \frac{dx}{2-x} = -\ln(2-x)\Big|_0^x = -\ln(2-x) + \ln 2$$

$$= -\ln\left(1 - \frac{x}{2}\right)$$

so

$$S(x) = -\frac{1}{x}\ln\left(1 - \frac{x}{2}\right), x \in [-2,0) \cup (0,2)$$

When $x = 0$, it is obvious that

$$S(0) = \frac{1}{2}$$

To sum up

$$\sum_{n=1}^{\infty} \frac{1}{n2^n} x^{n-1} = \begin{cases} -\frac{1}{x}\ln\left(1-\frac{x}{2}\right), \text{when } x \in [-2,0) \cup (0,2) \\ \frac{1}{2}, \text{when } x = 0 \end{cases}$$

Example 15 Compute the sum of the number series $\sum_{n=0}^{\infty} \frac{(n+1)^2}{n!}$.

Solution This number series is the series corresponds to the power series $\sum_{n=0}^{\infty} \frac{(n+1)^2}{n!} x^n$ at $x = 1$. It is obvious that the convergent set of the power series is $(-\infty, +\infty)$. Firstly, compute the sum function of this power series, because

$$S(x) = \sum_{n=0}^{\infty} \frac{(n+1)^2}{n!} x^n = \sum_{n=0}^{\infty} \frac{n(n-1) + 3n + 1}{n!} x^n$$

$$= \sum_{n=2}^{\infty} \frac{x^n}{(n-2)!} + 3\sum_{n=1}^{\infty} \frac{x^n}{(n-1)!} + \sum_{n=0}^{\infty} \frac{x^n}{n!}$$

$$= x^2 \sum_{k=0}^{\infty} \frac{x^k}{k!} + 3x \sum_{k=0}^{\infty} \frac{x^k}{k!} + \sum_{k=0}^{\infty} \frac{x^n}{n!} = (x^2 + 3x + 1)e^x$$

Here the Taylor series of e^x is used, so we have

$$\sum_{n=0}^{\infty} \frac{(n+1)^2}{n!} = S(1) = 5e$$

Chapter 11 Infinite Series

11.8 Some Applications of the Power Series

Using the expansion of functions as power series, some problems about the polynomial approximation of functions can be solved. At the same time, because some sum function of power series is not elementary, so it can be used in some integration and differential equations.

11.8.1 Approximation of Function Value

Example 1 Compute the value of e, let the error $r_n < 0.0001$.

Solution Because

$$e^x = 1 + x + \frac{x^2}{2!} + \cdots + \frac{x^n}{n!} + \cdots, x \in (-\infty, +\infty)$$

So, when $x = 1$, we have

$$e = 1 + 1 + \frac{1}{2!} + \cdots + \frac{1}{n!} + \cdots$$

If we select the first $n + 1$ terms to approximate e, then the error

$$|r_n| = \left| \frac{1}{(n+1)!} + \frac{1}{(n+2)!} + \cdots \right|$$

$$< \frac{1}{(n+1)!}\left[1 + \frac{1}{n+1} + \frac{1}{(n+1)^2} + \cdots \right]$$

$$= \frac{1}{(n+1)!} \cdot \frac{1}{1 - \frac{1}{n+1}} = \frac{1}{n! \, n}$$

To make $\dfrac{1}{n! \, n} < 0.0001$, we need to select $n = 7$, so

$$e \approx 2 + \frac{1}{2!} + \cdots + \frac{1}{7!} = \frac{1370}{504} \approx 2.7183$$

Example 2 Compute the approximation of $\ln 2$, accurate to the fourth decimal point.

Solution By the Maclaurin series of $\ln(1 + x)$

$$\ln 2 = 1 - \frac{1}{2} + \frac{1}{3} - \frac{1}{4} + \cdots + (-1)^{n-1}\frac{1}{n} + \cdots$$

This is an alternating series, its error

$$|r_n| < \frac{1}{n+1}$$

To make $|r_n| < 0.0001$, we select $n = 9999$ at least. It looks like this series converges too slowly, the computation is too much. We look for another quicker power series to compute $\ln 2$.

By (15) in Section 11.7

$$\ln\frac{1+x}{1-x} = 2\left(x + \frac{1}{3}x^3 + \frac{1}{5}x^5 + \cdots + \frac{1}{2n+1}x^{2n+1} + \cdots\right), x \in (-1, 1)$$

Let $\frac{1+x}{1-x} = 2$ and get $x = \frac{1}{3}$, substitute it into the above formula we get

$$\ln 2 = 2\left(\frac{1}{3} + \frac{1}{3}\cdot\frac{1}{3^3} + \frac{1}{5}\cdot\frac{1}{3^5} + \cdots + \frac{1}{2n+1}\cdot\frac{1}{3^{2n+1}} + \cdots\right)$$

If select the former n terms, the error is

$$|r_n| = 2\left[\frac{1}{2n+1}\cdot\frac{1}{3^{2n+1}} + \frac{1}{2n+3}\cdot\frac{1}{3^{2n+3}} + \frac{1}{2n+5}\cdot\frac{1}{3^{2n+5}} + \cdots\right]$$

$$< \frac{2}{(2n+1)3^{2n+1}}\left[1 + \frac{1}{9} + \left(\frac{1}{9}\right)^2 + \cdots\right]$$

$$= \frac{2}{(2n+1)3^{2n+1}}\cdot\frac{1}{1-\frac{1}{9}} = \frac{1}{4(2n+1)3^{2n-1}}$$

To make $|r_n| < 0.0001$, we only need to select $n = 4$, so

$$\ln 2 \approx 2\left[\frac{1}{3} + \frac{1}{3\cdot 3^3} + \frac{1}{5\cdot 3^5} + \frac{1}{7\cdot 3^7}\right]$$

Considering the error, compute the fifth after decimal point

$$\frac{1}{3} \approx 0.33333, \frac{1}{3\cdot 3^3} \approx 0.01235$$

$$\frac{1}{5\cdot 3^5} \approx 0.00082, \frac{1}{7\cdot 3^7} \approx 0.00007$$

So

$$\ln 2 \approx 0.6931$$

11.8.2 Application in Integration

Example 3 Compute the approximation of $\int_0^1 \frac{\sin x}{x}dx$, accurate to 10^{-4}.

Solution Because

$$\sin x = x - \frac{x^3}{3!} + \frac{x^5}{5!} - \frac{x^7}{7!} + \cdots + (-1)^n \frac{x^{2n+1}}{(2n+1)!} + \cdots$$

so

$$\frac{\sin x}{x} = 1 - \frac{x^2}{3!} + \frac{x^4}{5!} - \frac{x^6}{7!} + \cdots + (-1)^n \frac{x^{2n}}{(2n+1)!} + \cdots$$

$$x \in (-\infty, +\infty), x \neq 0$$

By $\lim\limits_{x \to 0} \frac{\sin x}{x} = 1$, so $\int_0^x \frac{\sin x}{x} dx$ is the normal definite integral

$$\int_0^x \frac{\sin x}{x} dx = \int_0^x \left[1 - \frac{x^2}{3!} + \frac{x^4}{5!} - \frac{x^6}{7!} + \cdots + (-1)^n \frac{x^{2n}}{(2n+1)!} + \cdots \right] dx$$

$$= x - \frac{x^3}{3! \cdot 3} + \frac{x^5}{5! \cdot 5} - \frac{x^7}{7! \cdot 7} + \cdots +$$

$$(-1)^n \frac{x^{2n+1}}{(2n+1)!(2n+1)} + \cdots, x \in (-\infty, +\infty)$$

Let $x = 1$, we get

$$\int_0^1 \frac{\sin x}{x} dx = 1 - \frac{1}{3! \cdot 3} + \frac{1}{5! \cdot 5} - \frac{1}{7! \cdot 7} + \cdots + (-1)^n \frac{1}{(2n+1)!(2n+1)} + \cdots$$

This is an alternating series, if select the former three terms, then the error

$$|r_3| < \frac{1}{7! \cdot 7} = \frac{1}{35\,280} < 10^{-4}$$

so

$$\int_0^1 \frac{\sin x}{x} dx \approx 1 - \frac{1}{3! \cdot 3} + \frac{1}{5! \cdot 5} = 1 - \frac{97}{1\,800} \approx 0.946\,11 \approx 0.946\,1$$

11.8.3 Solving Equations from Power Series

Example 4 The equation

$$xy - e^x + e^y = 0 \tag{1}$$

determines y as a function of x, express y as power series of x.

Solution Suppose

$$y = a_0 + a_1 x + a_2 x^2 + a_3 x^3 + \cdots$$

then by equation (1) we get, when $x = 0$, $y = 0$. Thus $a_0 = 0$. Then

$$xy = a_1 x^2 + a_2 x^3 + a_3 x^4 + \cdots \tag{2}$$

$$e^x = 1 + x + \frac{x^2}{2!} + \frac{x^3}{3!} + \frac{x^4}{4!} + \cdots \tag{3}$$

$$e^y = 1 + y + \frac{y^2}{2!} + \frac{y^3}{3!} + \frac{y^4}{4!} + \cdots$$

$$= 1 + (a_1 x + a_2 x^2 + a_3 x^3 + \cdots) + \frac{1}{2!}(a_1 x + a_2 x^2 + a_3 x^3 + \cdots)^2 +$$

$$\frac{1}{3!}(a_1 x + a_2 x^2 + a_3 x^3 + \cdots)^3 + \frac{1}{4!}(a_1 x + a_2 x^2 + a_3 x^3 + \cdots)^4 + \cdots$$

$$= 1 + a_1 x + \left(\frac{a_1^2}{2} + a_2\right) x^2 + \left(\frac{a_1^3}{6} + a_1 a_2 + a_3\right) x^3 +$$

$$\left(\frac{a_1^4}{24} + \frac{1}{2} a_1^2 a_2 + \frac{1}{2} a_2^2 + a_1 a_3 + a_4\right) x^4 + \cdots \tag{4}$$

Substitute (2), (3), (4) into equation (1), we get

$$\begin{cases} -1 + a_1 = 0 \\ a_1 - \frac{1}{2} + \frac{1}{2} a_1^2 + a_2 = 0 \\ a_2 - \frac{1}{6} + \frac{1}{6} a_1^3 + a_1 a_2 + a_3 = 0 \\ a_3 - \frac{1}{24} + \frac{1}{24} a_1^4 + \frac{1}{2} a_1^2 a_2 + \frac{1}{2} a_2^2 + a_1 a_3 + a_4 = 0 \\ \vdots \end{cases}$$

Solving this equation system, we get

$$a_1 = 1, a_2 = -1, a_3 = 2, a_4 = -4, \cdots$$

Then the expansion of the implicit function y determined by equation (1) is

$$y = x - x^2 + 2x^3 - 4x^4 + \cdots$$

Example 5 Solve the Bessel equation with order zero

$$xy'' + y' + xy = 0$$

Solution Suppose the equation has solutions as

$$y = \sum_{n=0}^{\infty} a_n x^n$$

Since

$$y' = \sum_{n=1}^{\infty} n a_n x^{n-1}, y'' = \sum_{n=2}^{\infty} n(n-1) a_n x^{n-2}$$

substitute into the equation, combine the terms with the same degree of x together, then

Chapter 11 Infinite Series

the coefficient of x^n is
$$a_{n-1} + (n+1)a_{n+1} + (n+1)na_{n+1} = a_{n-1} + (n+1)^2 a_{n+1}, n = 1,2,\cdots$$
the constant term is a_1. By comparison of coefficients
$$a_1 = 0, a_{n+1} = -\frac{a_{n-1}}{(n+1)^2}, n = 1,2,\cdots$$
so
$$a_1 = 0, a_3 = a_5 = a_7 = \cdots = a_{2k+1} = \cdots = 0$$
$$a_2 = -\frac{a_0}{2^2}, a_4 = -\frac{a_2}{4^2} = \frac{(-1)^2 a_0}{2^4 (2!)^2}, \cdots, a_{2k} = \frac{(-1)^k a_0}{2^{2k} (k!)^2}, \cdots$$

So the power series solution of the Bessel equation is
$$y = a_0 \left[1 - \frac{x^2}{2^2} + \frac{x^4}{2^4 (2!)^2} - \cdots + (-1)^k \frac{x^{2k}}{2^{2k} (k!)^2} + \cdots \right]$$

If we take $a_0 = 1$, we get a particular solution of the equation, and it's called the Bessel function, denoted by $J_0(x)$, i. e.
$$J_0(x) = 1 + \sum_{k=1}^{\infty} \frac{(-1)^k}{(k!)^2} \left(\frac{x}{2} \right)^{2k}$$

11.9 Fourier Series

In the nature and human practice, there are many back and forth and periodic things. For example, the turning around of planets, rotating of flywheels, up and down moving of plungers of steamers, vibrations of objects, flowing of sound, heat, light and electricity and so on. In mathematics, we can use the periodic functions to describe them. The simplest and basic one is the sine function
$$A\sin(\omega x + \varphi)$$
also called the harmonic function. Its period $T = \frac{2\pi}{\omega}$, the maximum value is A, called the amplitude, ω is called the frequency, φ is called the initial phase. The harmonic function is determined by these three variables. Besides the sine function, we often meet non-sine periodic functions, they indicate more complicate periodic phenomena. For example the rectangular wave in the electric technique (Fig. 11.5).

From the view of mathematics, we should divide a periodic functions into the sum of different frequency sine functions, i. e. denote the function with period T as

$$A_0 + \sum_{n=1}^{\infty} A_n \sin(n\omega x + \varphi_n), \omega = \frac{2\pi}{T} \qquad (1)$$

where $A_0, A_n, \varphi_n (n = 1, 2, \cdots)$ are all constants. Using the trigonometric formula

$$\sin(n\omega x + \varphi_n) = \sin \varphi_n \cos n\omega x + \cos \varphi_n \sin n\omega x$$

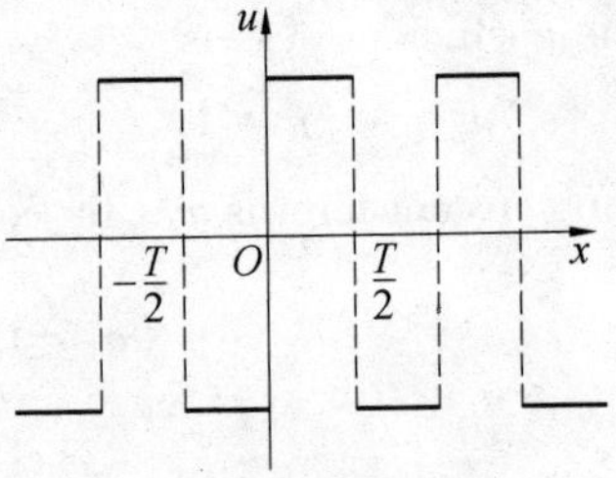

and letting $a_0 = 2A_0, a_n = A_n \sin \varphi_n, b_n = A_n \cos \varphi_n (n = 1, 2, \cdots)$, then (1) becomes

Fig. 11.5

$$\frac{a_0}{2} + \sum_{n=1}^{\infty} (a_n \cos n\omega x + b_n \sin n\omega x) \qquad (2)$$

A functional series expressed as equation (2) is called a trigonometric series.

The natural questions are: What is the condition for the function $f(x)$ being expandable as trigonometric series (2)? How to determine the coefficients a_0, a_n, b_n? For convenient, firstly we discuss the function with the period 2π, at this moment, $\omega = 1$. To solve the above questions, we introduce the orthogonality of trigonometric function system which plays the key role.

11.9.1 Orthogonality of Trigonometric Function System

The trigonometric function system

$$1, \cos x, \sin x, \cos 2x, \sin 2x, \cdots, \cos nx, \sin nx, \cdots$$

has the following two properties:

1. (Orthogonality) The integral of the product of any two different functions in the interval of a periodic long, $[-\pi, \pi]$, is zero.

2. The integral of the square of any function in $[-\pi, \pi]$ is not zero.

That is

$$\int_{-\pi}^{\pi} 1 dx = 2\pi$$

$$\int_{-\pi}^{\pi} \sin nx dx = \int_{-\pi}^{\pi} \cos nx dx = 0$$

$$\int_{-\pi}^{\pi} \cos nx \cos mx dx = \begin{cases} 0, \text{when } m \neq n \\ \pi, \text{when } m = n \end{cases}$$

Chapter 11 Infinite Series

$$\int_{-\pi}^{\pi} \sin nx \sin mx \, dx = \begin{cases} 0, & \text{when } m \neq n \\ \pi, & \text{when } m = n \end{cases}$$

$$\int_{-\pi}^{\pi} \cos nx \sin mx \, dx = 0$$

where $m, n = 1, 2, \cdots$. Using the product to sum formula of trigonometric functions, it is not difficult to check the above formulas. For example, when $m \neq n$

$$\int_{-\pi}^{\pi} \cos nx \cos mx \, dx = \frac{1}{2} \int_{-\pi}^{\pi} [\cos(m-n)x + \cos(m+n)x] \, dx$$

$$= \frac{1}{2} \left[\frac{\sin(m-n)x}{m-n} + \frac{\sin(m+n)x}{m+n} \right] \bigg|_{-\pi}^{\pi} = 0$$

When $m = n$

$$\int_{-\pi}^{\pi} \cos^2 nx \, dx = \int_{-\pi}^{\pi} \frac{1 + \cos 2nx}{2} \, dx = \left[\frac{x}{2} + \frac{\sin 2nx}{4n} \right] \bigg|_{-\pi}^{\pi} = \pi$$

11.9.2 Fourier Series

Theorem 11.15 If the function with the period 2π can be expanded as trigonometric series which is integrable term by term in the interval $[-\pi, \pi]$

$$f(x) = \frac{a_0}{2} + \sum_{n=1}^{\infty} [a_n \cos nx + b_n \sin nx] \tag{3}$$

then its coefficients formula is

$$\begin{cases} a_0 = \dfrac{1}{\pi} \displaystyle\int_{-\pi}^{\pi} f(x) \, dx \\ a_n = \dfrac{1}{\pi} \displaystyle\int_{-\pi}^{\pi} f(x) \cos nx \, dx \\ b_n = \dfrac{1}{\pi} \displaystyle\int_{-\pi}^{\pi} f(x) \sin nx \, dx \, (n = 1, 2, \cdots) \end{cases} \tag{4}$$

Proof Integrate the two sides of (3) in the interval $[-\pi, \pi]$, using the orthogonality of trigonometric function system, we have

$$\int_{-\pi}^{\pi} f(x) \, dx = \int_{-\pi}^{\pi} \frac{a_0}{2} \, dx + \sum_{n=1}^{\infty} \left[a_n \int_{-\pi}^{\pi} \cos nx \, dx + b_n \int_{-\pi}^{\pi} \sin nx \, dx \right] = a_0 \pi$$

so

$$a_0 = \frac{1}{\pi} \int_{-\pi}^{\pi} f(x) \, dx$$

Multiply both sides of (3) by $\cos kx$ and integrate from $-\pi$ to π, we get

$$\int_{-\pi}^{\pi} f(x)\cos kx dx$$
$$= \frac{a_0}{2}\int_{-\pi}^{\pi}\cos kx dx + \sum_{n=1}^{\infty}\left[a_n\int_{-\pi}^{\pi}\cos nx\cos kx dx + b_n\int_{-\pi}^{\pi}\sin nx\cos kx dx\right]$$
$$= a_k\int_{-\pi}^{\pi}\cos^2 kx dx = a_k\pi$$

so
$$a_k = \frac{1}{\pi}\int_{-\pi}^{\pi} f(x)\cos kx dx$$

Similarly, if we multiply the two sides of (3) by $\sin kx$, then integrate it from $-\pi$ to π, using the orthogonality of trigonometric function system, we have

$$b_k = \frac{1}{\pi}\int_{-\pi}^{\pi} f(x)\sin kx dx$$

From the coefficients formula (4), as long as the function $f(x)$ is integrable on the interval $[-\pi,\pi]$, then no matter if $f(x)$ can be expanded as trigonometric series (3), which is integrable term by term, the coefficient $a_0, a_n, b_n (n = 1,2,\cdots)$ in (4) can be computed, called the Fourier coefficients of the function $f(x)$. The trigonometric series constituted by these coefficients

$$\frac{a_0}{2} + \sum_{n=1}^{\infty}(a_n\cos nx + b_n\sin nx)$$

is called the Fourier series of the function $f(x)$, denoted by

$$f(x) \sim \frac{a_0}{2} + \sum_{n=1}^{\infty}(a_n\cos nx + b_n\sin nx)$$

Notice The Fourier series of $f(x)$ is not always convergent everywhere. Even if it is convergent, it may not be convergent to $f(x)$. So we cannot change the symbol " $\sim$ " into " $=$ " freely. For what kind of functions, their Fourier series converge to themselves? That is, what conditions must the functions satisfy such that they can be expanded as Fourier series? In the following we will introduce a convergent theorem.

Theorem 11.16 (Sufficient Conditions for Convergence) If the function $f(x)$ with the period 2π satisfies the Dirichlet conditions in the interval $[-\pi,\pi]$:

(i) Continuous everywhere except for finitely many first kind discontinuities;
(ii) Piecewise monotone, the number of monotonic intervals are finite;
Then the Fourier series of $f(x)$ converges everywhere on $[-\pi,\pi]$, and

Chapter 11 Infinite Series

$$\frac{a_0}{2} + \sum_{n=1}^{\infty}(a_n\cos nx + b_n\sin nx)$$

$$= \begin{cases} f(x), \text{when } x \text{ is the continuous point of } f(x) \\ \frac{1}{2}[f(x^-) + f(x^+)], \text{when } x \text{ is the discontinuous point of } f(x) \\ \frac{1}{2}[f(-\pi^+) + f(\pi^-)], \text{when } x = \pm\pi \end{cases}$$

This theorem explains that the Fourier series of a function satisfied the Dirichlet conditions converges to $f(x)$ at the point where $f(x)$ is continuous; At the discontinuous points, it converges to the arithmetic average of the left and right limit; at the point $x = \pm\pi$, it converges to the arithmetic average of the right limit of the left endpoint and the left limit of the right endpoint.

By the way, the trigonometric series expansion of a periodic functions is unique, which is its Fourier series. Its constant term $\frac{a_0}{2}$ is the average value of the function in a period.

Example 1 Suppose the period of the function $f(x)$ is 2π, and

$$f(x) = \begin{cases} -1, \text{when } -\pi < x \leq 0 \\ x^2, \text{when } 0 < x \leq \pi \end{cases}$$

The sum function of its Fourier series is denoted by $S(x)$, compute $S(0), S(1), S(\pi), S(2\pi)$.

Solution Because $f(x)$ satisfies the Dirichlet conditions in $[-\pi, \pi]$, we can expand $f(x)$ as the Fourier series. And

$$S(0) = -\frac{1}{2}, S(1) = 1$$

$$S(\pi) = \frac{\pi^2 - 1}{2}, S(2\pi) = -\frac{1}{2}$$

Example 2 Suppose $f(x)$ is a function with the period 2π, its expression in $[-\pi, \pi]$ is

$$f(x) = \begin{cases} -1, \text{when } -\pi \leq x < 0 \\ 1, \text{when } 0 \leq x < \pi \end{cases}$$

Try to expand $f(x)$ as the Fourier series.

Solution Firstly we compute the Fourier coefficients, noticing that $f(x)$ is an odd function

$$a_0 = \frac{1}{\pi} \int_{-\pi}^{\pi} f(x) \, dx = 0$$

$$a_n = \frac{1}{\pi} \int_{-\pi}^{\pi} f(x) \cos nx \, dx = 0 \, (n = 1, 2, \cdots)$$

$$b_n = \frac{1}{\pi} \int_{-\pi}^{\pi} f(x) \sin nx \, dx$$

$$= \frac{2}{\pi} \int_{0}^{\pi} \sin nx \, dx$$

$$= -\frac{2}{\pi} \frac{\cos nx}{n} \bigg|_0^{\pi} = \frac{2}{n\pi}[1 - \cos n\pi]$$

$$= \frac{2}{n\pi}[1 - (-1)^n] = \begin{cases} \dfrac{4}{n\pi}, & \text{when } n = 1,3,5,\cdots \\ 0, & \text{when } n = 2,4,6,\cdots \end{cases}$$

So the Fourier series of $f(x)$ is

$$f(x) \sim \frac{4}{\pi} \sum_{n=1}^{\infty} \frac{1}{2n-1} \sin(2n-1)x = \frac{4}{\pi}\left(\sin x + \frac{1}{3}\sin 3x + \frac{1}{5}\sin 5x + \cdots\right)$$

Because $f(x)$ satisfies the Dirichlet conditions, by Theorem 11.16

$$\frac{4}{\pi} \sum_{n=1}^{\infty} \frac{1}{2n-1} \sin(2n-1)x = \begin{cases} f(x), & x \in (-\pi, 0) \cup (0, \pi) \\ 0, & 0, \pm\pi \end{cases}$$

A group of graphs of Fig. 11.6 explain how the series converges to $f(x)$.

Example 3 If $f(x)$ satisfies $f(x + \pi) = -f(x)$ on $[-\pi, \pi]$, try to prove that the Fourier coefficients of $f(x)$: $a_0 = a_{2n} = b_{2n} = 0$.

Solution
$$a_{2n} = \frac{1}{\pi} \int_{-\pi}^{\pi} f(x) \cos 2nx \, dx$$

$$= \frac{1}{\pi} \int_{-\pi}^{0} f(x) \cos 2nx \, dx + \frac{1}{\pi} \int_{0}^{\pi} f(x) \cos 2nx \, dx$$

Let
$$x = \pi + t$$

then

$$\int_{0}^{\pi} f(x) \cos 2nx \, dx = \int_{-\pi}^{0} f(\pi + t) \cos 2nt \, dt$$

$$= -\int_{-\pi}^{0} f(t) \cos 2nt \, dt$$

So
$$a_{2n} = 0$$

Chapter 11 Infinite Series

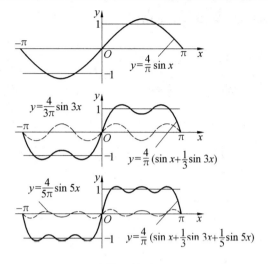

Fig. 11.6

Similarly $\qquad b_{2n} = 0$

Example 4 The period of $f(x)$ is 2π, and

$$f(x) = \begin{cases} x, & \text{when } -\pi \leq x < 0 \\ 0, & \text{when } 0 \leq x < \pi \end{cases}$$

Try to expand $f(x)$ as a Fourier series. (Fig. 11.7)

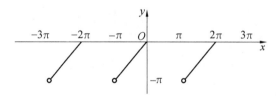

Fig. 11.7

Solution Firstly we compute the Fourier coefficients

$$a_0 = \frac{1}{\pi}\int_{-\pi}^{\pi} f(x)\,dx = \frac{1}{\pi}\int_{-\pi}^{0} x\,dx = -\frac{\pi}{2}$$

$$a_n = \frac{1}{\pi}\int_{-\pi}^{\pi} f(x)\cos nx\,dx = \frac{1}{\pi}\int_{-\pi}^{0} x\cos nx\,dx$$

$$= \frac{1}{\pi}\left[\frac{x\sin nx}{n} + \frac{\cos nx}{n^2}\right]\bigg|_{-\pi}^{0}$$

$$= \frac{1}{n^2\pi}[1-(-1)^n]$$

$$= \begin{cases} \frac{2}{n^2\pi}, n=1,3,5,\cdots \\ 0, n=2,4,6,\cdots \end{cases}$$

$$b_n = \frac{1}{\pi}\int_{-\pi}^{\pi} f(x)\sin nx dx = \frac{1}{\pi}\int_{-\pi}^{0} x\sin nx dx$$

$$= \frac{1}{\pi}\left[-\frac{x\cos nx}{n} + \frac{\sin nx}{n^2}\right]\Big|_{-\pi}^{0} = -\frac{\cos n\pi}{n}$$

$$= \frac{(-1)^{n+1}}{n}$$

So the Fourier series of $f(x)$

$$f(x) \sim -\frac{\pi}{4} + \sum_{n=1}^{\infty}\left\{\frac{1}{n^2\pi}[1-(-1)^n]\cos nx + \frac{(-1)^{n+1}}{n}\sin nx\right\}$$

$$= -\frac{\pi}{4} + \frac{2}{\pi}\left(\cos x + \frac{1}{3^2}\cos 3x + \frac{1}{5^2}\cos 5x + \cdots\right) +$$

$$\left(\sin x - \frac{1}{2}\sin 2x + \frac{1}{3}\sin 3x - \cdots\right)$$

Because $f(x)$ satisfies the Dirichlet conditions, by Theorem 11.16 we have

$$-\frac{\pi}{4} + \sum_{n=1}^{\infty}\left\{\frac{1}{n^2\pi}[1-(-1)^n]\cos nx + \frac{(-1)^{n+1}}{n}\sin nx\right\}$$

$$= \begin{cases} f(x), -\pi < x < \pi \\ -\frac{\pi}{2}, x = \pm\pi \end{cases}$$

11.9.3 Sine Series and Cosine Series

By the integral properties of the odd functions and the even functions, it is easy to get the following conclusions.

1. When $f(x)$ is an odd function with the period 2π, its Fourier coefficients are

$$\begin{cases} a_n = 0, n = 0,1,2,\cdots \\ b_n = \frac{2}{\pi}\int_0^{\pi} f(x)\sin nx dx, n = -1,2,\cdots \end{cases} \quad (5)$$

At this moment, the Fourier series of $f(x)$ only includes sine terms, that is

Chapter 11 Infinite Series

$$f(x) \sim \sum_{n=1}^{\infty} b_n \sin nx \tag{6}$$

It is called a sine series.

2. When $f(x)$ is an even function with the period 2π, its Fourier coefficients are

$$\begin{cases} a_0 = \dfrac{2}{\pi} \int_0^{\pi} f(x) \, dx \\ a_n = \dfrac{2}{\pi} \int_0^{\pi} f(x) \cos nx \, dx, n = 1,2,\cdots \\ b_n = 0, n = 1,2,\cdots \end{cases} \tag{7}$$

At this moment, the Fourier series of $f(x)$ only includes cosine terms and constant terms, that is

$$f(x) \sim \frac{a_0}{2} + \sum_{n=1}^{\infty} a_n \cos nx \tag{8}$$

It is called a cosine series.

When expand the functions as Fourier series, it is benefit to consider the odd and even of the functions firstly.

Example 5 Try to expand the function with the period 2π

$$f(x) = \begin{cases} -x, & -\pi \le x < 0 \\ x, & 0 \le x < \pi \end{cases}$$

as a Fourier series.

Solution The figure of the function is as Fig. 11.8 (In electricity it is called the sawtooth wave), and it is an even function

$$a_0 = \frac{2}{\pi} \int_0^{\pi} f(x) \, dx = \frac{2}{\pi} \int_0^{\pi} x \, dx = \pi$$

$$a_n = \frac{2}{\pi} \int_0^{\pi} f(x) \cos nx \, dx = \frac{2}{\pi} \int_0^{\pi} x \cos nx \, dx$$

$$= \frac{2}{\pi} \left[\frac{x \sin nx}{n} + \frac{\cos nx}{n^2} \right]_0^{\pi} = \frac{2}{n^2 \pi} [(-1)^n - 1]$$

$$= \begin{cases} -\dfrac{4}{n^2 \pi}, n = 1,3,5,\cdots \\ 0, n = 2,4,6,\cdots \end{cases}$$

Because $f(x)$ is continuous everywhere, so

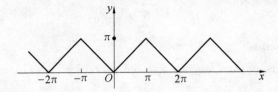

Fig. 11.8

$$f(x) = \frac{\pi}{2} - \frac{4}{\pi} \sum_{n=1}^{\infty} \frac{1}{(2n-1)^2} \cos(2n-1)x$$

$$= \frac{\pi}{2} - \frac{4}{\pi}\left(\cos x + \frac{1}{3^2}\cos 3x + \frac{1}{5^2}\cos 5x + \cdots\right), x \in (-\infty, +\infty)$$

By this expanding expression, it is easy to get several interesting conclusions of number series. Let $x = 0$, we get

$$\frac{\pi^2}{8} = 1 + \frac{1}{3^2} + \frac{1}{5^2} + \cdots$$

suppose

$$\sigma = 1 + \frac{1}{2^2} + \frac{1}{3^2} + \frac{1}{4^2} + \cdots$$

$$\sigma_1 = 1 + \frac{1}{3^2} + \frac{1}{5^2} + \cdots = \frac{\pi^2}{8}$$

$$\sigma_2 = \frac{1}{2^2} + \frac{1}{4^2} + \frac{1}{6^2} + \cdots$$

$$\sigma_3 = 1 - \frac{1}{2^2} + \frac{1}{3^2} - \frac{1}{4^2} + \cdots$$

Because $\sigma_2 = \frac{\sigma}{4} = \frac{\sigma_1 + \sigma_2}{4}$, so

$$\sigma_2 = \frac{\sigma_1}{3} = \frac{\pi^2}{24}, \sigma = \sigma_1 + \sigma_2 = \frac{\pi^2}{6}, \sigma_3 = \sigma_1 - \sigma_2 = \frac{\pi^2}{12}$$

11.9.4 The Fourier Series of The Function with Period 2*l*

In above we discussed the Fourier series of the function with period 2π, now we discuss the Fourier series of more general periodic functions. Suppose $2l(l > 0)$ is the period of $f(x)$, making transformation, let

$$t = \frac{\pi}{l}x$$

Chapter 11 Infinite Series

Then when x varies in $[-l,l]$, t varies in $[-\pi,\pi]$, the function $f(x)$ changes to the function of t with period 2π. Denote

$$f(x) = f\left(\frac{l}{\pi}t\right) = g(t)$$

As long as $f(x)$ is integrable on $[-l,l]$, $g(t)$ is integrable on $[-\pi,\pi]$, so, $g(t)$ has the Fourier series

$$g(t) \sim \frac{a_0}{2} + \sum_{n=1}^{\infty}(a_n\cos nt + b_n\sin nt)$$

Its Fourier coefficients are

$$\begin{cases} a_0 = \dfrac{1}{\pi}\displaystyle\int_{-\pi}^{\pi} g(t)\,dt \\ a_n = \dfrac{1}{\pi}\displaystyle\int_{-\pi}^{\pi} g(t)\cos nt\,dt \\ b_n = \dfrac{1}{\pi}\displaystyle\int_{-\pi}^{\pi} g(t)\sin nt\,dt, n = 1,2,\cdots \end{cases}$$

Substitute $t = \dfrac{\pi}{l}x$, we get the Fourier series of the function with period $2l$

$$f(x) \sim \frac{a_0}{2} + \sum_{n=1}^{\infty}\left(a_n\cos\frac{n\pi x}{l} + b_n\sin\frac{n\pi x}{l}\right) \qquad (9)$$

and the Fourier coefficients

$$\begin{cases} a_0 = \dfrac{1}{l}\displaystyle\int_{-l}^{l} f(t)\,dx \\ a_n = \dfrac{1}{l}\displaystyle\int_{-l}^{l} f(x)\cos\dfrac{n\pi x}{l}\,dx \\ b_n = \dfrac{1}{l}\displaystyle\int_{-l}^{l} f(x)\sin\dfrac{n\pi x}{l}\,dx, n = 1,2,\cdots \end{cases} \qquad (10)$$

When $f(x)$ satisfies the Dirichlet condtions on $[-l,l]$, the Fourier series of $f(x)$ (9), converges to $f(x)$ at the continuous point of $f(x)$, converges to $\dfrac{1}{2}[f(x_0^-) + f(x_0^+)]$ at the discontinuous point x_0; converges to $\dfrac{1}{2}[f(-l^+) + f(l^-)]$ at $\pm l$.

If $f(x)$ is an odd function with period $2l$, its Fourier series is a sine series

$$f(x) \sim \sum_{n=1}^{\infty} b_n\sin\frac{n\pi x}{l} \qquad (11)$$

The coefficients are

$$b_n = \frac{2}{l}\int_0^l f(x)\sin\frac{n\pi x}{l}\mathrm{d}x, n = 1,2,\cdots \tag{12}$$

If $f(x)$ is an even function with period $2l$, its Fourier series is a cosine series

$$f(x) \sim \frac{a_0}{2} + \sum_{n=1}^{\infty} a_n \cos\frac{n\pi x}{l} \tag{13}$$

The coefficients are

$$\begin{cases} a_0 = \dfrac{2}{l}\displaystyle\int_0^l f(x)\,\mathrm{d}x \\ a_n = \dfrac{2}{l}\displaystyle\int_0^l f(x)\cos\dfrac{n\pi x}{l}\mathrm{d}x, n = 1,2,\cdots \end{cases} \tag{14}$$

Example 6 Compute the Fourier coefficients b_n of the function with period $T = \dfrac{2\pi}{\omega}$

$$f(x) = \begin{cases} 0, -\dfrac{T}{2} \leqslant x < 0 \\ E\sin\omega x, 0 \leqslant x < \dfrac{T}{2} \end{cases}$$

Solution See Fig. 11.9(In the electricity it is called the half-wave rectifier type)

$$l = \frac{\pi}{\omega}, \frac{n\pi x}{l} = n\omega x$$

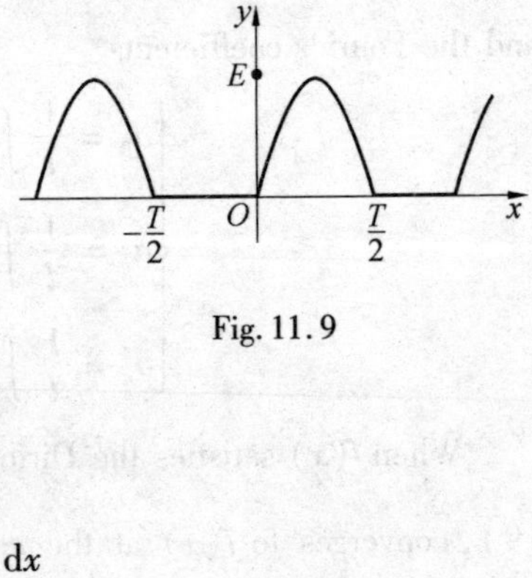

Fig. 11.9

so

$$\begin{aligned} b_n &= \frac{1}{l}\int_{-l}^{l} f(x)\sin\frac{n\pi x}{l}\mathrm{d}x \\ &= \frac{\omega}{\pi}\int_{-\pi/\omega}^{\pi/\omega} f(x)\sin n\omega x\mathrm{d}x \\ &= \frac{\omega}{\pi}\int_0^{\pi/\omega} E\sin\omega x\sin n\omega x\mathrm{d}x \\ &= \frac{\omega E}{2\pi}\int_0^{\pi/\omega}[\cos(n-1)\omega x - \cos(n+1)\omega x]\mathrm{d}x \\ &= \frac{\omega E}{2\pi}\left[\frac{\sin(n-1)\omega x}{(n-1)\omega} - \frac{\sin(n+1)\omega x}{(n+1)\omega}\right]\Bigg|_0^{\pi/\omega} = 0, n = 2,3,\cdots \end{aligned}$$

Demanding $n \neq 1$, so b_1 must be computed separately

$$b_1 = \frac{\omega E}{2\pi}\int_0^{\pi/\omega}(1 - \cos 2\omega x)\,\mathrm{d}x = \frac{\omega E}{2\pi}\left[x - \frac{\sin 2\omega x}{2\omega}\right]\Bigg| = \frac{E}{2}$$

Example 7 The period of the function $f(x)$ is 6, and when $-3 \leqslant x < 3$, $f(x) = x$, determine the Fourier series of $f(x)$.

Solution Here $l = 3$, $f(x)$ is an odd function, so $a_n = 0, n = 0,1,2,\cdots$. By (12)

$$b_n = \frac{2}{l}\int_0^l f(x)\sin\frac{n\pi x}{l}dx = \frac{2}{3}\int_0^3 x\sin\frac{n\pi x}{3}dx$$

$$= -\frac{2}{n\pi}\left[x\cos\frac{n\pi x}{3} - \frac{3}{n\pi}\sin\frac{n\pi x}{3}\right]\Big|_0^3 = (-1)^{n+1}\frac{6}{n\pi}, n = 1,2,\cdots$$

And $f(x)$ satisfies the Dirichlet conditions, so

$$\frac{6}{\pi}\left(\sin\frac{\pi x}{3} - \frac{1}{2}\sin\frac{2\pi x}{3} + \frac{1}{3}\sin\frac{3\pi x}{3} - \cdots\right) = \begin{cases} x, & -3 < x < 3 \\ 0, & x = \pm 3 \end{cases}$$

Example 8 Find the Fourier series of the function with period 1

$$f(x) = \frac{1}{2}e^x, 0 \leqslant x < 1$$

Solution By the integral properties, the integral interval in (10) just need to be kept for a interval of one period long. The graph of the function is as Fig. 11.10. So the coefficients

$$a_0 = \frac{1}{l}\int_0^{2l} f(x)dx = 2\int_0^1 \frac{1}{2}e^x dx = e - 1$$

$$a_n = \frac{1}{l}\int_0^{2l} f(x)\cos\frac{n\pi x}{l}dx$$

$$= 2\int_0^1 \frac{1}{2}e^x\cos 2n\pi x dx = \frac{e - 1}{1 + (2n\pi)^2}$$

$$b_n = \frac{1}{l}\int_0^{2l} f(x)\sin\frac{n\pi x}{l}dx$$

$$= 2\int_0^1 \frac{1}{2}e^x\sin 2n\pi x dx = \frac{-2n\pi(e - 1)}{1 + (2n\pi)^2}, n = 1,2,\cdots$$

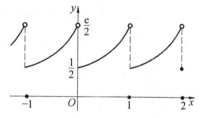

Fig. 11.10

Since $f(x)$ satisfies the Dirichlet conditions, so

$$\frac{e - 1}{2} + (e - 1)\sum_{n=1}^{\infty}\frac{1}{1 + (2n\pi)^2}(\cos 2n\pi x - 2n\pi\sin 2n\pi x)$$

$$= \begin{cases} \frac{1}{2}e^x, & 0 < x < 1 \\ \frac{e + 1}{4}, & x = 0, 1 \end{cases}$$

11.9.5 The Fourier Expansion for Functions in Finite Interval

In this part we will study how to expand the functions defined in the finite interval $[a,b]$ to sine and cosine series. The method is as following: If want to expand as sine series, firstly expand the definition of the function out of $[a,b]$, to expand it as an odd periodic function $F(x)$; If want to expand as cosine series, expand it to an even periodic function $F(x)$, then the Fourier series of $F(x)$ restricted on $x \in [a,b]$ is the Fourier series of $f(x)$. In fact, when computing in practice, what is needed is to apply the formula not the procedure.

Example 9 Expand the function
$$f(x) = x + 1, 0 \leqslant x \leqslant \pi$$
as sine series or cosine series.

Solution Firstly we find the sine series, at this moment, the period is 2π (Fig. 11.11), the coefficients
$$b_n = \frac{2}{\pi}\int_0^\pi (x+1)\sin nx\,dx = \frac{2}{\pi}\left[-\frac{(x+1)\cos nx}{n} + \frac{\sin nx}{n^2}\right]\Big|_0^\pi$$
$$= \frac{2}{n\pi}[1 - (\pi+1)\cos n\pi] = \frac{2}{n\pi}[1 + (-1)^{n+1}(\pi+1)], n = 1,2,\cdots$$

Because $f(x)$ satisfies the Dirichlet conditions, so the sine series is
$$\frac{2}{\pi}\left[(\pi+2)\sin x - \frac{\pi}{2}\sin 2x + \frac{1}{3}(\pi+2)\sin 3x - \frac{\pi}{4}\sin 4x + \cdots\right]$$
$$= \begin{cases} x+1, 0 < x < \pi \\ 0, t = 0, \pi \end{cases}$$

Similarly the cosine series of $f(x) = x+1$ is (Fig. 11.12)
$$1 + \frac{\pi}{2} - \frac{4}{\pi}\left(\cos x + \frac{1}{3^2}\cos 3x + \frac{1}{5^2}\cos 5x + \cdots\right) = x + 1, 0 \leqslant x \leqslant \pi$$

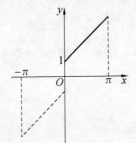

Fig. 11.11

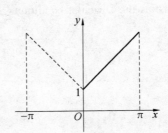

Fig. 11.12

Chapter 11 Infinite Series

Example 10 It is known that the sine and cosine series of the function $f(x)=x^2$ in $[0,1]$

$$S(x) = \sum_{n=1}^{\infty} b_n \sin n\pi x$$

Compute the period of $S(x)$ and the value of $S\left(-\dfrac{1}{2}\right)$.

Solution Because when $n=1$, the period of

$$\sin n\pi x$$

is 2, so the period of $S(x)$ is 2.

Here the sine series is the expansion to expand $f(x)$ as an odd function

$$F(x) = \begin{cases} -x^2, & -1 < x < 0 \\ x^2, & 0 \leqslant x \leqslant 1 \end{cases}$$

But $x = -\dfrac{1}{2}$ is the continuous point of $F(x)$, by Theorem 11.16

$$S\left(-\dfrac{1}{2}\right) = F\left(-\dfrac{1}{2}\right) = -\dfrac{1}{4}$$

11.9.6 The Complex Form of Fourier Series

We can often use the complex number form of the Fourier series in the electricity technique. Using the Euler formula

$$\cos x = \dfrac{e^{ix} + e^{-ix}}{2}, \quad \sin x = \dfrac{e^{ix} - e^{-ix}}{2i}$$

then

$$\dfrac{a_0}{2} + \sum_{n=1}^{\infty} \left[a_n \cos \dfrac{n\pi x}{l} + b_n \sin \dfrac{n\pi x}{l} \right]$$

$$= \dfrac{a_0}{2} + \sum_{n=1}^{\infty} \left[\dfrac{a_n}{2}(e^{i\frac{n\pi x}{l}} + e^{-i\frac{n\pi x}{l}}) - \dfrac{ib_n}{2}(e^{i\frac{n\pi x}{l}} - e^{-i\frac{n\pi x}{l}}) \right]$$

$$= \dfrac{a_0}{2} + \sum_{n=1}^{\infty} \left[\dfrac{a_n - ib_n}{2} e^{i\frac{n\pi x}{l}} + \dfrac{a_n ib_n}{2} e^{-i\frac{n\pi x}{l}} \right]$$

If denote

$$c_0 = \dfrac{a_0}{2}, \quad c_n = \dfrac{a_n - ib_n}{2}, \quad c_{-n} = \dfrac{a_n + ib_n}{2} \quad (n = 1, 2, \cdots)$$

the series above becomes

$$c_0 + \sum_{n=1}^{\infty}(c_n e^{i\frac{n\pi x}{l}} + c_{-n} e^{-i\frac{n\pi x}{l}}) = (c_n e^{i\frac{n\pi x}{l}})\big|_{n=0} + \sum_{n=1}^{\infty} c_n e^{i\frac{n\pi x}{l}} + \sum_{n=-\infty}^{-1} c_n e^{i\frac{n\pi x}{l}}$$

Write the final expression together, we get the complex number form of the Fourier series

$$\sum_{n=-\infty}^{\infty} c_n e^{i\frac{n\pi x}{l}} \qquad (15)$$

The coefficients c_n are

$$c_0 = \frac{a_0}{2} = \frac{1}{2l}\int_{-l}^{l} f(x)\,dx$$

$$c_n = \frac{1}{2}(a_n - ib_n) = \frac{1}{2}\left[\frac{1}{l}\int_{-l}^{l} f(x)\cos\frac{n\pi x}{l}dx - \frac{i}{l}\int_{-l}^{l} f(x)\sin\frac{n\pi x}{l}dx\right]$$

$$= \frac{1}{2l}\int_{-l}^{l} f(x)\left[\cos\frac{n\pi x}{l} - i\sin\frac{n\pi x}{l}\right]dx = \frac{1}{2l}\int_{-l}^{l} f(x)e^{-i\frac{n\pi x}{l}}dx\,(n=1,2,\cdots)$$

Similarly

$$c_{-n} = \frac{1}{2}(a_n + ib_n) = \frac{1}{2l}\int_{-l}^{l} f(x)e^{i\frac{n\pi x}{l}}dx\,(n=1,2,\cdots)$$

The above all coefficients can be expressed by one formula, that is

$$c_n = \frac{1}{2l}\int_{-l}^{l} f(x)e^{-i\frac{n\pi x}{l}}dx\,(n=0,\pm 1,\pm 2,\cdots) \qquad (16)$$

The two forms of the Fourier series are the same in fact, but the complex form (15) is simpler, and the coefficients can be combined as (16).

Example 11 Expand the rectangular wave (Fig. 11.13) with width τ and height h as a complex Fourier series.

Solution The wave function in a period $\left[-\frac{T}{2}, \frac{T}{2}\right]$ is

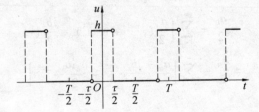

Fig. 11.13

Chapter 11 Infinite Series

$$u(t) = \begin{cases} 0, & -\dfrac{T}{2} \le t < -\dfrac{\tau}{2} \\ h, & -\dfrac{\tau}{2} \le t < \dfrac{\tau}{2} \\ 0, & \dfrac{\tau}{2} \le t < \dfrac{T}{2} \end{cases}$$

By (16), we have

$$c_0 = \frac{1}{T}\int_{-\frac{T}{2}}^{\frac{T}{2}} u(t)\,dt$$

$$= \frac{1}{T}\int_{-\frac{\tau}{2}}^{\frac{\tau}{2}} h\,dt = \frac{h\tau}{T}$$

$$c_n = \frac{1}{T}\int_{-\frac{T}{2}}^{\frac{T}{2}} u(t)\,e^{-i\frac{2n\pi t}{T}}\,dt = \frac{1}{T}\int_{-\frac{\tau}{2}}^{\frac{\tau}{2}} h\,e^{-i\frac{2n\pi t}{T}}\,dt$$

$$= \frac{h}{T}\left[-\frac{T}{i2n\pi}e^{-i\frac{2n\pi t}{T}}\right]\Big|_{-\frac{\tau}{2}}^{\frac{\tau}{2}} = \frac{h}{n\pi}\sin\frac{n\pi\tau}{T}\quad (n = \pm1, \pm2, \cdots)$$

Since $u(t)$ satisfies the Dirichlet conditions, so the complex Fourier series of $u(t)$ is

$$u(t) = \frac{h\tau}{T} + \frac{h}{\pi}\sum_{\substack{n=-\infty \\ n \ne 0}}^{\infty}\frac{1}{n}\sin\frac{n\pi\tau}{T}e^{i\frac{2n\pi t}{T}},\ t \ne \pm kT \pm \frac{\tau}{2},\ k = 0,1,2,\cdots$$

11.10 Examples

Example 1 Suppose $f(x)$ have the second continuous derivative at some neighborhood of the point $x = 0$, and $\lim\limits_{x\to 0}\dfrac{f(x)}{x} = 0$. Show that the series $\sum\limits_{n=1}^{\infty} f\left(\dfrac{1}{n}\right)$ is absolute convergent.

Solution 1 By $\lim\limits_{x\to 0}\dfrac{f(x)}{x} = 0$ and the Maclaurin formula of degree 1, we have

$$f(x) = \frac{1}{2}f''(\theta x)x^2\ (0 < \theta < 1)$$

Since $f''(x)$ is continuous at some neighborhood of $x = 0$, $f''(x)$ is bounded at a closed interval contained in the neighborhood, that is, there exists a constant $M > 0$, such that $|f''(x)| \le M$ whenever x belongs to the closed interval. Hence, $|f(x)| \le \dfrac{M}{2}x^2$ for x

sufficiently small. Let $x = \dfrac{1}{n}$, then

$$\left|f\left(\dfrac{1}{n}\right)\right| \leq \dfrac{M}{2}\dfrac{1}{n^2} \text{ for sufficiently large } n$$

Since $\sum\limits_{n=1}^{\infty}\dfrac{1}{n^2}$ is convergent, $\sum\limits_{n=1}^{\infty}f\left(\dfrac{1}{n}\right)$ is absolute convergent.

Solution 2 By the first half of Solution 1, we know $f(0) = f'(0) = 0$. It follows from L'Hospital's Rule that

$$\lim_{x \to 0}\dfrac{f(x)}{x^2} = \lim_{x \to 0}\dfrac{f'(x)}{2x} = \lim_{x \to 0}\dfrac{f''(x)}{2} = \dfrac{1}{2}f''(0)$$

Let $x = \dfrac{1}{n}$, then

$$\lim_{n \to \infty}\dfrac{\left|f\left(\dfrac{1}{n}\right)\right|}{\dfrac{1}{n^2}} = \dfrac{1}{2}|f''(0)|$$

Therefore, by the Limit Comparison Test, $\sum\limits_{n=1}^{\infty}f\left(\dfrac{1}{n}\right)$ is absolute convergent.

Example 2 Suppose that $\sum\limits_{n=1}^{\infty}(-1)^n a_n 2^n$ is convergent. Prove that $\sum\limits_{n=1}^{\infty}a_n$ is absolute convergent.

Solution 1 Since $\sum\limits_{n=1}^{\infty}(-1)^n a_n 2^n$ is convergent, the general term approach zero, i.e.

$$\lim_{n \to \infty}(-1)^n a_n 2^n = 0$$

Therefore

$$\lim_{n \to \infty}\dfrac{|a_n|}{\dfrac{1}{2^n}} = \lim_{n \to \infty}|a_n| 2^n = 0$$

As $\sum\limits_{n=1}^{\infty}\dfrac{1}{2^n}$ is a convergent geometric series, by the Limit Comparison Test $\sum\limits_{n=1}^{\infty}|a_n|$ converges, so $\sum\limits_{n=1}^{\infty}a_n$ absolutely converges.

Solution 2 By the given conditions, the power series

$$\sum_{n=1}^{\infty} a_n x^n$$

converges at $x = -2$, and it absolutely converges everywhere in the interval $(-2, 2)$, by Abel Lemma. In particular, $\sum_{n=1}^{\infty} a_n$ is absolute convergent.

Example 3 Determine the sum of the series $\sum_{n=0}^{\infty} \dfrac{2n+1}{n!}$.

Solution

$$\sum_{n=0}^{\infty} \frac{2n+1}{n!} = 2\sum_{n=0}^{\infty} \frac{n}{n!} + \sum_{n=0}^{\infty} \frac{1}{n!} = 2\sum_{n=1}^{\infty} \frac{1}{(n-1)!} + \sum_{n=0}^{\infty} \frac{1}{n!}$$

$$= 2\sum_{n=0}^{\infty} \frac{1}{n!} + \sum_{n=0}^{\infty} \frac{1}{n!} = 3\sum_{n=0}^{\infty} \frac{1}{n!} = 3e$$

Example 4 Determine the sum function for the power series

$$1 + \frac{x^2}{2!} + \frac{x^4}{4!} + \frac{x^6}{6!} + \cdots$$

Solution Here

$$\lim_{n \to \infty} \left| \frac{u_{n+1}(x)}{u_n(x)} \right| = \lim_{n \to \infty} \left| \frac{x^{2(n+1)}}{[2(n+1)]!} \cdot \frac{(2n)!}{x^{2n}} \right| = 0$$

It follows from the ratio test that the convergence interval of the power series is $(-\infty, +\infty)$. Suppose that the sum function is

$$S(x) = 1 + \frac{x^2}{2!} + \frac{x^4}{4!} + \frac{x^6}{6!} + \cdots, \quad -\infty < x < +\infty$$

then

$$S'(x) = x + \frac{x^3}{3!} + \frac{x^5}{5!} + \cdots, \quad -\infty < x < +\infty$$

Hence

$$S'(x) + S(x) = 1 + x + \frac{x^2}{2!} + \frac{x^3}{3!} + \cdots = e^x, \quad -\infty < x + \infty$$

Obviously, $S(x)$ is the solution to the following initial value problem

$$\begin{cases} S'(x) + S(x) = e^x \\ S(0) = 1 \end{cases}$$

Solving the initial value equation yields

$$S(x) = \frac{1}{2}(e^x + e^{-x}), \quad -\infty < x < +\infty$$

Example 5 Determine the convergence set of the series of functions $\sum_{n=1}^{\infty} \dfrac{(n+x)^n}{n^{n+x}}$.

Solution For any fixed x and sufficiently large n, we have $u_n(x) = \dfrac{(n+x)^n}{n^{n+x}} > 0$ with

$$u_n(x) = \dfrac{(n+x)^n}{n^{n+x}} = \dfrac{1}{n^x}\left(1 + \dfrac{x}{n}\right)^n$$

So

$$\lim_{n\to\infty} \dfrac{u_n(x)}{\dfrac{1}{n^x}} = \lim_{n\to\infty}\left(1 + \dfrac{x}{n}\right)^n = e^x$$

However, $\sum_{n=1}^{\infty} \dfrac{1}{n^x}$ is a P-series where $P = x$, therefore, it is convergent when $x > 1$ and it is divergent when $x \leqslant 1$. By the Limit Comparison Test, we conclude that the convergence set of the series $\sum_{n=1}^{\infty} \dfrac{(n+x)^n}{n^{n+x}}$ is $x > 1$.

Example 6 Find an expansion as power series in x for the function

$$f(x) = \dfrac{1}{4}\ln\dfrac{1+x}{1-x} + \dfrac{1}{2}\arctan x - x$$

Solution Since

$$f'(x) = \dfrac{1}{4}\left(\dfrac{1}{1+x} + \dfrac{1}{1-x}\right) + \dfrac{1}{2}\dfrac{1}{1+x^2} - 1$$

$$= \dfrac{1}{1-x^4} - 1 = \sum_{n=0}^{\infty} x^{4n} - 1 = \sum_{n=1}^{\infty} x^{4n}, \quad -1 < x < 1$$

Integrating two sides of the above equation from 0 to x with $f(0) = 0$ yields

$$f(x) = \int_0^x \sum_{n=1}^{\infty} x^{4n} dx = \sum_{n=1}^{\infty} \dfrac{1}{4n+1}x^{4n+1}, \quad -1 < x < 1$$

Example 7 Verify the identity

$$\sum_{n=1}^{\infty} \dfrac{(-1)^{n-1}}{n^2}\cos nx = \dfrac{\pi^2}{12} - \dfrac{x^2}{4} \text{ where } x \in [-\pi, \pi]$$

And find the sum of the series $\sum_{n=1}^{\infty}(-1)^{n-1}\dfrac{1}{n^2}$.

Solution The equality which we will prove is equivalent to

Chapter 11 Infinite Series

$$x^2 = 4\left(\frac{\pi^2}{12} - \sum_{n=1}^{\infty} \frac{(-1)^{n-1}}{n^2} \cos nx\right) = \frac{\pi^2}{3} + 4\sum_{n=1}^{\infty} \frac{(-1)^n}{n^2} \cos nx$$

Expand the function x^2 as a Fourier series in the interval $[-\pi, \pi]$, and consider that x^2 is an even function, we obtain

$$b_n = 0 \ (n = 1, 2, \cdots)$$

$$a_0 = \frac{2}{\pi} \int_0^{\pi} x^2 \, dx = \frac{2}{\pi} \frac{x^3}{3} \Big|_0^{\pi} = \frac{2}{3}\pi^2$$

$$a_n = \frac{2}{\pi} \int_0^{\pi} x^2 \cos nx \, dx = (-1)^n \frac{4}{n^2}$$

Thus

$$x^2 = \frac{1}{3}\pi^2 + 4\sum_{n=1}^{\infty} \frac{(-1)^n}{n^2} \cos nx, \quad -\pi \leq x \leq \pi$$

Let $x = 0$, we have

$$\sum_{n=1}^{\infty} \frac{(-1)^{n-1}}{n^2} = \frac{\pi^2}{12}$$

Exercises 11

11.1

1. Write out the general term u_n of each of the following series.

(1) $-\frac{1}{2} + 0 + \frac{1}{4} + \frac{2}{5} + \frac{3}{6} + \cdots$;

(2) $\frac{1}{2} + \frac{2}{5} + \frac{3}{10} + \frac{4}{17} + \cdots$;

(3) $\frac{\sqrt{3}}{2} + \frac{3}{2 \cdot 4} + \frac{3\sqrt{3}}{2 \cdot 4 \cdot 6} + \frac{3^2}{2 \cdot 4 \cdot 6 \cdot 8} + \cdots$.

2. Using the fact that $S_n = \frac{2n}{n+1} (n = 1, 2, \cdots)$ is the partial sums of the series $\sum_{n=1}^{\infty} u_n$,

(1) Find the general term u_n; (2) Determine whether the series is convergent or divergent.

3. Use the definition of the series to determine which series converge, and find the

sum if the series converges.

(1) $\sum_{n=1}^{\infty} \frac{1}{2^n}$;

(2) $\sum_{n=1}^{\infty} \sin \frac{n\pi}{2}$;

(3) $\sum_{n=1}^{\infty} \frac{1}{(5n-4)(5n+1)}$;

(4) $\sum_{n=1}^{\infty} \frac{1}{n(n+1)(n+2)}$;

(5) $\sum_{n=1}^{\infty} \frac{2n-1}{2^n}$;

(6) $\sum_{n=1}^{\infty} (\sqrt{n+2} - 2\sqrt{n+1} + \sqrt{n})$;

(7) $\frac{1}{3} + \frac{1}{15} + \frac{1}{35} + \frac{1}{63} + \cdots$;

(8) $\sum_{n=1}^{\infty} \arctan \frac{1}{n^2+n+1}$.

4. Suppose that the sequence $\{nu_n\}$ and the series $\sum_{n=1}^{\infty} n(u_n - u_{n-1})$ are both convergent. Verify that $\sum_{n=1}^{\infty} u_n$ is convergent.

5. Express the repeating decimal $0.\overline{73}$ as a fraction.

6. Use the properties of series to determine whether the following series converges.

(1) $\frac{1}{11} + \frac{2}{12} + \frac{3}{13} + \cdots$;

(2) $\frac{1}{4} + \frac{1}{5} + \frac{1}{6} + \frac{1}{7} + \cdots$;

(3) $\left(\frac{1}{6} + \frac{8}{9}\right) + \left(\frac{1}{6^2} + \frac{8^2}{9^2}\right) + \left(\frac{1}{6^3} + \frac{8^3}{9^3}\right) + \cdots$;

(4) $\frac{1}{2} + \frac{1}{10} + \frac{1}{4} + \frac{1}{20} + \cdots + \frac{1}{2^n} + \frac{1}{10n} + \cdots$;

(5) $\sum_{n=1}^{\infty} \ln \frac{n+1}{n}$;

(6) $\sum_{n=1}^{\infty} \frac{2^n + 3^n}{6^n}$;

(7) $\sum_{n=1}^{\infty} \frac{(-1)^n n^2}{2n^2 + n}$.

7. Discuss the convergence or divergence of the following series as the series $\sum_{n=1}^{\infty} u_n$ converges and diverges, respectively.

(1) $\sum_{n=1}^{\infty} (u_n + 0.0001)$;

(2) $1\,000 + \sum_{n=1}^{\infty} u_n$;

Chapter 11　Infinite Series

(3) $\sum_{n=1}^{\infty} \dfrac{1}{u_n}$.

8. Find the sum of the series $\sum_{n=1}^{\infty} \dfrac{1}{(2n-1)^2}$, given $\sum_{n=1}^{\infty} \dfrac{1}{n^2} = \dfrac{\pi^2}{6}$.

9. A type of chronic patients need to take a drug every day, and according to the pharmacology the general dosages in patients per day be required generally to maintain between 20~25mg. Suppose that 80% of the drug in vivo are excreted every day, find the daily dose of the patients.

10. It is known that the data in the computer are binary. Find the value in the decimal system for the infinite repeating decimal $(110.110\ 110\cdots)_2$ in binary system.

11.2

1. Use the Comparison test to determine whether the following series converges.

(1) $\dfrac{1}{2} + \dfrac{1}{5} + \dfrac{1}{10} + \dfrac{1}{17} + \cdots$;

(2) $1 + \dfrac{1+2}{1+2^2} + \dfrac{1+3}{1+3^2} + \cdots$;

(3) $\sum_{n=1}^{\infty} \sin \dfrac{\pi}{2^n}$;

(4) $\sum_{n=1}^{\infty} \left[\dfrac{1}{n} - \ln\left(1 + \dfrac{1}{n}\right)\right]$.

2. Use the Ratio test to determine whether the following series converges.

(1) $\sum_{n=0}^{\infty} \dfrac{n!}{n^n}$;

(2) $\sum_{n=0}^{\infty} \dfrac{5^n}{n!}$;

(3) $\sum_{n=1}^{\infty} \dfrac{2^n n!}{n^n}$;

(4) $\sum_{n=1}^{\infty} \dfrac{2 \cdot 5 \cdot \cdots \cdot (3n-1)}{1 \cdot 5 \cdot \cdots \cdot (4n-3)}$.

3. Use the Root test to determine whether the following series converges.

(1) $\sum_{n=1}^{\infty} \left(\dfrac{n}{3n+1}\right)^n$;

(2) $\dfrac{3}{1 \cdot 2} + \dfrac{3^2}{2 \cdot 2^2} + \dfrac{3^3}{3 \cdot 2^3} + \cdots$.

4. Use the Integral test to determine whether the following series converges.

(1) $\sum_{n=1}^{\infty} \dfrac{n+1}{n(n+2)}$;

(2) $\sum_{n=3}^{\infty} \dfrac{\ln n}{n^p}$.

5. Determine whether the given series is convergent or divergent.

(1) $\sum_{n=0}^{\infty} \dfrac{[(2n)!!]^2}{(4n)!!}$;

(2) $\sum_{n=1}^{\infty} \sqrt{\dfrac{n+1}{n}}$;

(3) $\sum_{n=2}^{\infty} \dfrac{1}{n \ln^{1+\sigma} n} (\sigma > 0)$;

(4) $\sum_{n=1}^{\infty} \dfrac{1}{\sqrt{n}} \arcsin \dfrac{1}{n}$;

(5) $\sum_{n=1}^{\infty} \dfrac{1}{1+a^n}(a>0)$;

(6) $\sum_{n=1}^{\infty} \ln\left(1+\dfrac{1}{n}\right)$;

(7) $\sum_{n=1}^{\infty} n\tan\dfrac{\pi}{2^{n+1}}$;

(8) $\sum_{n=1}^{\infty} \dfrac{n^2}{\left(2+\dfrac{1}{n}\right)^n}$;

(9) $\sum_{n=1}^{\infty} \dfrac{(n!)^2}{(2n)!}x^n\;(x>0)$;

(10) $\sum_{n=3}^{\infty} \dfrac{1}{n\ln n \cdot \ln\ln n}$;

(11) $\sum_{n=0}^{\infty} \left(\dfrac{b}{a_n}\right)^n$, given $\lim\limits_{n\to\infty} a_n = a, b \neq a, a>0, b>0$.

6. Discuss whether the series $\sum_{n=1}^{\infty} n^\alpha \beta^n$ is convergent or divergent for $\forall \alpha > 0, \beta > 0$.

7. Suppose that $\sum_{n=1}^{\infty} \dfrac{a^n n!}{n^n}$ is convergent and $\sum_{n=2}^{\infty} \dfrac{\sqrt{n+2} - \sqrt{n-2}}{n^a}$ is divergent where $\alpha > 0$, then( ).

(A) $a > e$ (B) $a = e$ (C) $\dfrac{1}{2} < a < e$ (D) $a \leqslant \dfrac{1}{2}$

8. Prove that:

(1) If the series of positive terms $\sum_{n=1}^{\infty} u_n$ is convergent, then so is $\sum_{n=1}^{\infty} u_n^2$;

(2) If the series of positive terms $\sum_{n=1}^{\infty} u_n$ and $\sum_{n=1}^{\infty} v_n$ are both convergent, then so are $\sum_{n=1}^{\infty} u_n v_n$, $\sum_{n=1}^{\infty} \sqrt{\dfrac{v_n}{n^p}}, p > 1$;

(3) If the series of positive terms $\sum_{n=1}^{\infty} u_n$ is divergent where $S_n = u_1 + \cdots + u_n$, then $\sum_{n=1}^{\infty} \dfrac{u_n}{S_n^2}$ is convergent;

(4) Let $u_n, v_n > 0$, with $\dfrac{u_{n+1}}{u_n} \leqslant \dfrac{v_{n+1}}{v_n}$, then $\sum_{n=1}^{\infty} u_n$ converges as $\sum_{n=1}^{\infty} v_n$ converges, and $\sum_{n=1}^{\infty} v_n$ diverges as $\sum_{n=1}^{\infty} u_n$ diverges.

9. Use properties of convergent series to prove that:

(1) $\lim\limits_{n\to\infty} \dfrac{n^n}{(2n)!} = 0$; (2) $\lim\limits_{n\to\infty} \dfrac{a^n}{n!} = 0 \, (a > 1)$.

10. Let $\sum\limits_{n=1}^{\infty} a_n$ and $\sum\limits_{n=1}^{\infty} b_n$ converge with $a_n < b_n$ for any positive integer n. Show that $\sum\limits_{n=1}^{\infty} c_n$ is convergent.

11. Given $\lim\limits_{n\to+\infty} na_n = a \neq 0$, then $\sum\limits_{n=1}^{\infty} a_n$ diverges.

12. Let $a_n = \int_0^{\pi/4} \tan^n x \, dx$. Verify that $\sum\limits_{n=1}^{\infty} \dfrac{a_n}{n^\lambda}$ converges for $\forall \lambda > 0$.

13. Suppose that the sequence $\{u_n\}$ satisfies $u_{n+1} = \dfrac{1}{2} u_n(u_n^2 + 1), n = 1, 2, \cdots$. Discuss the convergence or divergence of the series $\sum\limits_{n=1}^{\infty} u_n$ as the initial term $u_1 = \dfrac{1}{2}$ and $u_1 = 2$, respectively.

14. Determine whether the given series $\sum\limits_{n=1}^{\infty} \left(\int_0^n \sqrt[3]{1+x^2} \, dx \right)^{-1}$ converges or diverges.

11.3

1. Determine whether each of the following series converges or diverges. If it converges, is it conditionally convergent or absolute convergent?

(1) $1 - \dfrac{1}{\sqrt{2}} + \dfrac{1}{\sqrt{3}} - \dfrac{1}{\sqrt{4}} + \cdots + (-1)^{n-1} \dfrac{1}{\sqrt{n}} + \cdots$;

(2) $\sum\limits_{n=2}^{\infty} (-1)^n \dfrac{1}{n - \ln n}$; (3) $\sum\limits_{n=1}^{\infty} (-1)^{n-1} \dfrac{1}{3 \cdot 2^n}$;

(4) $\sum\limits_{n=1}^{\infty} (-1)^{n-1} \dfrac{1}{n^{p+\frac{1}{n}}}$; (5) $\sum\limits_{n=1}^{\infty} \dfrac{n! \, 2^n \sin \frac{n\pi}{5}}{n^n}$;

(6) $\sum\limits_{n=1}^{\infty} \left[\dfrac{\sin n\alpha}{n^2} - \dfrac{1}{\sqrt{n}} \right]$;

(7) $\sum\limits_{n=1}^{\infty} (-1)^n \left(1 - \cos \dfrac{\alpha}{n} \right)$ (the constant $\alpha > 0$);

(8) $\sum\limits_{n=1}^{\infty} \dfrac{(-\alpha)^n \cdot n!}{n^n}$ ($\alpha > 0$ is a constant);

(9) $\sum_{n=2}^{\infty} \sin\left(n\pi + \dfrac{1}{\ln n}\right)$.

2. Let $\sum_{n=1}^{\infty} a_n^2, \sum_{n=1}^{\infty} b_n^2$ be convergent, show that $\sum_{n=1}^{\infty} a_n b_n, \sum_{n=1}^{\infty} \dfrac{a_n}{n}$ and $\sum_{n=1}^{\infty} (-1)^{\frac{n+(n+1)}{2}} \cdot (a_n + b_n)^2$ are all convergent.

3. If the constant $k > 0$, then $\sum_{n=1}^{\infty} (-1)^n \dfrac{k+n}{n^2}$ ( ).

(A) diverges

(B) absolutely converges

(C) conditionally converges

(D) converges or diverges is independent of the value of k

4. Let $\sum_{n=1}^{\infty} u_n$ conditionally converges with $u_n^* = \dfrac{u_n + |u_n|}{2}, u_n^{**} = \dfrac{u_n - |u_n|}{2}$, then ( ).

(A) $\sum_{n=1}^{\infty} u_n^*$ and $\sum_{n=1}^{\infty} u_n^{**}$ converge

(B) $\sum_{n=1}^{\infty} u_n^*$ and $\sum_{n=1}^{\infty} u_n^{**}$ diverge

(C) $\sum_{n=1}^{\infty} u_n^*$ converges, but $\sum_{n=1}^{\infty} u_n^{**}$ diverges

(D) $\sum_{n=1}^{\infty} u_n^*$ diverges, but $\sum_{n=1}^{\infty} u_n^{**}$ converges

5. Determine whether the given series $\sum_{n=2}^{\infty} \dfrac{(-1)^n}{\sqrt{n}+(-1)^n}$ converges or diverges.

6. Let the partial sums $S_n = \sum_{k=1}^{n} u_k$, then the condition that the sequence $\{S_n\}$ is bounded is ( ) such that $\sum_{n=1}^{\infty} u_n$ is convergent.

(A) the sufficient but not necessary condition

(B) the necessary but not sufficient condition

(C) the necessary and sufficient condition

(D) neither a necessary nor a sufficient condition

Chapter 11 Infinite Series

7. Let $\sum_{n=1}^{\infty} u_n$, $\sum_{n=1}^{\infty} v_n$ be divergent. Determine whether the series $\sum_{n=1}^{\infty} [|u_n| + |v_n|]$ converges.

8. Suppose that the positive term sequence $\{a_n\}$ is monotonically decreasing, and $\sum_{n=1}^{\infty} (-1)^n a_n$ is divergent. Determine whether the series $\sum_{n=1}^{\infty} \left(\dfrac{1}{a_n + 1}\right)^n$ converges and state the reason.

9. For the infinite sequence $\{u_n\}$ ($u_n \neq 0$), if the notion of infinite product is introduced as follows

$$\prod_{n=1}^{\infty} u_n = u_1 \cdot u_2 \cdot \cdots \cdot u_n \cdot \cdots$$

then what is the initial question that should be discussed?

11.4

1. Determine whether the improper integral converges or diverges.

(1) $\displaystyle\int_0^{+\infty} \dfrac{x^2}{x^4 + x^2 + 1} dx$;

(2) $\displaystyle\int_1^{+\infty} \dfrac{dx}{x^3 \sqrt{x^2 + 1}}$;

(3) $\displaystyle\int_1^2 \dfrac{dx}{(\ln x)^3}$;

(4) $\displaystyle\int_1^2 \dfrac{dx}{\sqrt[3]{x^2 - 3x + 2}}$;

(5) $\displaystyle\int_2^{+\infty} \dfrac{dx}{x^3 \sqrt{x^2 - 3x + 2}}$;

(6) $\displaystyle\int_0^{\frac{\pi}{2}} \dfrac{dx}{\sin^p x \cos^q x} (p, q > 0)$.

11.5

1. Discuss the convergent set and the sum function of the series of functions $\sum_{n=1}^{\infty} u_n(x)$.

(1) Given $u_1 = \dfrac{x}{2}$, $u_n = \dfrac{x^n}{2^n} - \dfrac{x^{n-1}}{2^{n-1}}$, $n \geq 2$;

(2) Given $u_1 = \dfrac{x}{2}$, $u_n = \dfrac{nx}{n+1} - \dfrac{(n-1)x}{n}$, $n \geq 2$.

2. Determine the set of convergence of the following series of functions.

(1) $\displaystyle\sum_{n=1}^{\infty} \dfrac{1}{2n+1} \left(\dfrac{1-x}{1+x}\right)^n$;

(2) $x - \dfrac{x^3}{3 \cdot 3!} + \dfrac{x^5}{5 \cdot 5!} - \cdots$;

(3) $x + x^4 + x^9 + x^{16} + x^{25} + \cdots$.

3. Determine whether the following series is uniformly convergent.

(1) $\sum_{n=0}^{\infty} (1-x)x^n, 0 \leq x \leq 1$;

(2) $\sum_{n=1}^{\infty} \dfrac{\sin nx}{\sqrt[3]{n^4+x^4}}, |x| < +\infty$;

(3) $\sum_{n=1}^{\infty} \dfrac{\ln(1+nx)}{nx^n}, x \in [1+\alpha, +\infty)(\alpha > 0)$;

(4) $\sum_{n=1}^{\infty} \dfrac{1}{(x+n)(x+n+1)}, 0 < x < +\infty$.

11.6

1. Find the radius of convergence and interval of convergence of each of the following series.

(1) $\sum_{n=1}^{\infty} n! \left(\dfrac{x}{n}\right)^n$;

(2) $\sum_{n=1}^{\infty} \dfrac{1}{3^n + (-2)^n + 3 \cdot 2^n} x^n$;

(3) $\sum_{n=1}^{\infty} \dfrac{a^n - b^n}{a^n + b^n} (x - x_0)^n (0 < a < b)$.

2. Suppose that the radii of convergence of $\sum_{n=1}^{\infty} a_n x^n$ and $\sum_{n=1}^{\infty} b_n x^n$ are R_1 and R_2 respectively, and that $R_1 < R_2$. Verify that the radius of convergence of $\sum_{n=1}^{\infty} (a_n + b_n)x^n$ is R_1.

3. Find the set of convergence of the following series.

(1) $\sum_{n=1}^{\infty} \dfrac{2^n}{n^2+1} x^n$;

(2) $\sum_{n=1}^{\infty} \left(\dfrac{x}{n}\right)^n$;

(3) $\sum_{n=1}^{\infty} \dfrac{x^n}{(n+1)^p}$;

(4) $\sum_{n=1}^{\infty} \dfrac{2^n + 3^n}{n} x^n$;

(5) $\sum_{n=0}^{\infty} \dfrac{3^{-\sqrt{n}} x^n}{\sqrt{n^2+1}}$;

(6) $\sum_{n=1}^{\infty} \left(\dfrac{a^n}{n} + \dfrac{b^n}{n^2}\right) x^n (a > 0, b > 0)$;

(7) $\sum_{n=1}^{\infty} (-1)^n \left(1 + \dfrac{1}{2} + \dfrac{1}{3} + \cdots + \dfrac{1}{n}\right) x^n$;

(8) $\sum_{n=1}^{\infty} \dfrac{\ln(n+1)}{n+1} x^{n+1}$;

(9) $\sum_{n=1}^{\infty} \dfrac{(x-5)^n}{\sqrt{n}}$;

(10) $\sum_{n=1}^{\infty} (-1)^{n-1} \frac{(x-1)^n}{5n}$; (11) $\sum_{n=0}^{\infty} \frac{(x-3)^n}{n-3^n}$;

(12) $\sum_{n=1}^{\infty} \frac{(2x+1)^n}{n}$; (13) $\sum_{n=1}^{\infty} (-1)^n \frac{x^{2n+1}}{2n+1}$;

(14) $1 + \frac{x^2}{2} + \frac{x^4}{4} + \frac{x^6}{6} + \cdots$; (15) $\sum_{n=0}^{\infty} \frac{x^{n^2}}{2^n}$.

4. Find the set of convergence of the following series of functions.

(1) $\sum_{n=1}^{\infty} \frac{2^n}{n^2+1} x^n$; (2) $\sum_{n=1}^{\infty} \left(\frac{x}{n}\right)^n$;

(3) $\sum_{n=1}^{\infty} (\lg x)^n$; (4) $\sum_{n=1}^{\infty} \frac{n^2}{x^n}$;

(5) $\sum_{n=1}^{\infty} \frac{(x^2+x+1)^n}{n(n+1)}$; (6) $\sum_{n=1}^{\infty} \frac{1}{x^n} \sin \frac{\pi}{2^n}$.

5. Suppose that the power series $\sum_{n=0}^{\infty} a_n (x+1)^n$ converges conditionally at $x = 3$. Determine the interval of convergence and state the reason.

6. Suppose that $\sum_{n=1}^{\infty} (-1)^n a_n (a_n > 0)$ is conditionally convergent, determine the set of convergence of the power series $\sum_{n=0}^{\infty} a_n x^n$ and state the reason.

7. Let the series $\sum_{n=0}^{\infty} a_n x^n$ converge conditionally at $x = -3$ with its coefficients $a_n > 0 (n = 1, 2, \cdots)$. Find the set of convergence of the series and state the reason.

8. Determine the interval of convergence of the power series
$$1 + \frac{(x-1)^2}{1 \cdot 3^2} + \frac{(x-1)^4}{2 \cdot 3^4} + \cdots + \frac{(x-1)^{2n}}{n \cdot 3^{2n}} + \cdots$$

9. Given
$$\frac{1}{1-x} = 1 + x + x^2 + \cdots + x^n + \cdots, x \in (-1, 1)$$
find the power series expansion for functions $\ln(1-x)$ and $\frac{1}{(1-x)^2}$, respectively.

10. Use a fifth-degree polynomial in powers of x to approximate the expression of $\tan x$ around the origin. Use a forth-degree polynomial in powers of x to approximate the

expression of $e^{\sin x}$.

11.7

1. Use the direct expansion method to express $f(x) = a^x (a > 0, a \neq 1)$ as a power series in x.

2. Let $f(x) = \sum\limits_{n=0}^{\infty} a_n x^n$, prove that:
   (1) as $f(x)$ is odd, $a_{2k} = 0, k = 0, 1, 2, \cdots$;
   (2) as $f(x)$ is even, $a_{2k+1} = 0, k = 0, 1, 2, \cdots$.

3. Use the indirect expansion method to find a power series in x for each of the following functions.

   (1) $\sin^2 x$;
   (2) $\sin\left(x + \dfrac{\pi}{4}\right)$;

   (3) $\dfrac{x}{\sqrt{1 - 2x}}$;
   (4) $\ln(1 + x - 2x^2)$;

   (5) $\int_0^x \dfrac{\sin x}{x} dx$;
   (6) $\dfrac{d}{dx}\left(\dfrac{e^x - 1}{x}\right)$;

   (7) $\arcsin x$;
   (8) $\dfrac{1}{4}\ln\dfrac{1 + x}{1 - x} + \dfrac{1}{2}\arctan x - x$;

   (9) $\dfrac{1}{(x^2 + 1)(x^4 + 1)(x^8 + 1)}$.

4. Let $f(x) = \sum\limits_{n=0}^{\infty} a_n x^n$. Find the expansion of the power series (a Maclaurin series) for $g(x) = \dfrac{f(x)}{1 - x}$.

5. Find the series in powers of $x - x_0$ for the following functions about the given point x_0.

   (1) $\sqrt{x^3}, x_0 = 1$;
   (2) $\cos x, x_0 = -\dfrac{\pi}{3}$;

   (3) $\dfrac{x}{x^2 - 5x + 6}, x_0 = 5$.

6. Let $f(x) = (\arctan x)^2$, find $f^{(n)}(0)$.

7. Suppose that $f(x)$ can be expanded as a Maclaurin series, $g(x) = f(x^2)$ as $|x| < r$. Show that

Chapter 11 Infinite Series

$$g^{(n)}(0) = \begin{cases} 0, n = 2m+1 \\ \dfrac{(2m)!}{m!} f^{(m)}(0), n = 2m \end{cases} \quad (m = 1, 2, \cdots)$$

8. Find the sum function of each of the following series in the interval of convergence.

(1) $\sum\limits_{n=1}^{\infty} \dfrac{x^{3n}}{(3n)!}, \ |x| < +\infty$;

(2) $\sum\limits_{n=1}^{\infty} nx^{n-1}, \ |x| < 1$;

(3) $\sum\limits_{n=1}^{\infty} \dfrac{2n-1}{2^n} x^{2n-2}, \ |x| < \sqrt{2}$, and find $\sum\limits_{n=1}^{\infty} \dfrac{2n-1}{2^n}$;

(4) $\sum\limits_{n=1}^{\infty} (-1)^{n+1} n^2 x^n, \ |x| < 1$, and find $\sum\limits_{n=1}^{\infty} (-1)^n \dfrac{n^2}{2^n}$;

(5) $x - \dfrac{x^3}{3} + \dfrac{x^5}{5} + \cdots, \ |x| < 1$, and find $\sum\limits_{n=1}^{\infty} \dfrac{(-1)^n}{2n-1} \left(\dfrac{3}{4}\right)^n$;

(6) $\sum\limits_{n=0}^{+\infty} \dfrac{(n+1)x^n}{n!}, \ |x| < +\infty$.

9. Find the intervals of convergence and the sum functions of the following series.

(1) $\dfrac{x}{4} + \dfrac{x^2}{2 \cdot 4^2} + \cdots + \dfrac{x^n}{n \cdot 4^n} + \cdots$;

(2) $\sum\limits_{n=0}^{\infty} \dfrac{n^2+1}{2^n \cdot n!} x^n$;

(3) $\sum\limits_{n=1}^{\infty} (-1)^{n-1} \dfrac{x^{2n}}{n(2n-1)}$.

10. Use the series expansion for $\dfrac{d}{dx}\left(\dfrac{\cos x - 1}{x}\right)$ to find the sum of

$$\sum\limits_{n=1}^{\infty} (-1)^n \dfrac{2n-1}{(2n)!} \left(\dfrac{\pi}{2}\right)^{2n}$$

11. Find the sum of the series $\sum\limits_{n=0}^{\infty} \dfrac{(-1)^n (n^2 - n + 1)}{2^n}$.

12. Find the sum of the series $\sum\limits_{n=1}^{\infty} (-1)^{n-1} \dfrac{2n^2}{(2n)!} \dfrac{1}{2^n}$.

13. Suppose that the radius of convergence and the sum function of $\sum\limits_{n=1}^{\infty} a_n x^n$ are 3

Calculus(II)

and $S(x)$, respectively. Determine the intervals of convergence and the sum function of the power series $\sum_{n=1}^{\infty} na_n (x - 1)^{n+1}$.

14. Find the following limits.

(1) $\lim_{x \to 1^-} (1 - x^3) \sum_{n=1}^{\infty} n^2 x^n$;

(2) $\lim_{n \to \infty} \left(\dfrac{1}{a} + \dfrac{2}{a^2} + \cdots + \dfrac{n}{a^n} \right)$ $(a > 1)$;

(3) $\lim_{n \to \infty} \left(\dfrac{3}{2 \cdot 1} + \dfrac{5}{2^2 \cdot 2!} + \cdots + \dfrac{2n+1}{2^n \cdot n!} \right)$.

15. Let $f(x) = \int_0^{\sin x} \sin(t^2) \, dt$, $g(x) = \sum_{n=1}^{\infty} \dfrac{x^{2n+1}}{n^2 + 2}$, then as $x \to 0$, $f(x)$ is ( ).

(A) an equivalent infinitesimal of $g(x)$

(B) not an equivalent infinitesimal of $g(x)$, but an infinitesimal of the same order as $g(x)$

(C) an infinitesimal of lower order than $g(x)$

(D) an infinitesimal of higher order than $g(x)$

11.8

1. Find the approximate values of the following numbers (accurate to 4 decimal places).

(1) $\sqrt{e}$;

(2) $\sqrt[5]{245}$;

(3) $\ln 3$;

(4) $\cos 10°$;

(5) $\int_0^1 \dfrac{1 - \cos x}{x^2} dx$;

(6) $\int_0^{1/10} \dfrac{\ln(1 + x)}{x} dx$.

2. Solve the given differential equations by using power series.

(1) $\begin{cases} y' = y^2 + x^3, \\ y|_{x=0} = \dfrac{1}{2}; \end{cases}$

(2) $\begin{cases} (1 - x)y' + y = 1 + x, \\ y|_{x=0} = 0; \end{cases}$

(3) $y' + y = e^x$;

(4) $y'' = x^2 y$.

3. In the interval $[1, 2]$ use the function $\dfrac{2(x-1)}{x+1}$ to approximate the function $\ln x$ and estimate the error.

Chapter 11 Infinite Series

4. Let the series $\sum_{n=0}^{\infty} \dfrac{x^{3n}}{(3n)!}$ be a solution of the differential equation $y'' + y' + by = e^x$ in the interval $(-\infty, +\infty)$. Find the constant b and using this result to find the sum function of the given series.

11.9

1. Expand the function $f(x)$ of period 2π as a Fourier series, where $f(x)$ is defined in the interval $[-\pi, \pi]$ by the following expressions respectively.

(1) $f(x) = \dfrac{\pi}{4} - \dfrac{x}{2}$;

(2) $f(x) = e^x + 1$;

(3) $f(x) = 3x^2 + 1$;

(4) $f(x) = 2\sin\dfrac{x}{3}$;

(5) $f(x) = \begin{cases} x + 2\pi, & -\pi \leq x < 0 \\ x, & 0 \leq x < \pi \end{cases}$;

(6) $f(x) = \begin{cases} e^x, & -\pi \leq x < 0 \\ 1, & 0 \leq x < \pi \end{cases}$.

2. Let

$$\dfrac{a_0}{2} + \sum_{n=1}^{\infty} (a_n \cos nx + b_n \sin nx)$$

be the Fourier series of the function $f(x) = \pi x + x^2$ $(-\pi \leq x < \pi)$. Determine the coefficient b_3 and state the meaning of the constant $\dfrac{a_0}{2}$.

3. Express $f(x)$ defined in the interval $[0, \pi]$ by the following equations as a sine series.

(1) $f(x) = \dfrac{\pi - x}{2}$;

(2) $f(x) = \begin{cases} \dfrac{x}{\pi}, & 0 \leq x < \dfrac{\pi}{2} \\ 1 - \dfrac{x}{\pi}, & \dfrac{\pi}{2} \leq x \leq \pi \end{cases}$.

4. Express $f(x)$ defined in the interval $[0, \pi]$ by the following equations as a cosine series.

(1) $f(x) = \cos\dfrac{x}{2}$;

(2) $f(x) = \begin{cases} 1, & 0 \leq x < h \\ 0, & h \leq x \leq \pi \end{cases}$.

5. Expand the periodic functions $f(x)$ as a Fourier series, where the expression of $f(x)$ in a period is defined as follows:

(1) $f(x) = x^2 - x$, $-2 \leq x < 2$;

(2) $f(x) = \begin{cases} 2x+1, & -3 \leq x < 0 \\ x, & 0 \leq x < 3 \end{cases}$;

(3) $f(x) = \begin{cases} x, & 0 \leq x < 1 \\ 0, & 1 \leq x < 2 \end{cases}$, and using its Fourier series to prove the equality

$$1 + \frac{1}{3^2} + \frac{1}{5^2} + \frac{1}{7^2} + \cdots = \frac{\pi^2}{8}$$

6. Let $f(x)$ be a function of period 2 and defined in the interval $(-1, 1]$ by the equation

$$f(x) = \begin{cases} 2, & -1 < x \leq 0 \\ x^3, & 0 < x \leq 1 \end{cases}$$

Then what are the sums of the Fourier series for $f(x)$ at points $x = 0, \frac{1}{2}, 1$, respectively?

7. Expand the function

$$f(x) = \begin{cases} x, & 0 \leq x \leq \frac{l}{2} \\ l - x, & \frac{l}{2} < x \leq l \end{cases}$$

as a sine series.

8. Expand the function

$$f(x) = \begin{cases} \cos \frac{\pi x}{l}, & 0 \leq x \leq \frac{l}{2} \\ 0, & \frac{l}{2} < x < l \end{cases}$$

as a cosine series.

9. Expand the function $f(x) = x^2 \, (0 \leq x \leq 2)$ as a sine and cosine series respectively, and indicate their differences in convergence.

10. Let $f(x) = \begin{cases} e^x - 1, & -\pi \leq x < 0 \\ e^x + 1, & 0 \leq x < \pi \end{cases}$, and $a_0, a_n \, (n = 1, 2, \cdots)$ be the Flourier coefficients of $f(x)$, then the sum of the series $\frac{a_0}{2} + \sum_{n=1}^{\infty} a_n$ is _____.

11.10

1. Discuss the convergence of the series

Chapter 11 Infinite Series

$$\sum_{n=1}^{\infty} \frac{a^{\frac{n(n+1)}{2}}}{(1+a^0)(1+a^1)(1+a^2)\cdots(1+a^{n-1})}$$

given $a > 0$.

2. Discuss the convergence of the series $\sum_{n=1}^{\infty} \frac{(-1)^{n+1}}{\sqrt{n^{2k}+1}}$, where k is a real number.

3. Find if the series $\sum_{n=1}^{\infty} (-1)^{n+1} \left[e - \left(1 + \frac{1}{n}\right)^n \right]$ converges or divergence, if it is convergent, is it conditionally convergent or absolute convergent?

4. Let $\sum_{n=1}^{\infty} (-1)^{n-1} a_n = 2$, $\sum_{n=1}^{\infty} a_{2n-1} = 5$. Find the sum of the series $\sum_{n=1}^{\infty} a_n$.

5. Let $\sum_{n=1}^{\infty} \frac{1}{(2k-1)^2} = \frac{\pi^2}{8}$. Find the sum of the P-series $\sum_{n=1}^{\infty} \frac{1}{n^2}$ where $P = 2$.

6. Show that the series

$$\arctan \frac{1}{2} + \arctan \frac{1}{8} + \cdots + \arctan \frac{1}{2n^2} + \cdots$$

converges and find the sum S.

7. Verify that the power series $\sum_{n=1}^{\infty} \frac{(1!)^2 + (2!)^2 + \cdots + (n!)^2}{(2n)!} x^n$ converges absolutely in the interval $(-3,3)$.

8. Find the limit

$$\lim_{n \to \infty} \frac{1 + \frac{\pi^4}{5!} + \frac{\pi^8}{9!} + \cdots + \frac{\pi^{4(n-1)}}{(4n-3)!}}{\frac{1}{3!} + \frac{\pi^4}{7!} + \frac{\pi^8}{11!} + \cdots + \frac{\pi^{4(n-1)}}{(4n-1)!}}$$

9. Suppose than the power series $\sum_{n=0}^{\infty} a_n (x - x_0)^n (x_0 \neq 0)$ converges at $x = 0$ and diverges at $x = 2x_0$. Find the radius of convergence R and the interval of convergence, and state the reasons.

10. Find the radius of convergence and the interval of convergence of each of the given functions:

(1) $\sum_{n=1}^{\infty} \frac{4^{2n-1}}{n\sqrt{n}} (x-2)^{2n-1}$; (2) $\sum_{n=1}^{\infty} 8^n (2n-1)^{3n+1}$.

11. Determine the interval of convergence of the series

$$\sum_{n=1}^{\infty} \frac{x^n}{(1+x)(1+x^2)\cdots(1+x^n)}$$

12. Use the power series expansion to find the derivatives of the given order for the following functions at the point $x = 0$:

(1) $f(x) = \dfrac{x}{1+x^2}$, find $f^{(7)}(0)$; (2) $f(x) = x^6 e^x$, find $f^{(10)}(0)$.

13. Use the power series expansion for functions to calculate the following limits.

(1) $\lim\limits_{x \to \infty} \left[x - x^2 \ln\left(1 + \dfrac{1}{x}\right) \right]$; (2) $\lim\limits_{x \to 0} \dfrac{2(\tan x - \sin x) - x^3}{x^5}$.

14. Let $f(x) = \begin{cases} \dfrac{1+x^2}{x} \arctan x, & x \neq 0 \\ 1, & x = 0 \end{cases}$, expand $f(x)$ as a power series in powers of x and find the sum of $\sum\limits_{n=1}^{\infty} \dfrac{(-1)^n}{1-4n^2}$.

15. Suppose that $f(x)$ is a continuous function with 2π as a period, and that a_n, b_n are its Fourier coefficients. Find Fourier coefficients A_n, B_n of

$$F(x) = \frac{1}{\pi} \int_{-\pi}^{\pi} f(t) f(x+t) \, dt$$

and prove that

$$\frac{1}{\pi} \int_{-\pi}^{\pi} f^2(t) \, dt = \frac{a_0^2}{2} + \sum_{n=1}^{\infty} (a_n^2 + b_n^2)$$

16. Given $f(x) = \dfrac{\pi}{2} \cdot \dfrac{e^x + e^{-x}}{e^\pi - e^{-\pi}}$.

(1) Find the Fourier coefficients in $[-\pi, \pi]$;

(2) Find the sum of $\sum\limits_{n=1}^{\infty} \dfrac{(-1)^n}{1+(2n)^2}$.

17. Expand the function $f(x) = \arcsin(\sin x)$ as a Fourier series.

18. Let a_n, b_n be the Fourier coefficients of the integrable function $f(x)$ with 2π as a period. Compute the Fourier coefficients of the translation function $f(x+h)$, where h is a constant.

Appendix IV Change of Variables in Multiple Integrals

The formula of integration by substitution of definite integral is as follows

$$\int_a^b f(x)\,dx = \int_\alpha^\beta f(\varphi(t))\varphi'(t)\,dt$$

where the substitution $x = \varphi(t)$ is monotonic and has continuous derivative, and $a = \varphi(\alpha)$, $b = \varphi(\beta)$. In this formula we not only substitute $x = \varphi(t)$ into the integrand $f(x)$, but also consider the following relationship between infinitesimal interval elements dx and dt

$$dx = \varphi'(t)\,dt$$

here $\varphi'(t)$ is the ratio of dx to dt. Finally, we should convert the integral interval of x into the new integral interval of t.

Similarly, there exist formulas of integration by substitution for multiple integrals. For example, the transformation formula of the double integral from rectangular to polar coordinates is

$$\iint_\sigma f(x,y)\,dxdy = \iint_\sigma f(r\cos\theta, r\sin\theta)\,rdrd\theta$$

In fact, it is the result of the following substitution

$$\begin{cases} x = r\cos\theta \\ y = r\sin\theta \end{cases}$$

Furthermore, transformation formulas of triple integrals from rectangular to cylindrical or spherical coordinates can be obtained by changing variables. Next we will introduce change of variables in multiple integrals. In particular, we focus on double integrals.

For the double integral $\iint_{\sigma_{xy}} f(x,y)\,d\sigma_{xy}$, take a transformation T

$$\begin{cases} x = x(u,v) \\ y = y(u,v) \end{cases} \qquad (1)$$

Suppose the transformation T is a one-to-one mapping from the region σ_{uv} in the uv-plane to the region σ_{xy} in the xy-plane, and let $x = x(u,v)$, $y = y(u,v)$ have continuous first order partial derivatives, then what is the new double integral?

The key is the relationship between the area elements $d\sigma_{uv}$ of σ_{uv} and $d\sigma_{xy}$ of σ_{uv}. Hence, inside the region σ_{uv} take a small rectangle $ABCD$ whose area is denoted by $\Delta\sigma_{uv}$, where $A(u,v)$, $B(u+\Delta u,v)$, $C(u+\Delta u,v+\Delta v)$, $D(u,v+\Delta v)$, are as showed in Fig. IV.1.

The transformation T maps the small rectangle $ABCD$ to the small curved quadrilateral $A_1 B_1 C_1 D_1$ in the region σ_{xy}. Denote the area of $A_1 B_1 C_1 D_1$ by $\Delta\sigma_{xy}$ and the coordinates of its four vertices are as follows:

$A_1(x_1,y_1)$, where $x_1 = x(u,v)$, $y_1 = y(u,v)$;
$B_1(x_2,y_2)$, where $x_2 = x(u+\Delta u,v)$, $y_2 = y(u+\Delta u,v)$;
$C_1(x_3,y_3)$, where $x_3 = x(u+\Delta u,v+\Delta v)$, $y_3 = y(u+\Delta u,v+\Delta v)$;
$D_1(x_4,y_4)$, where $x_4 = x(u,v+\Delta v)$, $y_4 = y(u,v+\Delta v)$.

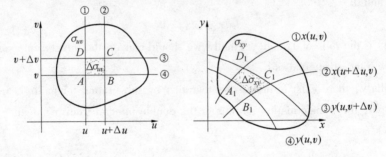

Fig. IV.1

Applying the Taylor formula of degree 1 for the function of two variables, yields

$$\begin{cases} x_2 = x(u,v) + \dfrac{\partial x}{\partial u}\Delta u + o(\Delta u) \\ y_2 = y(u,v) + \dfrac{\partial y}{\partial u}\Delta u + o(\Delta u) \end{cases}$$

$$\begin{cases} x_3 = x(u,v) + \dfrac{\partial x}{\partial u}\Delta u + \dfrac{\partial x}{\partial v}\Delta v + o(\sqrt{\Delta u^2 + \Delta v^2}) \\ y_3 = y(u,v) + \dfrac{\partial y}{\partial u}\Delta u + \dfrac{\partial y}{\partial v}\Delta v + o(\sqrt{\Delta u^2 + \Delta v^2}) \end{cases}$$

Appendix IV Change of Variables in Multiple Integrals

$$\begin{cases} x_4 = x(u,v) + \dfrac{\partial x}{\partial v}\Delta v + o(\Delta v) \\ y_4 = y(u,v) + \dfrac{\partial y}{\partial v}\Delta v + o(\Delta v) \end{cases}$$

If we omit the infinitesimal of higher order $o(\Delta u)$, $o(\Delta v)$ and $o(\sqrt{\Delta u^2 + \Delta v^2})$, then

$$x_2 - x_1 \approx x_3 - x_4 \approx \frac{\partial x}{\partial u}\Delta u, \; y_2 - y_1 \approx y_3 - y_4 \approx \frac{\partial y}{\partial u}\Delta u$$

$$x_4 - x_1 \approx x_3 - x_2 \approx \frac{\partial x}{\partial v}\Delta v, \; y_4 - y_1 \approx y_3 - y_2 \approx \frac{\partial y}{\partial v}\Delta v$$

Thus, the lengths of the two pairs of opposite sides of the small quadrilateral are approximately equal. So we regard $A_1B_1C_1D_1$ as a parallelogram with the approximating area as follows

$$|\overrightarrow{A_1B_1} \times \overrightarrow{A_1D_1}|$$

Because

$$\overrightarrow{A_1B_1} = (x_2 - x_1)\boldsymbol{i} + (y_2 - y_1)\boldsymbol{j} \approx \frac{\partial x}{\partial u}\Delta u \boldsymbol{i} + \frac{\partial y}{\partial u}\Delta u \boldsymbol{j}$$

$$\overrightarrow{A_1D_1} = (x_4 - x_1)\boldsymbol{i} + (y_4 - y_1)\boldsymbol{j} \approx \frac{\partial x}{\partial v}\Delta u \boldsymbol{i} + \frac{\partial y}{\partial v}\Delta v \boldsymbol{j}$$

It follows that

$$\Delta\sigma_{xy} \approx |\overrightarrow{A_1B_1} \times \overrightarrow{A_1D_1}|$$

$$\approx \begin{Vmatrix} \boldsymbol{i} & \boldsymbol{j} & \boldsymbol{k} \\ \dfrac{\partial x}{\partial u}\Delta u & \dfrac{\partial y}{\partial u}\Delta u & 0 \\ \dfrac{\partial x}{\partial v}\Delta v & \dfrac{\partial y}{\partial v}\Delta v & 0 \end{Vmatrix} = \begin{vmatrix} \dfrac{\partial x}{\partial u} & \dfrac{\partial y}{\partial u} \\ \dfrac{\partial x}{\partial v} & \dfrac{\partial y}{\partial v} \end{vmatrix} \Delta u \Delta v$$

Therefore, we obtain the area element

$$\mathrm{d}\sigma_{xy} = \begin{vmatrix} \dfrac{\partial x}{\partial u} & \dfrac{\partial y}{\partial u} \\ \dfrac{\partial x}{\partial v} & \dfrac{\partial y}{\partial v} \end{vmatrix} \mathrm{d}\sigma_{uv} = \left|\frac{\partial(x,y)}{\partial(u,v)}\right| \mathrm{d}\sigma_{uv} \tag{2}$$

where $\left|\dfrac{\partial(x,y)}{\partial(u,v)}\right|$ denotes the ratio of the area elements $\mathrm{d}\sigma_{xy}$ to $\mathrm{d}\sigma_{uv}$ under the transformation T.

Thus, under the transformation T the formula of integration by substitution of double integral is

$$\iint_\sigma f(x,y)\,dxdy = \iint_{\sigma_{u,v}} f(x(u,v),y(u,v)) \left|\frac{\partial(x,y)}{\partial(u,v)}\right| dudv \qquad (3)$$

By the way, in order to make the transformation T one to one we need to let its Jacobi determinant non-zero.

It is simple to calculate the Jacobi determinant

$$\begin{vmatrix} \dfrac{\partial x}{\partial r} & \dfrac{\partial y}{\partial r} \\ \dfrac{\partial x}{\partial \theta} & \dfrac{\partial y}{\partial \theta} \end{vmatrix} = \begin{vmatrix} \cos\theta & \sin\theta \\ -r\sin\theta & r\cos\theta \end{vmatrix} = r$$

of the transformation from rectangular to polar coordinates

$$x = r\cos\theta,\ y = r\sin\theta$$

Example 1 Compute $\iint_\sigma y^2 d\sigma$, where σ is the region enclosed by $x > 0, y > 0$, $1 \leqslant xy \leqslant 3$, $1 \leqslant \dfrac{y}{x} \leqslant 2$.

Solution As shown in Fig. IV. 2, sketch the region σ. Obviously, if we make a transformation $u = xy$, $v = \dfrac{y}{x}$, or

$$x = \sqrt{\dfrac{u}{v}},\ y = \sqrt{uv}$$

then the region σ can be expressed as

$$1 \leqslant u \leqslant 3, 1 \leqslant v \leqslant 2$$

Since the Jacobi determinant

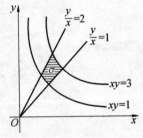

Fig. IV.2

$$\frac{\partial(x,y)}{\partial(u,v)} = \begin{vmatrix} \dfrac{1}{2\sqrt{uv}} & \dfrac{\sqrt{v}}{2\sqrt{u}} \\ -\dfrac{\sqrt{u}}{2v\sqrt{v}} & \dfrac{\sqrt{u}}{2\sqrt{v}} \end{vmatrix} = \frac{1}{2v}$$

so

$$\iint_\sigma y^2 d\sigma = \iint_\sigma uv \left|\frac{1}{2v}\right| dudv = \frac{1}{2}\int_1^3 u\,du \int_1^2 dv = 2$$

Appendix IV Change of Variables in Multiple Integrals

Example 2 Calculate $\iint_\sigma [(x+y)^2 + (x-y)^2] d\sigma$, where the region σ is a square with vertices $(0,0), (1,1), (2,0)$ and $(1,-1)$.

Solution Sketch the region σ, as shown in Fig. IV.3. Obviously, if let
$$u = x+y, v = x-y$$
or
$$x = \frac{1}{2}(u+v), y = \frac{1}{2}(u-v)$$

then the region σ can be expressed as
$$0 \leqslant u \leqslant 2, 0 \leqslant v \leqslant 2$$

Now

$$\frac{\partial(x,y)}{\partial(u,v)} = \begin{vmatrix} \frac{\partial x}{\partial u} & \frac{\partial y}{\partial u} \\ \frac{\partial x}{\partial v} & \frac{\partial y}{\partial v} \end{vmatrix} = \begin{vmatrix} \frac{1}{2} & \frac{1}{2} \\ \frac{1}{2} & -\frac{1}{2} \end{vmatrix} = -\frac{1}{2}$$

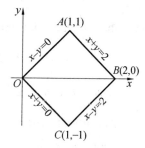

Fig. IV.3

Hence
$$\iint_\sigma [(x+y)^2 + (x-y)^2] d\sigma = \iint_\sigma [u^2 + v^2] \cdot \frac{1}{2} du dv$$
$$= \frac{1}{2} \int_0^2 du \int_0^2 (u^2 + v^2) dv$$
$$= \frac{16}{3}$$

Example 3 Applying the transformation $x = ar\cos\theta, y = br\sin\theta$ to calculate the area of the ellipse $\frac{x^2}{a^2} + \frac{y^2}{b^2} = 1$.

Solution By the given transformation we have $0 \leqslant r \leqslant 1, 0 \leqslant \theta \leqslant 2\pi$ and

$$\frac{\partial(x,y)}{\partial(r,\theta)} = \begin{vmatrix} \frac{\partial x}{\partial r} & \frac{\partial y}{\partial r} \\ \frac{\partial x}{\partial \theta} & \frac{\partial y}{\partial \theta} \end{vmatrix} = \begin{vmatrix} a\cos\theta & b\sin\theta \\ -ar\sin\theta & br\cos\theta \end{vmatrix} = abr$$

Thus, we obtain
$$\iint_\sigma 1 d\sigma = \iint_{\sigma_{r\theta}} abr\, dr d\theta = ab \int_0^{2\pi} d\theta \int_0^1 r dr = \pi ab$$

Similarly, if $x = x(u,v,w), y = y(u,v,w), z = z(u,v,w)$ have continuous first order partial derivatives and the Jacobi determinant

$$\frac{\partial(x,y,z)}{\partial(u,v,w)} = \begin{vmatrix} \frac{\partial x}{\partial u} & \frac{\partial y}{\partial u} & \frac{\partial z}{\partial u} \\ \frac{\partial x}{\partial v} & \frac{\partial y}{\partial v} & \frac{\partial z}{\partial v} \\ \frac{\partial x}{\partial w} & \frac{\partial y}{\partial w} & \frac{\partial z}{\partial w} \end{vmatrix} \neq 0$$

then the formula of integration by substitution of triple integral is

$$\iiint_{V_{xyz}} f(x,y,z) \, dxdydz$$
$$= \iiint_{V_{uvw}} f[x(u,v,w), y(u,v,w), z(u,v,w)] \mid \frac{\partial(x,y,z)}{\partial(u,v,w)} \mid dudvdw \qquad (4)$$

And it isn't hard to get the following results: for the transformation from rectangular to cylindrical coordinates

$$x = r\cos\theta, y = r\sin\theta, z = z$$

we have

$$\frac{\partial(x,y,z)}{\partial(r,\theta,z)} = r$$

for the transformation from rectangular to spherical coordinates

$$x = \rho\sin\varphi\cos\theta, y = \rho\sin\varphi\sin\theta, z = \rho\cos\varphi$$

we have

$$\frac{\partial(x,y,z)}{\partial(\rho,\varphi,\theta)} = \rho^2 \sin\varphi$$

Example 4 Evaluate the volume of the ellipsoid $\frac{x^2}{a^2} + \frac{y^2}{b^2} + \frac{z^2}{c^2} \leq 1$.

Solution Take a transformation

$$\begin{cases} x = a\rho\sin\varphi\cos\theta \\ y = b\rho\sin\varphi\sin\theta \\ z = c\rho\cos\varphi \end{cases}$$

Then

$$\mid \frac{\partial(x,y,z)}{\partial(\rho,\varphi,\theta)} \mid = abc\rho^2 \mid \sin\varphi \mid$$

and the ellipsoid is determined by the system of inequalities

$$0 \leq \theta \leq 2\pi, 0 \leq \varphi \leq \pi, 0 \leq \rho \leq 1$$

Appendix IV Change of Variables in Multiple Integrals

Hence, the volume of the ellipsoid is

$$\iiint_V dV = \iiint_V abc\rho^2 \sin\varphi\, d\rho\, d\varphi\, d\theta = abc \int_0^{2\pi} d\theta \int_0^{\pi} \sin\varphi\, d\varphi \int_0^1 \rho^2\, d\rho = \frac{4}{3}\pi abc$$

Appendix V Radius of Convergence of Power Series

There is a radius of convergence R for any power series

$$\sum_{n=0}^{\infty} a_n x^n = a_0 + a_1 x + a_2 x^2 + \cdots + a_n x^n + \cdots \quad (1)$$

such that the power series convergences absolutely in the interval of convergence $(-R, R)$. We will present a universal method to find the radius of convergence as follows.

Cauchy-Hadamard Theorem) If the upper limit

$$\varlimsup_{n \to +\infty} \sqrt[n]{|a_n|} = \rho$$

then as $\rho = 0, R = +\infty$, and as $\rho = +\infty, R = 0$.

Example Find the interval of convergence of the functional series $\sum_{n=1}^{\infty} \left(1 + \frac{1}{n}\right)^{-n^3} e^{-n^2 x}$.

Solution By setting $y = e^{-x} > 0$, we obtain a power series as follows

$$\sum_{n=1}^{\infty} \left(1 + \frac{1}{n}\right)^{-n^3} y^{n^2}$$

whose radius of convergence is

$$R = \varlimsup_{n \to +\infty} \frac{1}{\sqrt[n]{|a_n|}} = \varlimsup_{n \to +\infty} \frac{1}{\sqrt[n^2]{|a_n^2|}} = \varlimsup_{n \to +\infty} \frac{1}{\sqrt[n^2]{\left(1 + \frac{1}{n}\right)^{-n^3}}}$$

$$= \lim_{n \to +\infty} \left(1 + \frac{1}{n}\right)^n = e$$

Hence, as $0 < y < e$ this power series is convergent, and as $y = e$ the corresponding numerical series has a general term

$$\left(1 + \frac{1}{n}\right)^{-n^3} e^{n^2} = \left[\frac{e}{\left(1 + \frac{1}{n}\right)^n}\right]^{n^2} > 1$$

Appendix V Radius of Convergence of Power Series

By properties of series, we know the numerical series is divergent. In conclusion, the interval of convergence of the power series in y is $(0,e)$. Therefore, the interval of convergence of the series of functions $\sum_{n=1}^{\infty}\left(1+\dfrac{1}{n}\right)^{-n^3}e^{-n^2}$ is $(-1,+\infty)$.